T0313316

Low Power Circuit Design Using Advanced CMOS Technology

EDITORS

Milin Zhang
Tsinghua University, Beijing, China

Zhihua Wang
Tsinghua University, Beijing, China

Jan Van der Spiegel
University of Pennsylvania, Philadelphia, USA

Franco Maloberti
University of Pavia, Pavia, Italy

Tutorials in Circuits and Systems

For a list of other books in this series, visit www.riverpublishers.com

Series Editors

Peter (Yong) Lian

President IEEE
Circuits and Systems Society
York University, Canada

Franco Maloberti

Past President IEEE
Circuits and Systems Society
University of Pavia, Italy

Jan Van der Spiegel

Past-President,
IEEE Solid-State Circuits Society
University of Pennsylvania, USA

LONDON AND NEW YORK

Published 2018 by River Publishers

River Publishers

Alsbjergvej 10, 9260 Gistrup, Denmark

www.riverpublishers.com

Distributed exclusively by Routledge

4 Park Square, Milton Park, Abingdon, Oxon OX14 4RN

605 Third Avenue, New York, NY 10017, USA

Low Power Circuit Design Using Advanced CMOS Technology / by Milin Zhang, Zhihua Wang, Jan Van der Spiegel, Franco Maloberti.

Routledge is an imprint of the Taylor & Francis Group, an informa business

ISBN 978-87-7022-000-2 (print)

While every effort is made to provide dependable information, the publisher, authors, and editors cannot be held responsible for any errors or omissions.

Table of contents

Introduction

This book is a summary of lectures from the first Advanced CMOS Technology School (ACTS) summer 2017. The slides are selected from the handouts, while the text was edited according to the lecturers talk.

ACTS is a joint activity supported by the IEEE Circuit and System Society (CASS) and the IEEE Solid-State Circuits Society (SSCS). The goal of the school is to provide society members as well researchers and engineers from industry the opportunity to learn about new emerging areas from leading experts in the field. ACTS is an example of high-level continuous education for junior engineers, teachers in academe, and students. ACTS was the results of a successful collaboration between societies, the local chapter leaders, and industry leaders. This summer school was the brainchild of Dr. Zhihua Wang, with strong support from volunteers from both the IEE SCS and CASS. In addition, the local companies, Synopsys China and Beijing IC Park, provided financial support.

This first ACTS was held in the summer 2017 in Beijing. The lectures were given by academic researchers and industry experts, who presented each 6-hour long lectures on topics covering process technology, EDA skill, and circuit and layout design skills. The school was hosted and organized by the CASS Beijing Chapter, SSCS Beijing Chapter, and SSCS Tsinghua Student Chapter. The co-chairs of the first ACTS were Dr. Milin Zhang, Dr. Hanjun Jiang and Dr. Liyuan Liu. The first ACTS was a great success as illustrated by the many participants from all over China as well as by the publicity it has been received in various media outlets, including Xinhua News, one of the most popular news channels in China.

CHAPTER 01

Semiconductor Innovation – a Continuum of Opportunities

Rakesh Kumar

TCX Technology Connexions

Phenomenal growth in the use of mobile wireless, and ubiquitous sensor products has fueled a resurgence of excitement in the semiconductor industry. In spite of increasing investment costs and predictions of doom and gloom, these are exciting times in the semiconductor innovation pipeline for those that are willing to adapt and make adjustments.

This talk will present a discussion of the important technical and business challenges, and the myriad of opportunities. The focus will be on Innovation and Entrepreneurship opportunities. The semiconductor industry continues to serve as the foundation for continued development of new solutions in many applications spaces – IoT, Arificial Intelligence, brain-inspired chips, neuromorphic computing etc. These multi-disciplinary solutions offer new opportunities to the engineer that reaches out beyond their own specific expertise area. Leveraging the industry's use of collaborative research and development, and the Fabless Integrated Circuits model provides many opportunities for developing and scaling new ideas. Because the success rate of start-up companies remains relatively low, innovators need to pay attention to a broad spectrum of factors besides their technical idea. This course will provide guidelines for innovators to launch companies with increased probabilities of success, and will offer hope for entrepreneurs, researchers, and designers in fulfilling their dreams.

Some material and illustrations are from the author's book, *Fabless Semiconductor Implementation*, McGraw Hill, 2008

THE CHANGING WORLD

1

Smaller Number of Players for Leading Edge Nodes

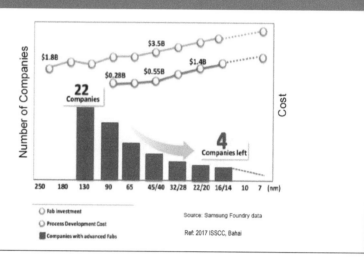

Some years ago, times were pretty bad for the semiconductor industry. It was a time for "doom and gloom". In 2003, there were 22 companies that were at 130nm node, but only 4 were left in 2013. Why? Cost of the wafer fab facilities, processing and design was going up significantly. Companies were shutting down because of much consolidation in the industry.

2

Theme of this Talk

However, today my perspective is that these are exciting times. You have to adapt to change, you have to innovate, you have to be entrepreneurial and you have to think through what your career is going to be, what you want to do in five years or ten years. Actually in my career I had no clue myself! However, I was constantly focusing on doing things that were exciting to me, and that I knew the industry would want. One of the principles I had was to always evaluate any new assignment and assess how that would look on my resume. If it was something that was not consistent with the direction that I wanted to plan for myself, I would try to get that adjusted.

3

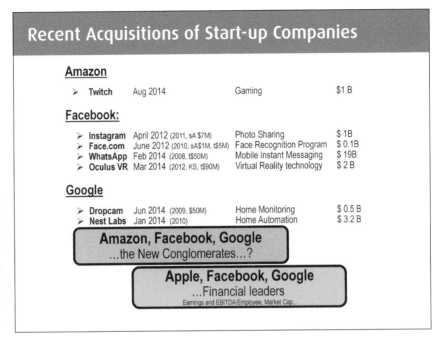

Recent Acquisitions of Start-up Companies

Amazon

➤ Twitch	Aug 2014	Gaming	$1 B

Facebook:

➤ **Instagram**	April 2012 (2011, sA $7M)	Photo Sharing	$ 1B
➤ **Face.com**	June 2012 (2010, sA$1M, t$5M)	Face Recognition Program	$ 0.1B
➤ **WhatsApp**	Feb 2014 (2008, t$50M)	Mobile Instant Messaging	$ 19B
➤ **Oculus VR**	Mar 2014 (2012, KS, t$90M)	Virtual Reality technology	$ 2 B

Google

➤ **Dropcam**	Jun 2014 (2009, $50M)	Home Monitoring	$ 0.5 B
➤ **Nest Labs**	Jan 2014 (2010)	Home Automation	$ 3.2 B

Amazon, Facebook, Google
...the New Conglomerates...?

Apple, Facebook, Google
...Financial leaders
Earnings and EBITDA/Employee, Market Cap...

As an entrepreneur you're always looking for ideas and looking for applications of what is a customer problem, if you can help solve the problem and you can create a product that actually solves that problem, you've got a really good chance to be successful. There are many examples of successful startups, and their acquisition.

Companies like Amazon, Facebook, Google were startups at the beginning, then they went public and have since been hugely successful. Five years ago, you would not even think about applying to Amazon, Google, Apple for a job, right? Today they are the leaders.

4

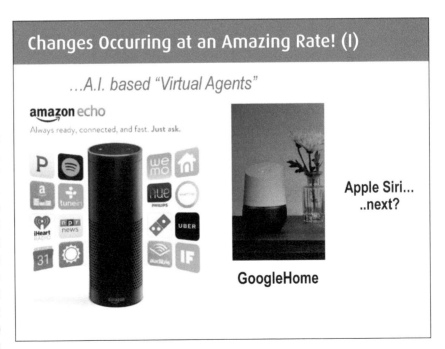

Changes Occurring at an Amazing Rate! (I)

...A.I. based "Virtual Agents"

amazon echo

Always ready, connected, and fast. Just ask.

Apple Siri...
..next?

GoogleHome

But you have to make a decision for yourself. You have to start somewhere. What I do warn my students is that while not everybody is going to be successful, the opportunities are very much there. You constantly have to look at where you can make a difference. Here are some products that these companies have introduced. In order to make these work, you got to have analog circuits for the human interfaces. And of course one has to leverage digital processors.. Successful development and shipment of a product requires the confluence of multiple disciplines.

As a student you have to be aware of the various disciplines and you really have to make the choice for yourself and decide whether you want to be the designer only.

5 Changes Occurring at an Amazing Rate! (II)

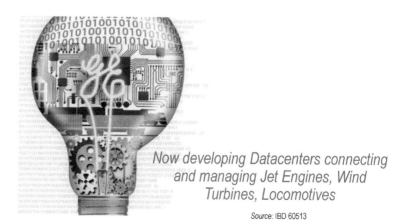

...GE transforming itself into "The Apple of the Industrial Internet"

Now developing Datacenters connecting and managing Jet Engines, Wind Turbines, Locomotives

Source: IBD 60513

Here is an example of another change that's happening in the industry. Take a company like General Electric, they are wanting to make a strategic change and become "the Apple of the industrial internet". They are making major investments in new product ideas for energy, jet engines wind turbines, locomotives. It turns out they don't actually sell engines anymore, they use a different business model and only lease the engines now. The world is changing. If the old model was GE built the engines and ship the engine to somebody, now they'll lease them out and every three years somebody buys a new engine. It also affects how the rest of the company operates.

6 Semiconductors Drive the Electronics Food Chain...

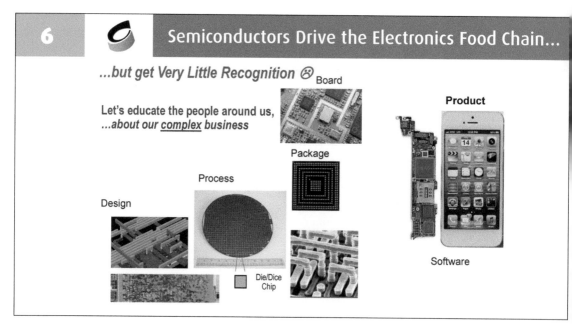

...but get Very Little Recognition ☹

Let's educate the people around us, ...about our _complex_ business

Board

Product

Package

Process

Design

Die/Dice Chip

Software

Today we all cannot live and survive it without the smartphone. There are over 5 billion plus worldwide. However, people using smartphones have no idea what's involved in building this device, or that you are involved in building this device? You, know, us engineers contribute the technology but are the absolute worst marketers! I always use this chart to show that smartphones require a combination of advanced software and hardware technologies. For example, the processors used in them have over a thousand pins.

INDUSTRY BACKGROUND AND CHALLENGES

7

The Electronics System Industry

~$ 30,000B	**SYSTEMS** Auto, Computing, Comms, Industrial, ...
~$ 1,500B	**ELECTRONICS**
~$ 300B Semiconductors	**SEMI**
~$ 80B Materials & Equipment	**M&E**

The semiconductor industry supports a very large electronics and systems industry. Systems could be automobiles computers, or communications related. Systems industry revenue is of the order of thirty trillion dollars annually. That is supported by electronics, which is about one point five trillion dollars. Systems can't do without electronics, but they make many times as much money. The semiconductor industry is more like 300 billion dollars, about five to one, right? For every dollar of the semiconductor industry ships into the electronics field, electronic manufacture makes five times as much money. And to continue the value chain here materials and the equipment is eighty billion dollars, is four to one.

8

Technology Scaling Basics

Now let us look at some basics of the very successful semiconductor industry. The basic switching element is a Transistor which has been systematically scaled to smaller dimensions per Moore's Law. This slide shows the layout of a basic transistor. By shrinking in both dimensions (X and Y) one is able to reduce chip area. By keeping the chip area the same, one can pack 2x the number of Transistors in the same area. And further increase the number of transistors by another 2x by increasing the chip size. This was the methodology commonly used in the industry down to the 28nm technology. This very simple concept was repeated over and over. Semiconductor Process people use design rules to represent limitations in the process area, as well as to guild the designer's layout design.

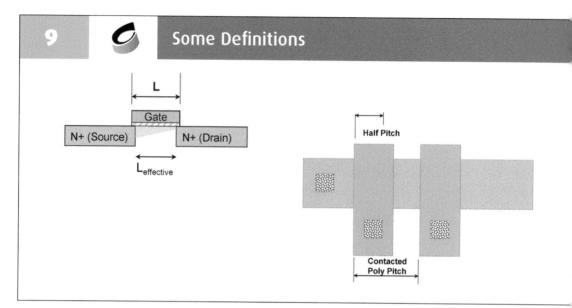

9 **Some Definitions**

This slide depicts the relationship between the layout view and the Transistor cross section. Just remember, there are equipment and other limitations to how much the physical dimensions can be scaled, and also how large a chip is feasible to make. Scaling laws also require careful adjustments to scaling the vertical dimension (Z). As the physical dimensions have been scaled down there have been many challenges that have been overcome. One example is that for advanced devices, the electrically effective Transistor gate length (Leff) is no longer the same as the physical gate length (L), and is actually less.

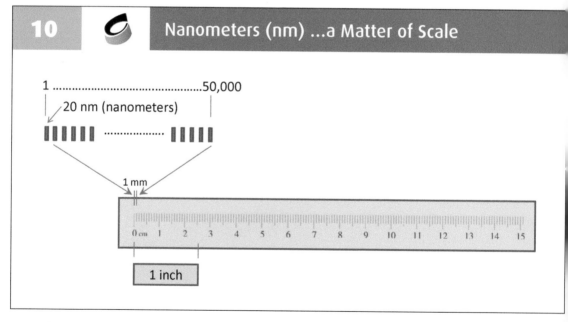

10 **Nanometers (nm) ...a Matter of Scale**

For advanced technologies the difference between the effective L and the visible L becomes quite significant. For maximizing the routing track on the chip what really matters is not just the length of gate poly but the poly pitch after adjacent contacts have been placed in there. Half that pitch, called "half pitch" then becomes a standard metric for the technology.

Let's look at an evolution of microprocessors. Historically, Intel introduced the first microprocessor, the 4004, consisting of 2300 transistors. It is a DIP (dual in line) package with 16 pins. Leading devices today have over 5 billion transistors. In Intel's core i7, there are over 1000 pins. There are complicated chips with quad-core, or four-cores, shared L3 cache, memory controller, IO and other functionality modules.

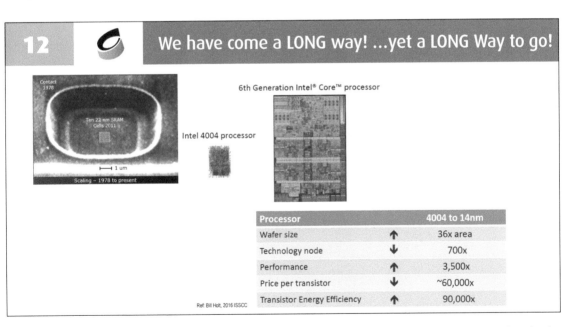

This is a comparison between 4004 and 14 nanometer devices, the Intel 6th generation processor. The wafer size increases by 36x. Technology node increases 700x. The performances are 3500 times. The price for single transistor is 60,000 times, that is the driving force while reusing it and the transistor energy efficiency is the other reason why they're using it is 90,000 times. Energy didn't used to be a big factor, initially it was all about performance. Now power dissipation is a very important consideration especially for mobile and IoT type applications. Energy consumption becomes a real strong consideration.

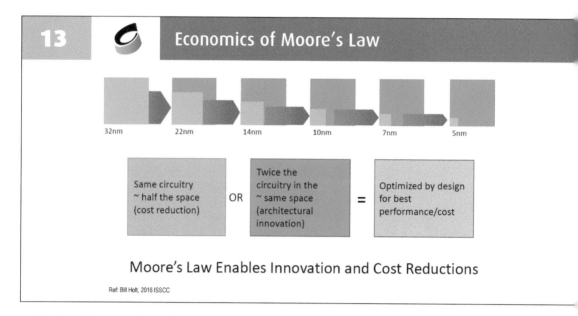

This chart illustrates the scaling principles outlined earlier. It illustrates the crux of Moor'e economics law. When 32nm chip, shrinking to the next generation technology, 22nm node, chip gets smaller, but designers actually raise chip size again back to the same to get to twice the number of transistors in the same area. Multiple is about 0.7x. By the way, this is all run by what's called Dennard's Scaling proposed in 1974, in a paper that talks about linear scaling, and that's been a big driver for Moore's law for over 30+ years.

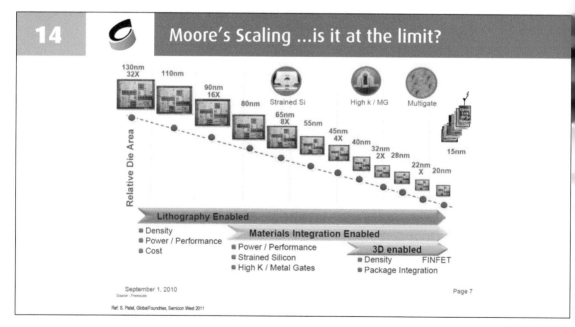

This chart again shows the relative die area with technology node shrinking from 130nm, to 20nm. For many years, the Moore's Law was really linear scaling about lithography and design rules. Around the 90nm node, changes had to be made to include new materials like stained silicon, and high K metal gates were introduced. And then at about 20 nm, people started to look at more type of technologies, i.e. 3-D, FinFET.

15

Voltage Scaling Evolution

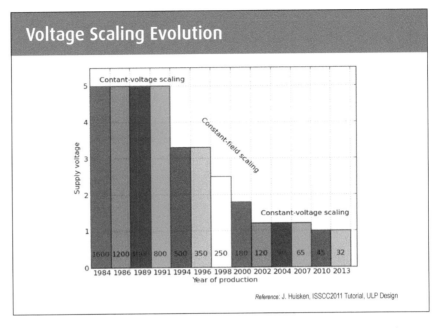

Reference: J. Huisken, ISSCC2011 Tutorial, ULP Design

Dennard's Scaling requires the reduction of the Power Supply voltage as the linear dimensions were reduced. Unfortunately the industry did not adhere to this as shown in this chart. This resulted in the increase of the electric field thereby causing increased stress. At the 500nm node, the voltage was reduced to 3.3 volts for the first time. Later, it went down to 2.5 volts and then 1.8 volts, then 1.1 volts, and now is <1 volt, maybe about 0.9 volts. Changing the power supply voltage is not easy because power sources constraints at the board level.

16

Previous Predictions of the End

The End of Scaling is Near?

"Optical lithography will reach its limits in the range of 0.75-0.50 microns"

"Minimum geometries will saturate in the range of 0.3 to 0.5 microns"

"X-ray lithography will be needed below 1 micron"

"Minimum gate oxide thickness is limited to ~2 nm"

"Copper interconnects will never work"

"Scaling will end in ~10 years"

Perceived barriers are meant to be

surmounted, circumvented or tunneled through

Source: M. Bohr, ISSCC 2009

As the industry progressed thru scaling devices there have been many technical hurdles and roadblocks, especially as the atomic dimensions are fast approaching. This has led to much discussion of the death of Moore's law, the end of scaling, and much doom and gloom. Why have people been talking about this Doom and Gloom? Here's a quote from Mark Bohr, the leading technologist at Intel, and he put on this paper at ISSCC, 7 years ago. Here's some of the reasons. Optical lithography reaches limit the 0.75 to 0.5 um. Remember that submicron can never be done. Minimum geometries will saturate in the range of a 0.3 to 0.5 um. X-ray lithography will be needed to below one micron, x-ray lithography still we're there. Minimum oxide, gate oxide thickness is limited to about 2 nanometers, copper interconnect will never work, scaling will end in about ten years.

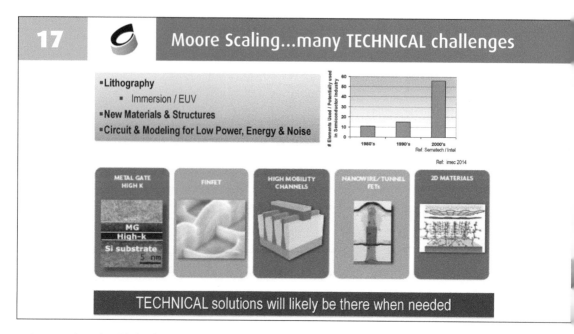

The wavelength of light that is used to pattern the features onto a silicon wafer is exactly 193nm and yet today we are actually printing 16nm! It's extremely challenging to print dimensions smaller than one micron. Yet people have found creative ways to do this. So, while technical solutions have been found, the fabrication cost has increased very rapidly.

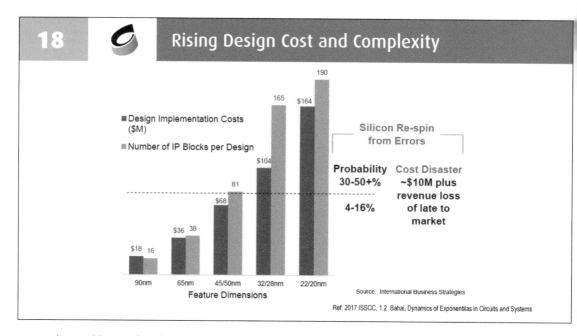

Another problem is that the cost of implementing large design has escalated very rapidly. Chips are complex, with millions/billions of transistors on it. Automated tools are needed for simulation to make sure that functionality and performance matches expectations. Automated tools are used to place and route the transistors, make the connections, and to verify the design timing, functionality and performance.

19

Economic Challenges are Threatening

Increased Cost of Capital, R&D, Design

EUV Lithography System

Costs associated with node progression have been rising significantly

| Fab cost $ Millions | Process development cost $ Millions | Chip design cost including fabless overhead costs* $ Millions |

6,700

>30%   1,850

1,300   150

Cost per Function must continue to REDUCE for continued Scaling

400
1,450   1,800   250   310   15   24   34

130nm 90nm 65nm 45nm 32nm 22nm   130nm 90nm 65nm 45nm 32nm 22nm   130nm 90nm 65nm 45nm 32nm 22nm

This chart shows how Fab facility, process development and design costs have increased. The cost of a leading edge wafer fab is around $10B. Process development costs go up every time a new element is added in the process flow. Historically, any given technology node starts up being pretty expensive, since it's not mature and the number of defects per square millimeter is high. Switching new designs over to the new technology becomes a trick. Whenever a new generation technology is ready for mass production, there's a huge amount of effort to determine when designers should switch to the next generation. The decision is primarily based on economics – to maximize revenue and margin. This is the cost effectiveness of the business.

20

Cost Reduction thru Technology Scaling

130 nm Technology

180 nm Technology   Source: Intel

Figure 23: Comparison of 180nm technology to 130nm technology

Wafer Cost of new technology wafer	~1.4x
Each side on new die	~0.7x
Area of new die	~0.49x
Die Cost on new technology	~0.7x

How does cost reduction thru technology scaling work? Here's an example for a 180 nm technology chip going to 130nm. The ratio is 0.7. The wafer cost actually goes up to 1.4x. Due to the scaling, each side of the die now goes from 1 to 0.7x, and the area of the new die is 0.49x, square this. Therefore, the wafer will have approximately 2x the number of possible die. This will result in a net die cost that will be 0.7x that in the old technology.

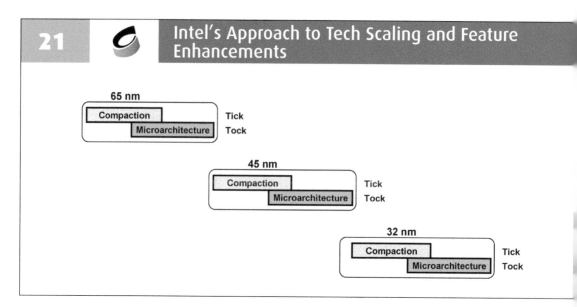

21 — **Intel's Approach to Tech Scaling and Feature Enhancements**

So Intel was actually very good at practising this approach and called it the Tick Tock methodology. The way this works is as follows. If they had an optimized design in any one technology ("Tick"), they would create an initial design in the new technology very quickly ("Tock"). Initially this was done via an "optical shrink" This would give them a first-pass design to check functionality and market presence. Then as the new technology was optimized for stability, yield and performance they would create an optimized design in the new technology The Tick and Tock were also called "Compaction" and "Microarchitecture".

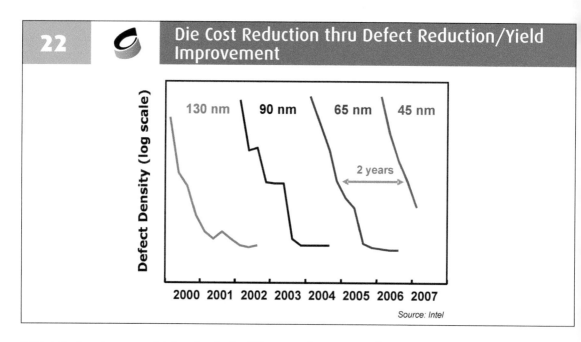

22 — **Die Cost Reduction thru Defect Reduction/Yield Improvement**

Source: Intel

This is the learning curve of defect density for different technologies. The whole idea was to minimize the time it took to go from high defect density to low defect density. It's all about the cost effective design shipments and maximizing profitability. It is a history of the maximum die size. Also, over the years the Silicon wafer size has increased from 50mm diameter to 300mm. The next step is 450mm.

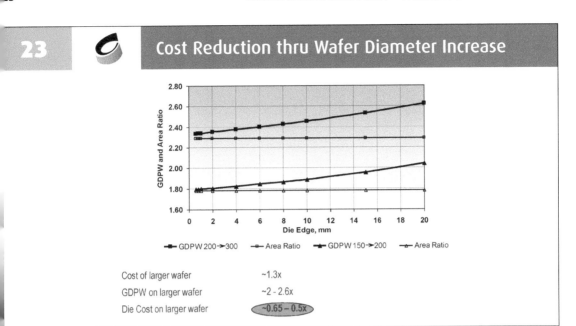

23   Cost Reduction thru Wafer Diameter Increase

Cost of larger wafer	~1.3x
GDPW on larger wafer	~2 - 2.6x
Die Cost on larger wafer	~0.65 – 0.5x

Why do we want to increase the wafer size? Here is a plot that shows the increase in maximum possible die on a wafer as as you increase the wafer size for any given chip size. It's all about cost effectiveness per die and profitability. It basically shows you that larger wafer cost 30% more to process. However, you can get anywhere from 2 to 2.6x the number of the die on it and therefore the die cost on the larger wafers is 50-30% better.

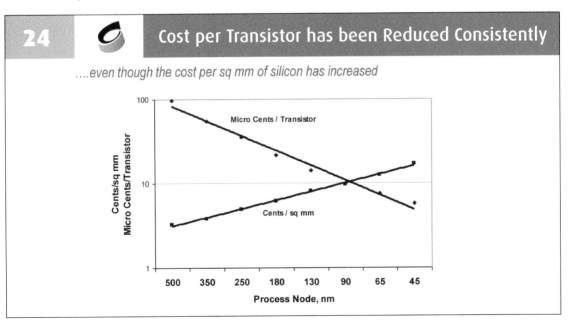

24   Cost per Transistor has been Reduced Consistently

....even though the cost per sq mm of silicon has increased

Overall this industry operates on reducing the cost per transistor even though the cost to build the silicon wafer and the cost /sq mm of silicon keeps going up. The cost/sq mm go up for new technologies and larger wafer diameter. But because there are more transistors on the same chip, the cents per transistors keep going down.

25 — Cost per Transistor Reduction by Wafer Size and Node

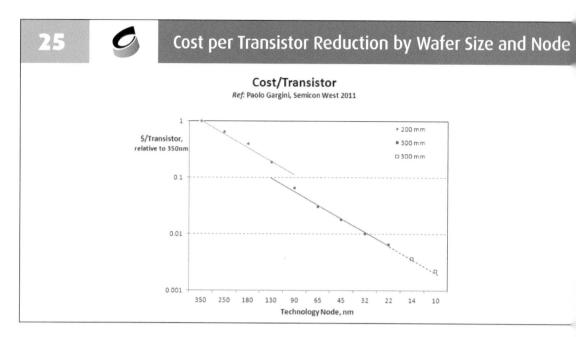

Many companies track the trend of Transistor cost/transistor. It is interesting to note Intel's prediction in 2011 that the industry can continue to progress down the curve at least thru the 10nm node. As mentioned earlier there will be changes in device structures, materials and device physics together with innovative circuit techniques to reduce power dissipation and energy consumption as the industry marches forward.

26 — "More than Moore" Activities

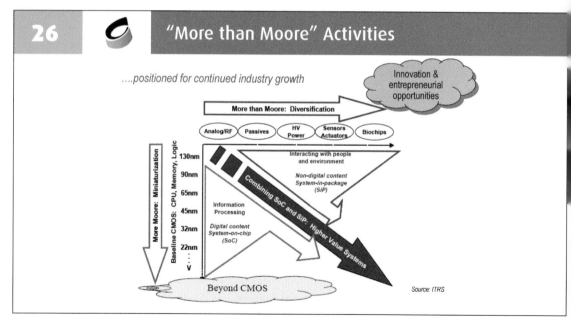

Many years ago the industry recognized limits of Moore's Law scaling. Therefore they have been looking at alternatives, called "More than Moore". As the industry was shrinking the base technologies, there was also an efforts to look beyond CMOS. There has also been much new innovation especially in the area of packaging. Another primary focus here was adding analog functions on the chip.

MARKET DRIVERS

27

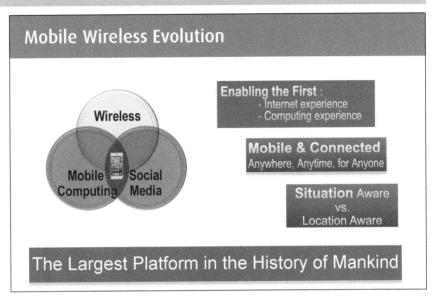

N ow let's discuss Market drivers for this industry. One of the big drivers in the last 5 to 10 years has been the smartphone. It enables more capability, a computer capability on a phone. Starting in ~2005, as the smartphone evolved to integrate wireless capabilities with mobile computing and social media, it has become the largest platform in the history of mankind. There are over 5 billion people in the world, and over 5 billion phones.

28 The "Internet of Things"

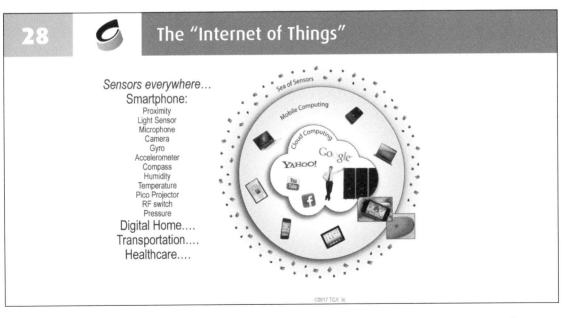

A nother big drive is the IoT (Internet of Things) platform. It is anticipated that the hub of mobile wireless devices connected to data centers will have a sea of sensors. The seas of sensors will connect almost every conceivable "thing" to the internet!

This chart is from a recent ISSCC keynote speech. This chart illustrates the types of things that can be connected - the market is huge. One word of caution, though for the semiconductor industry. While the unit volume of chips can go up, the total number of silicon wafer starts may not be that significant. This is because typical sensors and IoT chips are relatively small in size, and there could be large numbers of potential good dice on a single wafer.

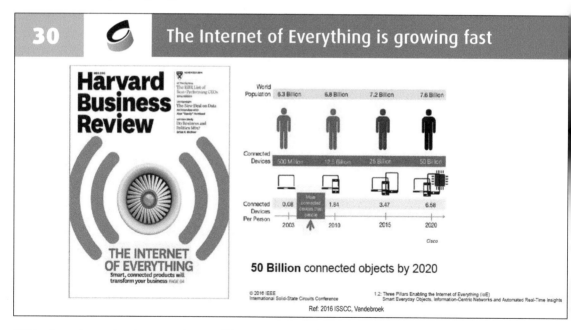

This chart shows that by 2020 there will be over 6 devices per person connected to the internet, for a total of over 50 billion devices!. This is the kind of information that gets me excited about opportunities in this industry. You can figure out where you want to play in this space. It's wide open. Here shows you the example of 50 billion connected devices by 2020.

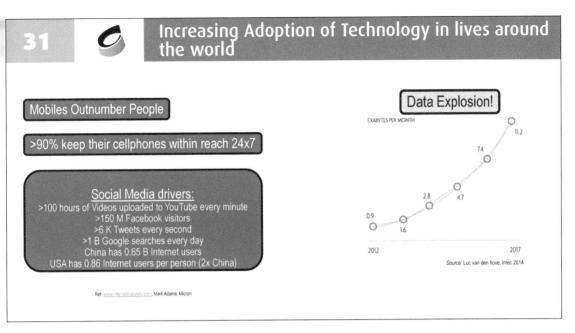

It is exciting to know that the mobile devices now outnumber people. Over 90% of people keep their cellphones within reach 24x7. Typically we need a new, smarter cellphone every two years. This kind of impetus keeps the industry growing. A long as we, as users, keep pushing for improved capability the demand will be there and hence the semiconductor volume. These devices must have increased compute capability at high speeds, manage increasing amounts of data, and yet dissipate less power and energy. There are significant challenges that will require research and innovation.

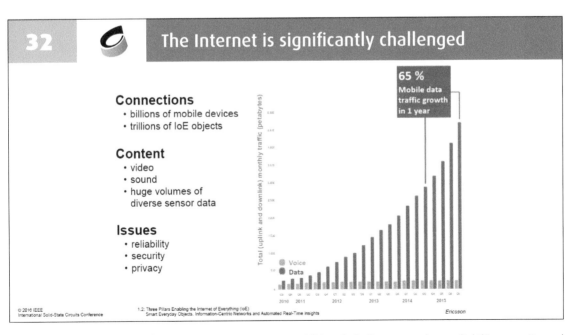

The internet of things phenomena is causing a tremendous explosion of Data that is generated and has to be managed. It also leads to many additional challenges such as reliability, security and privacy.

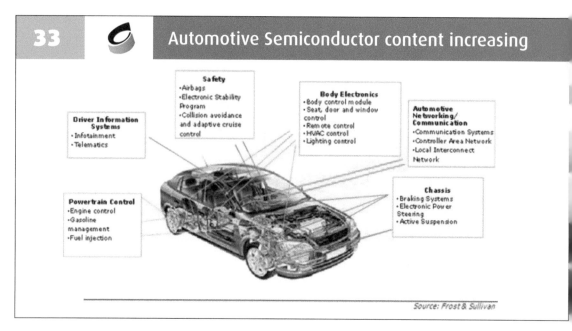

33   Automotive Semiconductor content increasing

Safety
- Airbags
- Electronic Stability Program
- Collision avoidance and adaptive cruise control

Body Electronics
- Body control module
- Seat, door and window control
- Remote control
- HVAC control
- Lighting control

Automotive Networking/Communication
- Communication Systems
- Controller Area Network
- Local Interconnect Network

Driver Information Systems
- Infotainment
- Telematics

Powertrain Control
- Engine control
- Gasoline management
- Fuel injection

Chassis
- Braking Systems
- Electronic Power Steering
- Active Suspension

Source: Frost & Sullivan

Automobiles have been another market driver for semiconductors. In the average vehicle, there is approximately $350 worth of electronics. For a hybrid vehicle, is about $600 dollars of semiconductor electronics. If you look at a luxury vehicle, Mercedes is 5,000 dollars worth of semiconductors!

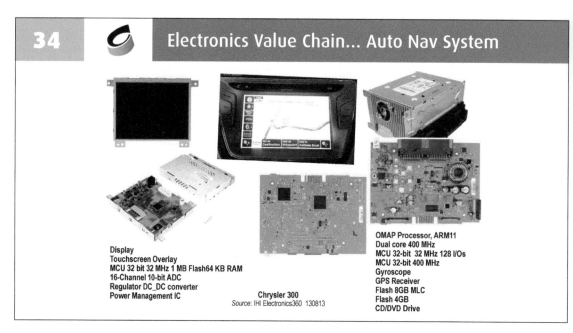

34   Electronics Value Chain... Auto Nav System

Display
Touchscreen Overlay
MCU 32 bit 32 MHz 1 MB Flash64 KB RAM
16-Channel 10-bit ADC
Regulator DC_DC converter
Power Management IC

Chrysler 300
Source: IHI Electronics360 130813

OMAP Processor, ARM11
Dual core 400 MHz
MCU 32-bit 32 MHz 128 I/Os
MCU 32-bit 400 MHz
Gyroscope
GPS Receiver
Flash 8GB MLC
Flash 4GB
CD/DVD Drive

Another example in the automobile is the GPS system. Behind the screen there are two sub-systems – the display and the power supply. Each has an electronics printed circuit board with many semiconductor components. Who makes the chips? - could be Texas Instruments(TI), could be Linear Tech for the power management IC. Now, if you develop a better power management chip and you find the customer really interested in yours and no one else can make it - that's an opportunity for you to form a startup Company!

35

Multi-Tiers of Value Chain and Ecosystems

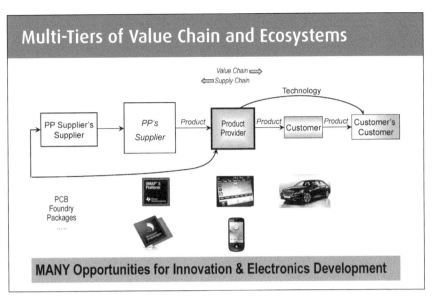

MANY Opportunities for Innovation & Electronics Development

Let's talk about a typical Product Value Chain. This chart shows the value chain for a GPS navigation system, and a Smartphone. The GPS supplier has a supplier. In this case, it is the OMAP Texas Instruments chip provider. On the other side, there's the customer. Mercedes might be the customer. You might be the customer's customer. Using this as an example, if you end up starting your own business and you're providing a product, you will need to identify your customer, and clearly define what you supply to the customer. And, relate your product to the customer's problem and identify how yours is the only solution to the customer's problem. This illustrates the importance of the value chain analysis. If you look at a smartphone supplied by HTC or Huawei, they are the phone provider. They buy the chips from a supplier such as Qualcomm. Qualcomm has their own suppliers. While this chart shows a 2 layer deep value chain, one could have value chains that are multiple layers deep.

36

Markets and Opportunities are exploding!

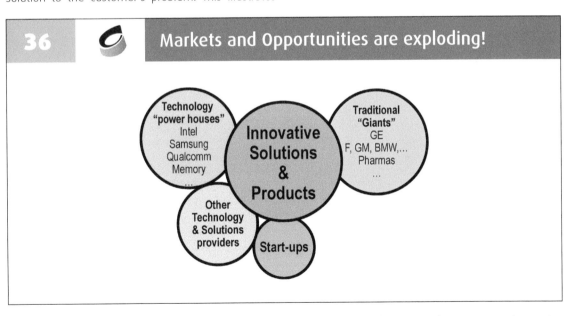

This chart shows that innovative solutions and products can come from many sources. Technology "power-houses" and Technology Giants develop automobiles, pharmaceuticals, systems and electronics products. Even if their main street business is not electronics, they may use electronics. Start-ups play an increasingly important role in this eco-system.

OPPORTUNITIES

37

Many Smartphone Features

We will now discuss opportunities in the semiconductor space. As a designer or a semiconductor technologist, if you look at a device like a phone, what are the things that really matter? Of primary importance these days is the user experience. You take a device like a smartphone, traditionally, everybody wants to put more features on it, we as users demand and they are competition in each of these areas, each of these providers wants to build more functionality.

>2 GHz processing

>32 GB FLASH Memory

Multi-core CPU

>1 GB DRAM

Multimedia A/V Gaming

GPS Navigation

Faster GPU

Integrated Wi-Fi

Source: imec

38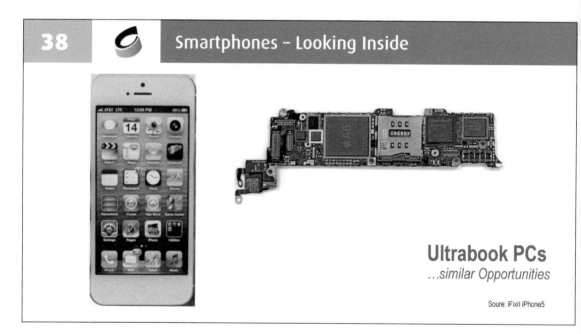

Smartphones – Looking Inside

Ultrabook PCs
...similar Opportunities

Soure: iFixit iPhone5

So, what are some of the opportunities? If you look inside an iPhone smartphone, there is a motherboard that is loaded with components, the CPU A10, a Qualcomm chip set, a Broadcom transceiver chip, a TI chip and a bunch of other support chips. Laptop PC's such as the Ultrabook have the same kinds of configuration.

39

Successful development and implementation of a smartphone or a Laptop PC requires integration of multiple disciplines is extremely important. Various components have to work together - software has to work with the display, the Antenna, the Touchscreen in the processor, etc. Co-design is extremely important. Silicon folks have to be talking to the packaging guys, and the test people, and the designers, and bringing the most leverage together. That's what made me motivated to work on product integration and cooperative activities. Whenever starting a new program, we would bring all the people together, all the different disciplines, plan out what your plan is going to be and make sure everybody knew they were going to do their piece, in order to get the device out the door.

MANY Levels of Innovation

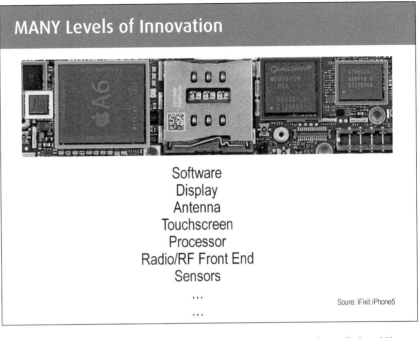

Software
Display
Antenna
Touchscreen
Processor
Radio/RF Front End
Sensors
...
...

Soure: iFixit iPhone5

40

The need for co-design was highlighted a couple of weeks ago when I went to imec and they presented this chart., They showed that while Moore's law will provide an increased number of transistors in the advanced technologies such as the Finfet, one really needs to have DTCO, Design Technology Co-Optimization. This means that the process and design people have to work together, not only to make sure what features are working together, but how you set up the design rules, so that you get the most number of routing tracks for instance to get the maximum packing density on the chip.

Leading Technology Advances from IMEC... July 2017

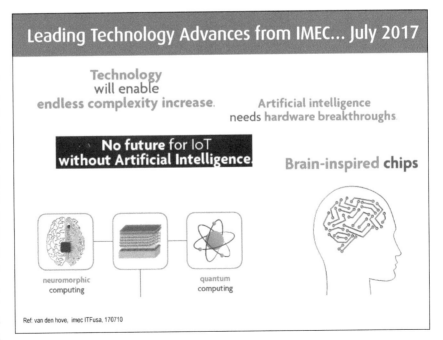

41 — Product Innovation Through Early Co-Design

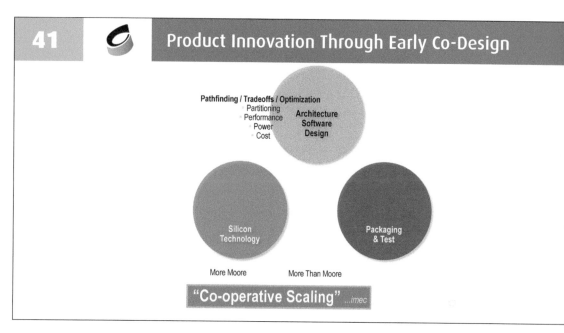

This is an example where by changing slight dimensions in the design rules, you can get much better packing density, that would be the co-optimization. System Technology Co-Optimization(STCO) could be similar examples where you can depend on the constraints, the system designer can get much better either access to lower leakage and improved models to predict the performance, leakage etc.

42 — HTC Desire Smartphone uses Qualcomm's MSM 8255

Here is an example again of co-optimization from imec. They created this sensor chip, that gets implanted into the prosthetic arm replacing an amputated arm. The sensor connects to the brain and helps the gentleman drive his fingers based on what his brain is thinking. That takes a lot of cooperative design

New implantable chip to hook up bionic arm to brain

Creating a more embodied experience for patients with artificial limbs

Ref: imec magazine 1707

between biomedical people, electronics engineers, packaging engineers, chip designers, all working together to get this implant to work. That is phenomenal. Reliability is an important consideration, packaging is an important consideration, circuit design and the system level, tremendous opportunity.

43

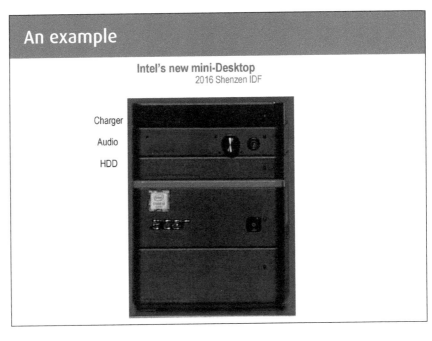

An example

Intel's new mini-Desktop
2016 Shenzen IDF

Charger

Audio

HDD

N ow Intel is investing to design a compact desktop. You will have huge competition if you want to launch a startup to develop a compact desktop, therefore this is not a good idea. You could go work for Intel, and work in one or more of these departments and do the circuit design that goes into the audio chips, as an example. If you have the passion to be doing a startup in 5-10 years, get the experience, learn all the things you can learn, while you are working at Intel and then go to your start up when the time is right. The big companies have a lot of resources.

44

Applying SoC technologies in "Adjacent" Markets

Automotive

Networking

Mobile compute

SoC

Wearables

Smart homes

Smart cities

Robotics and drones

Ref: Qualcomm Analyst Day 160211

Here's an example from Qualcomm's analyst day presentation from last year. , Qualcomm is known for their SoCs, right, System-on-Chips, that go into the phone. Looking forward they are also looking at adjacent areas for expanding their market opportunities. They are asking what can you do with this kind of a chip. In fact, if you think about it, Apple iWatch for instance, that connects to the phone, but it's not a standalone device. It needs the iPhone to be able to receive messages. The reason I brought it up is Qualcomm is looking to expand their market reach, not just into smart phones, smart phones are the base, they want to get into wearables, smart phones, smart cities, robotics, mobile computing, networking, automotive, and the whole space is getting very fuzzy. Intel is getting into mobile, Qualcomm is getting into computing, it's all kind of getting fuzzy together, but all that creates opportunities for the big companies especially as they create the demand. As they expand into new application areas they may need novel solutions which could increase opportunities for startups, and that's what you have to keep your eyes open for.

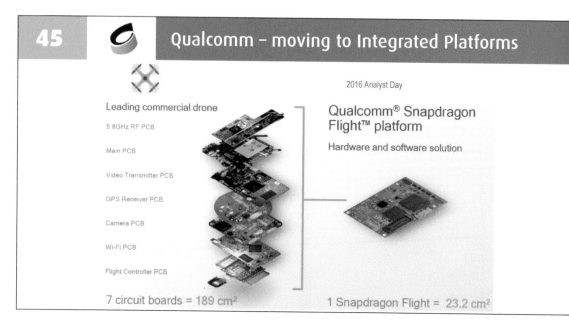

45

Qualcomm – moving to Integrated Platforms

2016 Analyst Day

Leading commercial drone

5.8GHz RF PCB

Main PCB

Video Transmitter PCB

GPS Receiver PCB

Camera PCB

Wi-Fi PCB

Flight Controller PCB

Qualcomm® Snapdragon Flight™ platform

Hardware and software solution

7 circuit boards = 189 cm²

1 Snapdragon Flight = 23.2 cm²

Here's an example how you could use the Snapdragon and a drone, so what they are saying is that, if somebody were to make a Drone, they have to have all these different PC boards, and the total area of the boards is about 190 square centimeters. However, if you use the Snapdragon reference board, it's only 23 sq cm. This is a significant form factor (size) reduction. Is that important for a drone? Absolutely, because that could determine how far the drone can go on its own. There are a lot of innovation opportunities related to Drones. If you have a neat way to do something that nobody else can, maybe that's a startup opportunity, right.

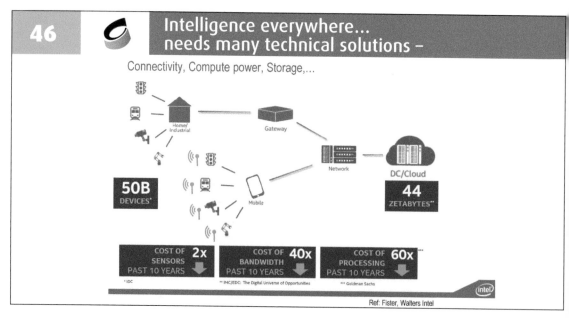

46

Intelligence everywhere... needs many technical solutions –

Connectivity, Compute power, Storage,...

Home/Industrial

Gateway

Network

DC/Cloud

50B DEVICES*

Mobile

44 ZETABYTES**

| COST OF SENSORS PAST 10 YEARS | **2x** | COST OF BANDWIDTH PAST 10 YEARS | **40x** | COST OF PROCESSING PAST 10 YEARS | **60x** *** |

* IDC   ** IMC/EDC: The Digital Universe of Opportunities   *** Goldman Sachs

(intel)

Ref: Fister, Walters Intel

Now this whole area of Internet of Things(IoT) and 5G is also very exciting. Increasing demand for semiconductors, but also creates all kinds of opportunities.

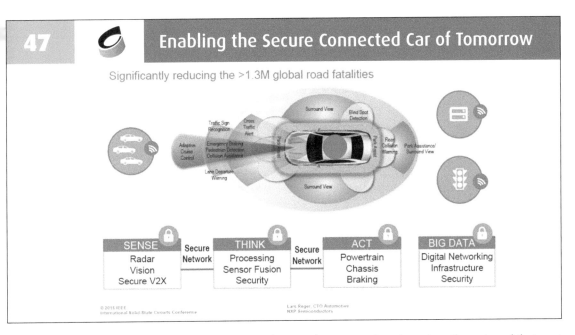

For example, security becomes a big concern with 5G. In the automobiles, all kinds of information that you got to have for the brake detection. It is easy to buy a sensor and put sensors on the side of the car, but integrating those into the automobile is not easy. It requires lots of data transmission, security, and sensitive information to be processed.

Today, lots of people have wearables devices, but they are really clumsy.

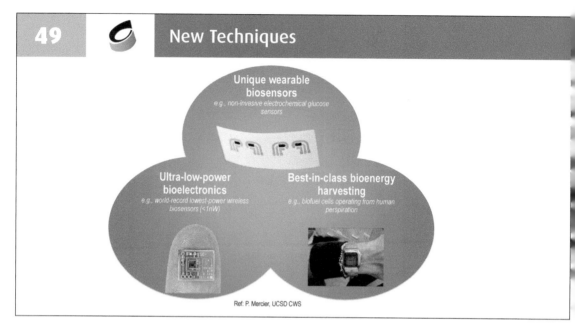

49 **New Techniques**

Unique wearable
biosensors
e.g., non-invasive electrochemical glucose
sensors

Ultra-low-power
bioelectronics
e.g., world-record lowest-power wireless
biosensors (<1nW)

Best-in-class bioenergy
harvesting
e.g., biofuel cells operating from human
perspiration

Ref: P. Mercier, UCSD CWS

One opportunity is to develop new techniques of doing biosensors, e.g., non-invasive electrochemical glucose sensors. It might require new technology, new packaging technology, new chemical technology, and new detection technology, all kinds of interesting stuff that could be conceived. Now whether there is a market or not again, that is determined by the potential customer.

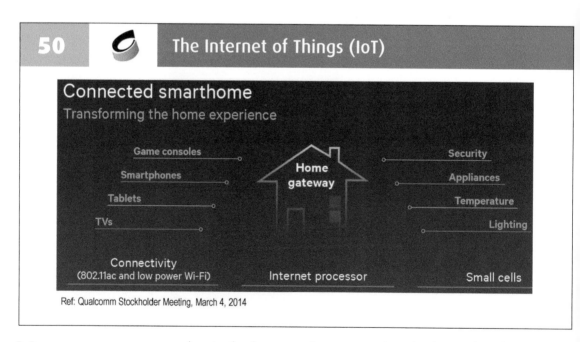

50 **The Internet of Things (IoT)**

Connected smarthome
Transforming the home experience

Game consoles

Smartphones

Tablets

TVs

Home
gateway

Security

Appliances

Temperature

Lighting

Connectivity
(802.11ac and low power Wi-Fi)

Internet processor

Small cells

Ref: Qualcomm Stockholder Meeting, March 4, 2014

Here are some more examples, in the home, there's all kinds of gateway, security, appliances. One of the start-ups in my class were for developing a camera that goes in the doorway that detects whoever come into the door and sends a message to your phone when there is somebody at the door. When you are not home, it may record that image and let you know that there's somebody at the door.

51 ## The Skyworks View...

Ref: Skyworks, Investor Presentation 2014

Here's an example for Skyworks as a company in California, there're the drawing pictures of what the home is going to be like, and how the appliances are going to link together. Now if you come up with a brilliant idea that plays into the home market, do your research and figure out whether it's worth doing a start-up.

52 ## Expand Focus on holistic SOLUTIONS

...not just the Technology and point solutions!

5G Technical Benefits:
➢ Latency
➢ Capacity
➢ Speed
➢ ...

➢ Does Customer Care?

Some Challenges:
➢ Security, Privacy, Compliance
➢ Interoperability and Standards
➢ Legal
➢ Reliability
➢ ...

Now 5G from a technical perspective, one gets lower latency, more capacity, more speed. However, there's all kinds of implications that are not technical and I think in the bigger scheme of things all these problems have to be addressed. In the end one has to explore if the customer really cares.

53 Technology Transitions driven by Power & Innovations

This is a chart from Bill Holt, that talks about how the industry has migrated feature sizes from the '60s down to now. The industry as overcome many hurdles over the years. I believe that CMOS is here to stay as the mainstream technology. I think that even if some new technology comes along, CMOS will still be the base technology in the future. I don't think CMOS is not going away anywhere, anytime soon. It's my opinion. There are still many opportunities for technology innovation.

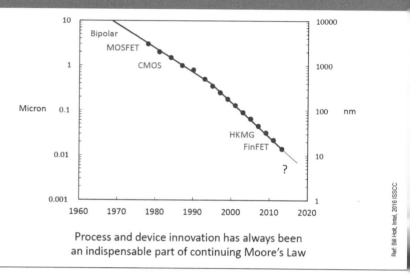

Process and device innovation has always been an indispensable part of continuing Moore's Law

Ref: Bill Holt, Intel, 2016 ISSCC

is the industry has a history of introducing new technologies such as copper, lo-k dielectrics, Finfets etc. from process technology perspective.

54 A history of Design & Technology Innovations

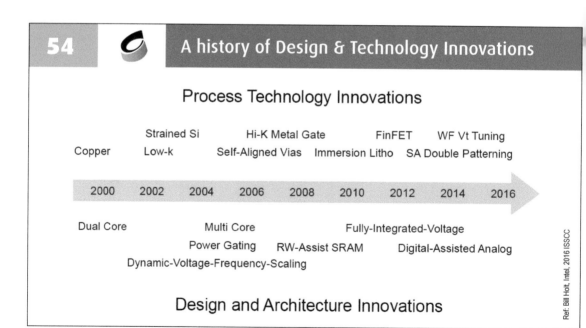

Ref: Bill Holt, Intel, 2016 ISSCC

This chart shows the history of Process and Design innovations. The upper line is Process Technology Innovations and the lower line is Design and Architecture Innovations. To reduce the capacitance and leakage, the Process comes from copper to Hi-K Metal Gates and now is Finfet. To reduce the power and speed, the Design and Architecture comes from dual core to multicore, Power Gating and DVFS and so on.

55 Beyond CMOS Computing

Hierarchical Level	CMOS	Beyond CMOS
Materials	Silicon	III-V , Correlated oxides, High-Z metals
Device	MOSFET	Tunneling-FET MESO (Magneto-Electric / Spin Orbit Torque)
Interconnect	Electronic	Electronic
Circuits	CMOS	Electronic, Spintronic
Architecture	von Neumann	von Neumann, Non- von Neumann
Memory	SRAM/DRAM	Electronic, Spintronic

- Beyond CMOS devices are <u>different</u>

- Beyond CMOS provides opportunities for active <u>research</u>

- Beyond CMOS will not replace CMOS, it will <u>augment</u> CMOS

Ref: Bill Holt, Intel, 2016 ISSCC

Beyond CMOS means beyond Silicon. There are few new kinds of structures, like tunneling-FETs, Magneto-Electric, Spintronic. They are in lab research now and can't replace CMOS. Maybe spintronic have the chance.

56 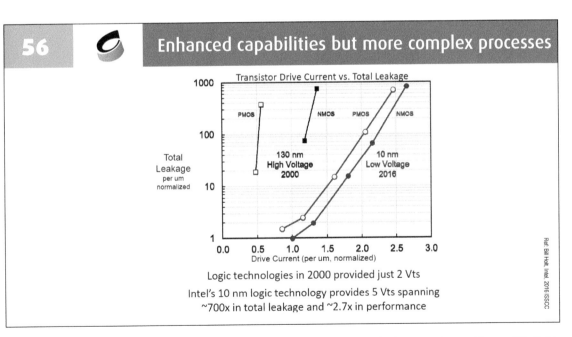 Enhanced capabilities but more complex processes

Logic technologies in 2000 provided just 2 Vts

Intel's 10 nm logic technology provides 5 Vts spanning ~700x in total leakage and ~2.7x in performance

Ref: Bill Holt, Intel, 2016 ISSCC

CMOS development is unfriendly for analog design because of the matching and leakage. This figure shows the leakage difference between 130nm and 10nm. You can find more about this on Bill Holt's research published in 2017.

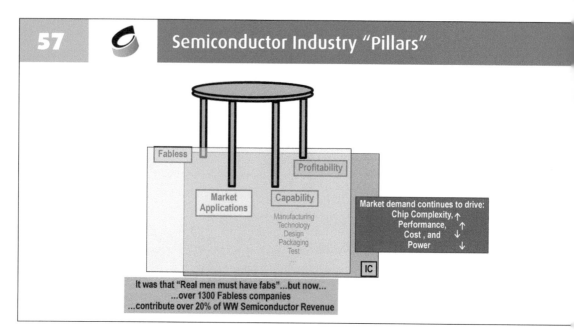

Now let's discuss the Fabless Semiconductor opportunities. Fabless is a good opportunity in the Industry. Now fabless contributes over 20% of worldwide Semiconductor Revenue.

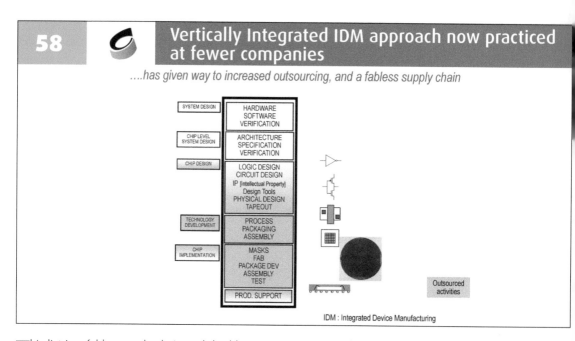

This list is a fables supply chain and the blue ones represents what the fabless companies might outsource and then they would keep ownership of the top part.

59 — Electronics Value Chain... Smartphones

BoM (Bill of Materials)...iPhone5 S

Table 1: Preliminary Teardown Bill of Materials and Manufacturing Cost Estimate for the Apple iPhone 5s (Cost in US Dollars)

Components / Hardware Elements	Details	16GB	32GB	64GB
	Pricing without contract	$649.00	$749.00	$849.00
Implied Margin		69%	72%	74%
Total BOM Cost		$190.70	$200.10	$210.30
Manufacturing Cost		$8.00	$8.00	$8.00
BOM + Manufacturing		$198.70	$208.10	$218.30
Major Cost Drivers				
Memory				
NAND Flash		$9.40	$18.80	$29.00
DRAM	1GB LPDDR3	$11.00	$11.00	$11.00
Display & Touch Sreen	4" Retina Display w/ Touch	$41.00	$41.00	$41.00
Processor	64-Bit A7 Processor + M7 Co-Processor	$19.00	$19.00	$19.00
Camera(s)	8MP (1.5-micron) + 1.2MP	$13.00	$13.00	$13.00
Wireless Section - BB/RF/PA	Qualcomm MDM9615M+WTR1605L+Front End	$32.00	$32.00	$32.00
User interface & Sensors	Includes fingerprint sensor assembly	$15.00	$15.00	$15.00
WLAN / BT / FM / GPS	Murata Dual-Band Wireless-N Module	$4.20	$4.20	$4.20
Power Management	Dialog + Qualcomm	$7.50	$7.50	$7.50
Battery	3.8V~1560mAh	$3.60	$3.60	$3.60
Mechanical / Electro-Mechanical		$28.00	$28.00	$28.00
Box Contents		$7.00	$7.00	$7.00

Source: IHS, September 2013

This is the BOM of iPhone5S. You can compare the red ones, and you will find something interesting.

60 — HTC Incredible 44%
IC Suppliers don't need their own Fabs

...Fabless IC Content Increasing

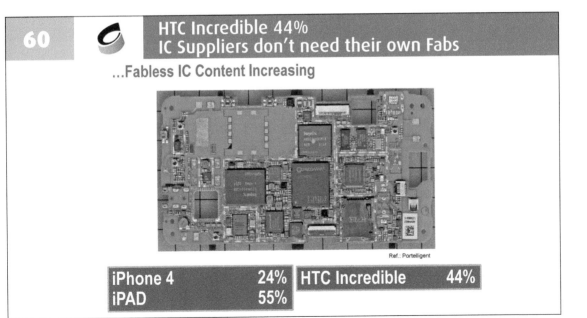

Ref.: Portelligent

iPhone 4	24%	HTC Incredible	44%
iPAD	55%		

This is data from a few years ago, but what was interesting is to look at the phones, that for an iPhone 4 20% of the components were made by fabless companies, or an iPad little over half, and the HTC was 44%.

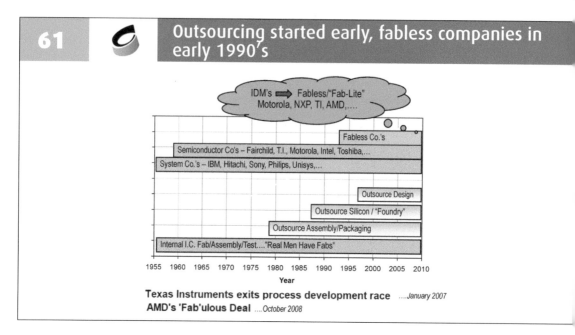

DM's transfer to Fabless companies in early 1990's.

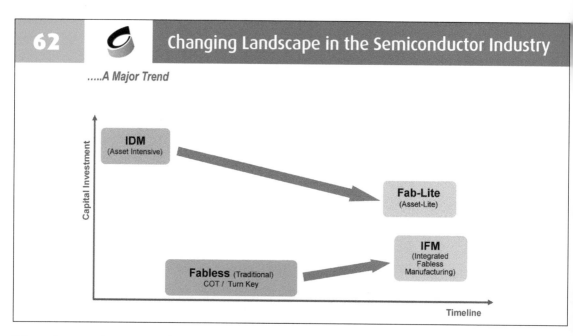

From the Capital Investment list, you will find why fabless. IDM needs the highest capital investment than others. And fabless is the lowest one in capital investment. So maybe IDM will be Fab-Lite and Fabless will be IFM in future.

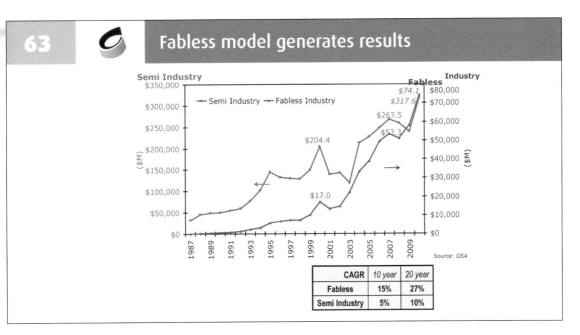

63 **Fabless model generates results**

This figure shows that the revenue in semiconductor. The red one is Semi Industry and the blue one is Fabless Industry. The fabless has more power in growth.

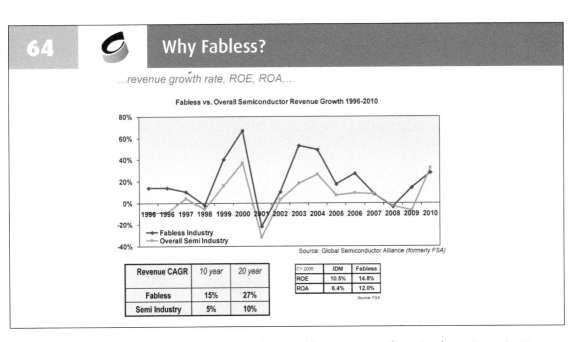

64 **Why Fabless?**

…revenue growth rate, ROE, ROA, …

We look at CAGR about Fabless and Semi Industry. Fabless is 15% and Semi Industry is 5% in 10 years. Fabless is 27% and Semi Industry is 10% in 20 years. So Fabless is consistently better financial results.

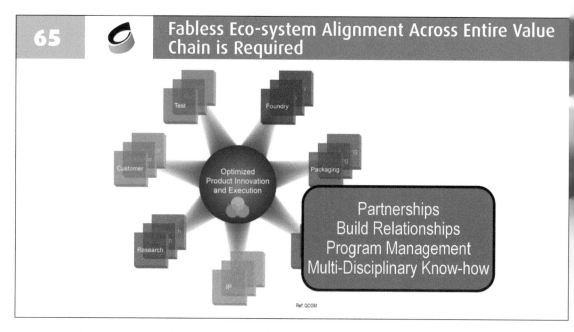

DM owns all and fabless needs an Eco-system alignment across entire value chain.

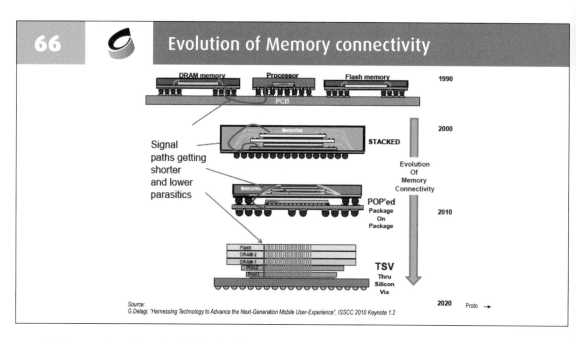

With the speedup of CPU's clk, the bottleneck is DRAM to Processor. So the evolution of memory connectivity is on the way. In 1990 we place all of them in pcb. In 2000, we stacked all of them and connect them in wire bonding. In 2010, we pop'd package on package. Signal paths getting shorter and parasitics getting lower. In 2020, we will use TSV to solve the problem.

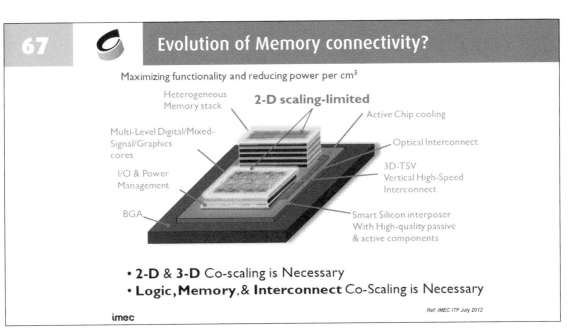

67   **Evolution of Memory connectivity?**

Maximizing functionality and reducing power per cm³

Heterogeneous Memory stack

2-D scaling-limited

Multi-Level Digital/Mixed-Signal/Graphics cores

Active Chip cooling

Optical Interconnect

I/O & Power Management

3D-TSV Vertical High-Speed Interconnect

BGA

Smart Silicon interposer With High-quality passive & active components

- **2-D & 3-D** Co-scaling is Necessary
- **Logic, Memory, & Interconnect** Co-Scaling is Necessary

Ref: IMEC ITF July 2012

imec

This is the super chip of future imaged by IMEC. High complexity means high power. So to achieve this goal, first have to reduce the power, or we need active chip cooling. Secondly is 3D packing should be co scaling with 2D. Thirdly high speed and low power data interconnect tech like optical interconnect will be taken.

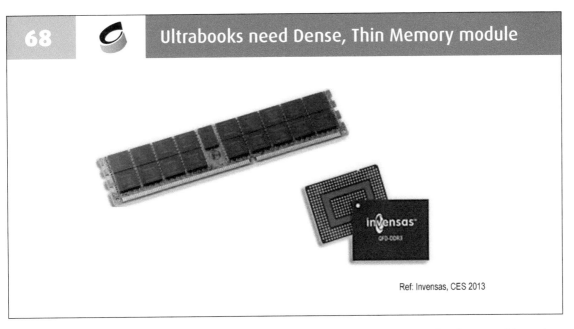

68   **Ultrabooks need Dense, Thin Memory module**

Ref: Invensas, CES 2013

Ultrabooks need dense and thin memory module. Invensas is a startup company but focus on this market.

69   **Micron's Hyper Memory Cube**

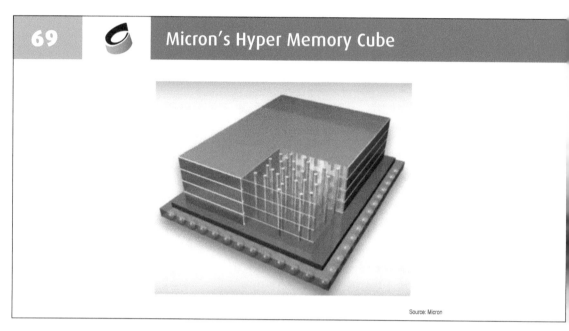

Source: Micron

Micron created what's called a Hyper Memory Cube, here is an example where you might have a stack of memories with via going all the way up and down making that connection and that's pretty impressive.

70   **Innovative Packaging Examples for Reducing Form Factor**

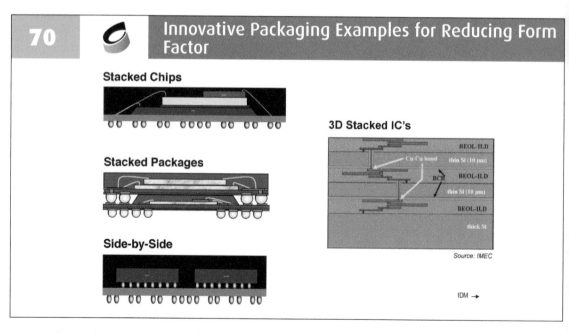

Now chip package is 2D or 2.5D. They are side-by-side, stacked packages and stacked chips. IMEC give an innovative packaging examples for reducing form factor. They connect each other using TSV's (though silicon via).

71 Next Generation Mobile User experience

...multiple connections to the applications processor

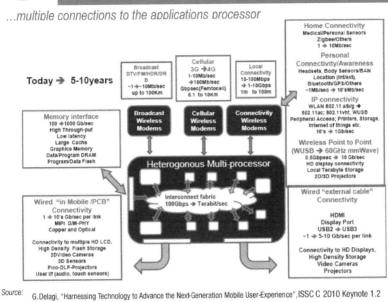

Source: G. Delagi, "Harnessing Technology to Advance the Next-Generation Mobile User-Experience", ISSC C 2010 Keynote 1.2

72 The Age of Entrepreneurship

➢ High growth Startup companies are a major source of job growth in the US

➢ Many Innovators

➢ Many new Opportunities for Entrepreneurship
  ➢ CloudComputing, Mobile Apps, Internet of Things,...

➢ "Easy" to start a new company!   After Ash Maurya, "Running Lean"

201 MILLION PEOPLE WERE UNEMPLOYED IN 2014

74 MILLION YOUTH WERE UNEMPLOYED IN 2014

Source: ILO 2014

49% OF MILLENNIALS WANT TO WORK AT A SUSTAINABLE COMPANY

60% OF GENERATION ZERS WANT JOBS THAT HAVE A SOCIAL IMPACT

72% OF GENERATION ZERS WANT TO START THEIR OWN BUSINESS

Source: Nielsen (2014), Sparks & Honey (2014)

About the your future career, entrepreneurship is an opportunity. From these numbers you will find that many in the younger generation want to pursue new opportunities in entrepreneurship.

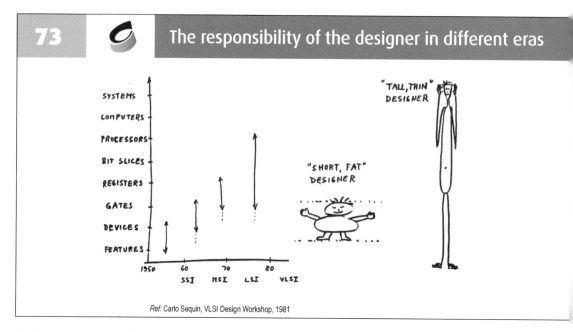

73   The responsibility of the designer in different eras

Ref: Carlo Sequin, VLSI Design Workshop, 1981

This figure shows that the designer can choose to be as short fat designers, focused in a small area. Or they can be tall thin designers, familiar with a breadth of disciplines. This chart is from 1981 and shows a very similar scenario of choices, then and now.

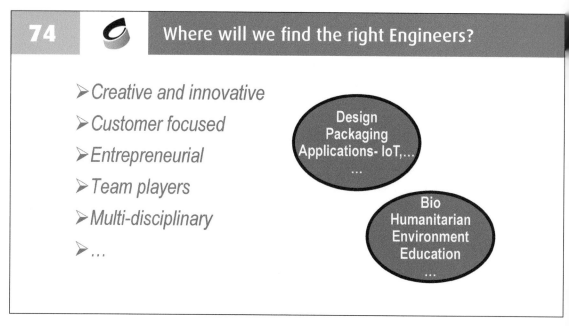

74   Where will we find the right Engineers?

➤ *Creative and innovative*

➤ *Customer focused*

➤ *Entrepreneurial*

➤ *Team players*

➤ *Multi-disciplinary*

➤ *...*

Design
Packaging
Applications- IoT,...
...

Bio
Humanitarian
Environment
Education
...

Here are some of the characteristics of a new employee, what the managers desire. Recognizing this I recommend moulding your career to pick up some or all of these traits.

So learn from what you're hearing and I would say do some soul-searching, figure out what your strengths are, what you really like to do, today and maybe in the future. Identify the opportunities will terminal long-term, focus on what you are doing.

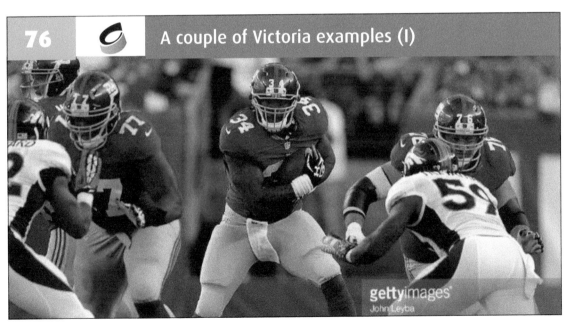

This is an example from American football. This guy is the ball carrier, he's got to find the right holes in the defense and make it through. One way to look at it, when there is an opportunity, find the right solution! Here is another example from a soccer game where, you know, you got one little angle here to get the ball in the goal, and that maybe how you career looks like, right.

77 A couple of Victoria examples (II)

So I would say overall, I think the industry has been good to me, and I think it's got a great future.

CHAPTER 02

Analog Front-End Design for Mobile and Multimedia SoC

Seng-Pan U

University of Macau & Synopsys

ncreasingly, SoCs that follow the trend are at the center of mobile communications and wireless connectivity applications in smartphones or tablets, as well as other multimedia applications in the home such as digital TVs and set-top-boxes. The SoCs digitally process modulated analog signals that can be the output of wireless radios, wireline transceivers, or sensors connecting to the physical worlds. These signals must be accurately digitized for internal processing by baseband or application processors in the SoCs through the analog front-ends (AFEs) which are mainly constituted by the analog signal conditioning circuitries and data converters. Typically, the AFE components are specified in terms of their electrical characteristics, whereas the system designer evaluates the system performance with a different set of metrics. Therefore, system designers must interpret the AFE components' electrical specifications and how they determine the overall performance in units that are meaningful to the system.

For broadband communication applications in the context of wireless or wireline connectivity, cellular communications and digital TV and radio broadcast, we will firstly explore tradeoffs between relative performance and operating modes of different components to find the optimal performance, power, area and cost for baseband SoCs. Then, a quick method will be introduced to determine if the electrical characteristics of any given AFE are adequate for their targeted application, thus avoiding over-specification and inefficient power consumption. Secondly, as the core of such broadband AFEs, various data converter architectures will be addressed with respect to the advanced process nodes.

For the portable consumer applications, such as smartphones, tablets, portable gaming systems, and portable media devices, the audio AFEs, namely, audio codecs (coder/decoders) that encode analog audio as digital signals and decode digital audio back into analog is a key component for multimedia SoCs. The main functionalities, features and performance index of the audio codec and its design considerations will be explained comprehensively.

Finally, the integration of AFE components in SoCs is sometimes perceived as complex because it requires careful custom place-and-route for dealing with the mixed-signal design challenges to avoid the pitfalls that can, sometimes, jeopardize overall system performance. The design techniques that address the common issues of integration in a methodical way will be introduced to help ensure the successful integration of high-performance AFEs in SoCs.

1. INTRODUCTION TO ANALOG FRONT-ENDS

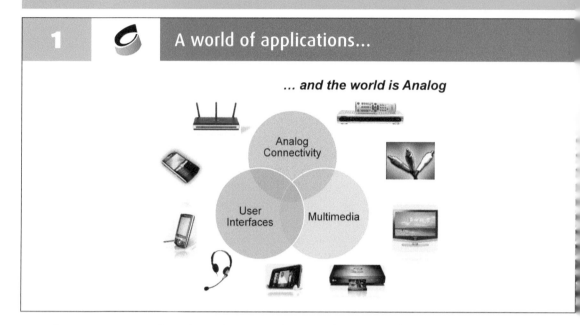

Analog connectivity, multimedia and user interface are the AFEs so these are very important block to connect the analog world.

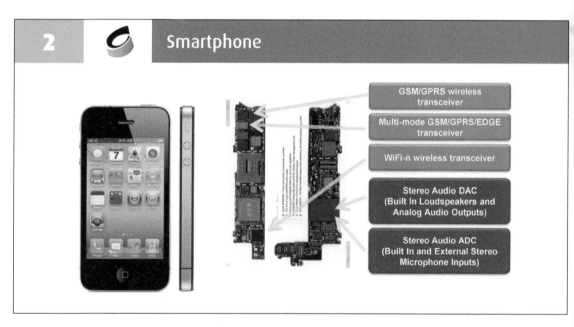

There are multiple SoCs inside of a cellphone, i.e. wireless transceivers, the audio codecs. These are called user interface. It can be found in lots of consumer devices. This talk will mainly focus on the wireless AFE, and the audio codecs.

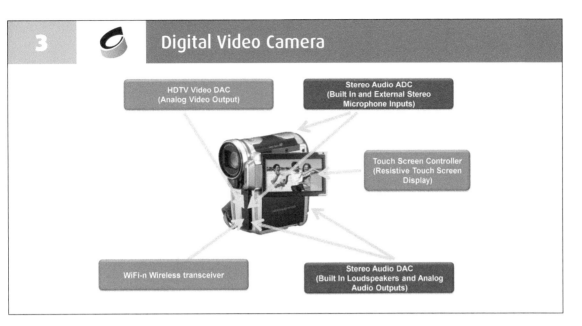

This is the digital video camera. There are analog video output, stereo audio ADCs, and also wireless transceivers, so basically the same content of that.

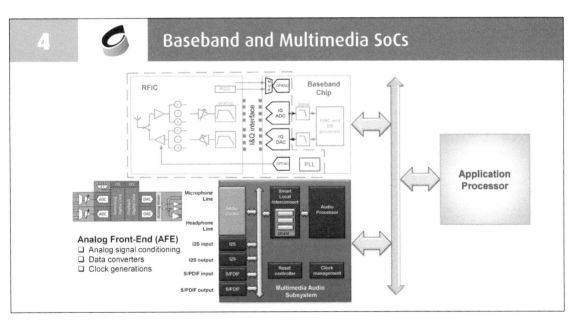

Basically, in modern SoC consumer electronics, the wireless RF IC is usually a separate IC. The baseband chip normally contains the MAC and baseband process. The MAC and baseband process will integrate together with the AFE. The AFE include the ADC, the DAC, the PLL and so on. There is also audio/voice interface. The voice recognition, all these functions actually can be found in every embedded portable device. The audio part which includes, we call multimedia SoC, is audio sub- system. It has audio AFE, which includes ADC and DAC, and also the drivers, in-PGAs, clock generation and interface, and then you have audio processor all together.

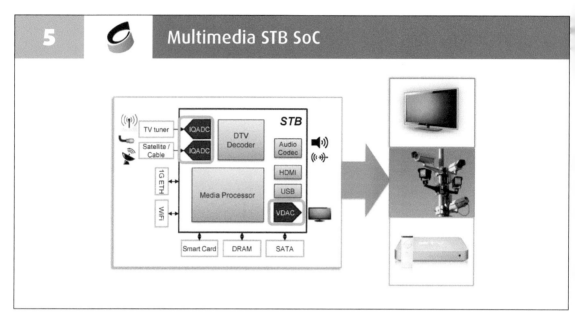

This is the SoC of STB (SET TOP BOX). This is the STB, the video and the audio codecs and so on.

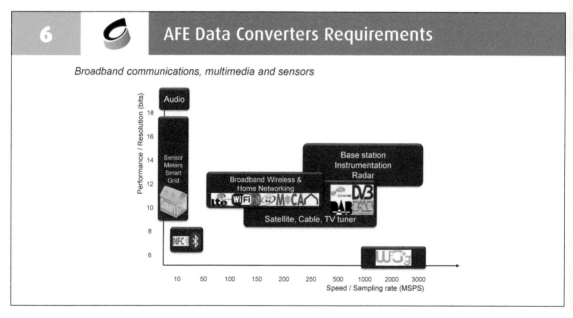

In AFE, data converter is the core. All the above applications require the converters up to 200MHz and the resolution is 10 to 12 bits. And communication at about 3GHz with around 7 bit resolutions. 5G requires around 1G Sample/s and 12 bit resolutions. Different applications require different resolutions.

2. AFES FOR COMMUNICATION SOCS

2.1 DETERMINATION OF AFE PERFORMANCE

7   **AFE requirements in communication SoCs**

• Market for mobile communication, wireless and wireline connectivity systems is huge

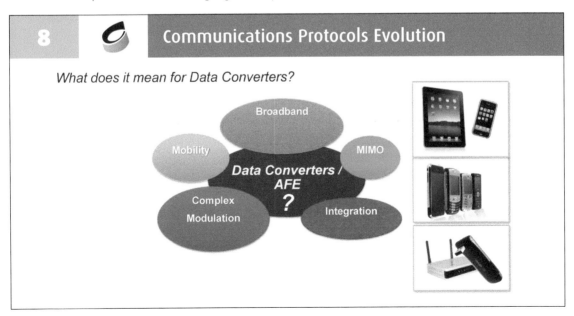

Source: ABI Research Dec 2012

The market for mobile communication, wireless and wireline connectivity system is huge. Look at the shipment of WiFi chipset these years, you will see the requirement is increasing significantly.

People nowadays travel without WiFi is not doable and it's not a good quality life without that. That's why lots of chips are required.

8   **Communications Protocols Evolution**

That's why wireless AFEs are very important. There are requirements of wide-band, motilities, compact baud rates, MIMOs and integration of these

many factors. This is from a system level perspective how we can divide that from the system level requirement.

9 **AFE Specification Challenge**

- Performance requirements for communication systems where the transmitted signal is modulated using complex modulation schemes, such as Orthogonal Frequency-Division Multiplexing (OFDM), are often defined in terms of Error Vector Magnitude (EVM) parameters.

- OFDM modulation is an efficient and robust method used to transmit data over a non-ideal channel that is subject to fading and multiple path interference, typically seen in wireless communication systems. In OFDM, data is encoded on multiple closely-spaced orthogonal carrier frequencies, each modulated using schemes such as Quadrature Amplitude Modulation (QAM).

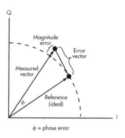

In nowadays communication systems, the transmission signals are normally modulated using the complex modulation schemes, i.e. OFDM due to its efficiency and robustness. The guide in system design is mainly the Error Vector Magnitude (EVM) parameters, which determine the magnitude errors around the reference for this.

 10 **AFE Specification Challenge**

The EVM parameter represents the deviation of each QAM symbol position in the constellation diagram, relative to their ideal position.

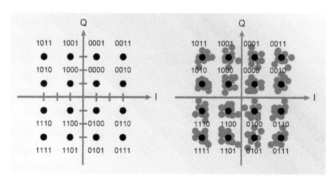

QAM16 modulation example

EVM parameters represent the deviations of each QAM symbol position in the constellation diagram, relative to their ideal position. It emerges with other positions then it is not good result you want. For the ideal case of QAM 16 modulation, the EVM parameter is to describe the deviations to the ideal course about the magnitude errors.

11 AFE Specification Challenge

- To verify a match between AFE performance (and its components) and specific system requirements, system designers must use the AFE electrical specifications to determine the AFE system level performance (e.g EVM contribution), while taking into account signal characteristics and many other factors.

- For systems where Gaussian noise sources are the main impairments to the system performance, the overall SNR is very well correlated to EVM:

$$EVM_{dB} = -SNR_{dB}$$

- To determine the contribution of the AFE to the overall system, SNR is enough to confirm if the AFE performance is adequate to meet system requirements.

To verify the match between the AFE and its component and a specific system requirements, the designers need to use AFE electrical specifications, i.e. SNR, ENoB, to determine the system level performance. But from the system level, only EVM matters. If the Gaussian noise is the main impairments to the system performance, it's not fully accurate but it's a very good match that is the overall SNR is very well correlated here to the EVM. However, there are multiple considerations of SNR.

12 AFE Performances Contribution

AFE performance contribution to the total transceiver SNR performance.

The key characteristics of the signal being processed:
- ➤ Input signal bandwidth
- ➤ Input signal amplitude
- ➤ Modulation scheme
- ➤ ADC SNR
- ➤ PLL clock jitter.

The key characteristics of the signal being processed in this AFE includes the input signal bandwidth, amplitude modulation, ADC SNR, PLL clock jitters and so on. All these factors combined together are correlated to the EVM.

13 Accounting for Input Signal Bandwidth (I)

To improve a data converter SNR performance is by spreading the total noise power it generates through a larger frequency spectrum thereby increasing the converter sampling rate beyond the minimum Nyquist limit.

$$SNR_{BW} = 3 \times log_2(OSR) \quad (dB)$$

$$OSR = \frac{(Fs/2)}{BW}$$

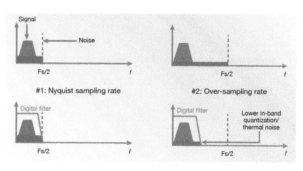

The quantization noise of the data converters is flat over the Fs/2. If the signal is oversampled, the actual SNR band of interest will be 3*log2(OSR), where OSR equals to (Fs/2)/BW. It means that although the noise spread whole spectrum, the bandwidth that system required should be cared, and the rest will be filtered out.

14 Accounting for Input Signal Bandwidth (II)

Oversampling simplifies the analog, anti-aliasing, filtering at the input of an ADC, or the reconstruction filter at the output of a DAC. This is due to the signal images, centered in multiples of the sampling frequency, having larger frequency separation and being easier to filter.

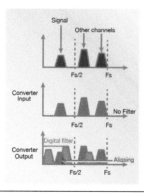

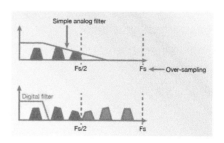

Another reason for oversample is the simplify anti-aliasing, filtering with reconstruction filters and so on. By over sampled, the design of analog anti-aliasing filter and digital filter, which is much more soft kit, adequate and effective compared to the analog filter, will be very simple.

15 Accounting for Input Signal Amplitude

The main contributors to a data converter intrinsic noise are quantization noise and thermal noise, typically assumed to be white noise with a uniform power distribution. The power of these noise components is mainly independent of the signal amplitude.

Any reduction (back-off) of full-scale signal amplitude leads to a reduction of the effective SNR.

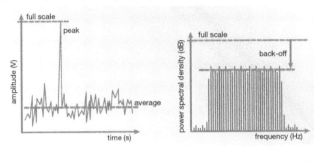

The main contributors to a data converter intrinsic noise are the quantization noise and thermal noise, which are normally right assumed to be like white noise uniformly distributed around the frequency band. Therefore any back off or reduction of the full scale signal amplitude will lead to a reduction of the effective SNR. If the signal is smaller than the full scale of the ADC, only a portion of the dynamic range will be used.

16 Accounting for Input Signal Amplitude

In communication systems, the signal often uses complex modulation schemes with a large Peak-to-Average Ratio (PAR). In order not to saturate an ADC, the signal has to be backed-off (attenuated) such that the peak of the signal falls within the ADC full scale range.

The amount of signal back-off that is implemented takes into account several impairments:
- Presence of strong out-of-band signals that were not adequately filtered
- Variations of the radio signal strength that are not compensated by the transceiver automatic gain control
- Gain inaccuracy in the analog signal chain due to Process-Voltage-Temperature (PVT) variations

$$SNR_A = -[(PAR - 3) + IBO] \quad \text{(dB)}$$

PAR: peak-to-average ratio in dB
IBO: input signal back-off in dB

Unfortunately it is necessary to back off or attenuate because in the communication system the compressed modulations usually have a very large peak to average ratios (PAR). The peak signals should be attenuated in order not to saturate the ADC at a very big PAR. Besides, in multi-channel communication system, the strong out of band signal which may not be filtered out accurately, should also be attenuated before entering an ADC. An additional IBO back-off accounting for the PAR should be added due to the back off the input signal. This 3dB is basically because of the normal situation is coming by a sine wave. The peak to average of sine wave is 3dB so that it should minus 3dB for peak to average. So these are the overall SNR of the system.

17 Accounting for Data Converter

- The overall contribution of the data converter to the system SNR is

$$SNR_{ADC} = SNR_{Nyq} + SNR_{BW} + SNR_A \qquad \text{(dB)}$$

$$SNR_{ADC} = SNR_{Nyq} + 3 \times log_2(OSR) - (PAR - 3) - IBO \qquad \text{(dB)}$$

So generally, the SNRADC will be basically equal to SNRnyq (which is the target SNR for an ideal ADC) plus some SNR giving from the effective bandwidth, and also get some SNR degradations because of the amplitude on that. So this is the discussed SNR, bandwidth, and peak average ratio and input back-offs.

18 Accounting for PLL jitter (I)

The clock jitter contribution to the SNR :

$$SNR_J = -20 \times log_{10}(f_{in} \times 2\pi\sigma_{LTJ}) \quad \text{(dB)}$$

f_{in} is the frequency of the (single tone) input signal in Hz units
σ_{LTJ} is long term jitter in sec-rms units.

Long term jitter is the integrated phase noise of the clock signal. In the frequency domain,

$$\sigma_{LTJ} = \frac{1}{2\pi F_S}\sqrt{2 \times \int_0^\infty 10^{\frac{L(f)}{10}}}$$

Nowadays, jitter is becoming a very critical issue of the system and especially the system with very high speed. The SNR of jitter is actually proportional to the fin (input frequency), so that the jitter becomes very sensitive due to the high frequency of the input signal. Unlike the kind of single tone input signal expression of SNRjitter, the long term jitter is integrated the phase noise of a clock signal in the frequency domains.

19 Accounting for PLL jitter (II)

Effect of clock jitter in OFDM modulation

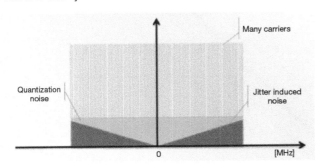

- 802.11ac and LTE use OFDM modulation
  - Large number of sub-carriers
- Band centered in baseband at the input of ADCs
- ADC quantization adds uniform power noise density in the full band
- Sampling clock Jitter intermodulates with each sub-carrier
- Clock phase noise is convolved with each sub-carrier scaled by the sub-carrier frequency

Clock jitter distribution can be simplified as: $SNR_J = -20 \times log_{10}(\frac{BW}{\sqrt{3}} \times 2\pi\sigma_{LTJ})$ (dB)

In the OFDM modulation system, there are large numbers of sub-carriers. So the band will be centered in baseband at the input of ADCs and the ADC quantization adds uniform power noise distributions in the full band. The jitter induced noise is not flat, which is proportional to the input signal frequency. The clock phase noise is convolved with each sub-carrier scaled by the sub-carrier frequency. So the clock jitter distribution can be simplified as the formula at the bottom.

 20 OFDM Signals – back to basics...

- 802.11ac and LTE use OFDM modulation
  - Large number of sub-carriers
- Band centered in baseband at the input of ADCs
- ADC quantization adds uniform power noise density in the full band
- Sampling clock Jitter intermodulates with each sub-carrier
- Clock phase noise is convolved with each sub-carrier scaled by the sub-carrier frequency

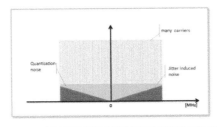

The jitter can also affect the system performance, because the very strong out of band interference will be modulated into the band if the clock is not pure.

21 — Accounting for PLL jitter (I)

Sampling clock phase noise can also impact system performance in the presence of strong out-of-band interferer.

➢ Impose large back-off attenuation to the signal to avoid ADC input saturation

➢ Impose additional clock jitter constrains to avoid SNR degradation

To avoid SNR degradation, additional clock jitter constrains will be imposed.

22 — Accounting for PLL jitter (II)

• In RX where input signal with back-off (IBO), the jitter-induced noise contribution will be less relevant, since it is proportional to the input strength.

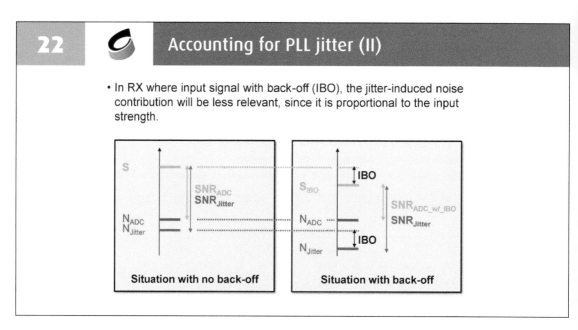

So, interesting at the back off of the input signals actually will be helpful for clock jitter. The jitter sensitivity becomes relatively small due to the smaller amplitude of signal. So it shows that with back-off, the jitter induced noise contribution will be relatively smaller, because it is proportional to the input strength as well.

23   Accounting for Other Error Sources

Modern demodulation architectures typically implement direct demodulation schemes. Any gain, phase or offset mismatch between I and Q channels also contribute to the total SNR degradation. Built-in calibration procedures are often implemented to reduce these impairments to manageable levels across the transceiver system.

For the purpose of budgeting SNR, an additional SNR margin of ~1 to 2 dB to take into account the residual effect of the calibrated mismatch on the SNR.

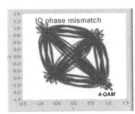

In complex modulations, I Q channel are used to obtain the analog signal. If the IQ channels are not fully matched, such as gain, phase, offset mismatch, the SNR will also degraded. When IQADC, IQDAC in AFE are not designed with good matching, an additional SNR degradation will occur. The points in the constellation are shifted due to the IQ phase mismatch in the left figure, and the gain mismatch in the right figure. Because the matching now is not so bad if designed carefully, an additional SNR margin of 1~2dB will be taken into account.

24   Calculating AFE Performance Contributions (I)

- **Step 1**: Calculate the contribution of the data converter to the system SNR

$$SNR_{ADC} = SNR_{Nyq} + 3 \times log_2(OSR) - (PAR - 3) - IBO$$

- **Step 2**: Calculate the contribution of the PLL clock jitter to the system SNR

$$SNR_J = -20 \times log_{10}(\frac{BW}{\sqrt{3}} \times 2\pi\sigma_{LTJ})$$

- **Step 3**: Assume a few dBs as the typical contribution to system SNR due to un-compensated IQ-imbalances

$$SNR_{IQ} = 1 \sim 2$$

- **Step 4**: The total contribution of the AFE to the overall system SNR is the sum of these three contributions

$$SNR_{total} = -10 \times log_{10}(10^{-\frac{SNR_{ADC}}{10}} + 10^{-\frac{SNR_J}{10}}) + SNR_{IQ}$$

- **Step 5**: Determining AFE contribution to system level EVM

$$EVM_{dB} = -SNR_{total}$$

All of the factors have been talking about. First of all, the SNRADC contain not only the pure SNRnyq, but also the bandwidth of interest, the amplitude, the input back-off and the PLL jitter. Additional few dB will also be added because of the other effect like un-compensated IQ-imbalance. Compare to see whether the SNRtotal is actually corresponding to the system requirement. It's a straight-forward way but very practical to get in so.

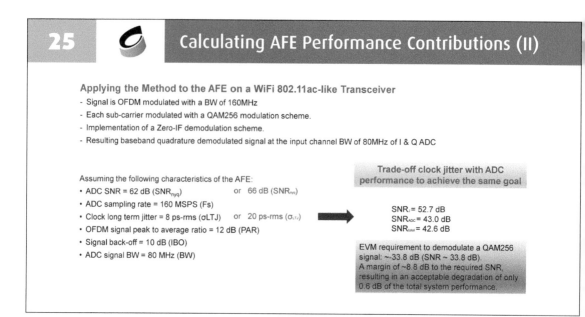

25 — Calculating AFE Performance Contributions (II)

Applying the Method to the AFE on a WiFi 802.11ac-like Transceiver
- Signal is OFDM modulated with a BW of 160MHz
- Each sub-carrier modulated with a QAM256 modulation scheme.
- Implementation of a Zero-IF demodulation scheme.
- Resulting baseband quadrature demodulated signal at the input channel BW of 80MHz of I & Q ADC

Assuming the following characteristics of the AFE:
- ADC SNR = 62 dB (SNR$_{nyq}$)   or   66 dB (SNR$_{rms}$)
- ADC sampling rate = 160 MSPS (Fs)
- Clock long term jitter = 8 ps-rms (σLTJ)   or   20 ps-rms (σ_{LTJ})
- OFDM signal peak to average ratio = 12 dB (PAR)
- Signal back-off = 10 dB (IBO)
- ADC signal BW = 80 MHz (BW)

Trade-off clock jitter with ADC performance to achieve the same goal

SNR$_j$ = 52.7 dB
SNR$_{ADC}$ = 43.0 dB
SNR$_{total}$ = 42.6 dB

EVM requirement to demodulate a QAM256 signal: ~-33.8 dB (SNR ~ 33.8 dB).
A margin of ~8.8 dB to the required SNR, resulting in an acceptable degradation of only 0.6 dB of the total system performance.

There is an example to give a concept how to determine the system requirement, which applying the method to the AFE on WiFi 802.11ac transceiver. The calculation means that there is a trade-off between clock jitter and ADC performance.

So with a margin of about 8.8dB, it will be designed either a relatively lower ADC but with a high requirement on the PLL clock or high requirement of ADC but a relatively relaxed PLL.

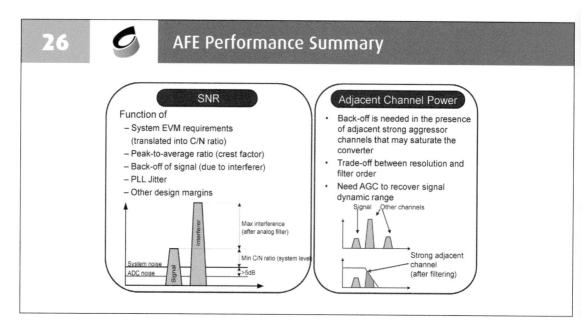

26 — AFE Performance Summary

SNR

Function of
- System EVM requirements (translated into C/N ratio)
- Peak-to-average ratio (crest factor)
- Back-off of signal (due to interferer)
- PLL Jitter
- Other design margins

Max interference (after analog filter)

Min C/N ratio (system level)

>5dB

Interferer

Signal

System noise
ADC noise

Adjacent Channel Power
- Back-off is needed in the presence of adjacent strong aggressor channels that may saturate the converter
- Trade-off between resolution and filter order
- Need AGC to recover signal dynamic range

Signal   Other channels

Strong adjacent channel (after filtering)

To summarize, the AFE performance is basically a kind of function of system EVM requirements (translated to SNR), PAR, signal back-offs because of interferences, and PLL jitters and other design margins. There is a tradeoff between the ADC performance and jitter performance as well.

2. AFES FOR COMMUNICATION SOCS

2.2 SCALING NYQUIST ADC ARCHITECTURES

The second portion is about the basic building block of data converters.

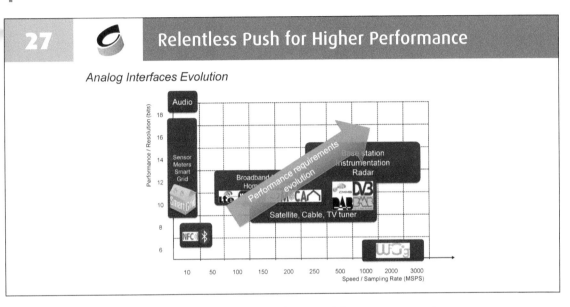

The analog interface evolution is actually pushing high point of data converter performance.

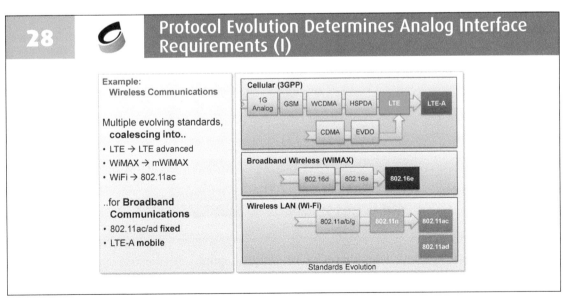

The protocol evolution is determined analog interface. The bandwidth becomes wider so that the speed is required higher as well as the resolutions.

29 · Protocol Evolution Determines Analog Interface Requirements (II)

Example: Wireless Communications

- Evolving to support higher data rates
  - Multi antenna arrays
  - Higher channel bandwidth
  - Complex modulation schemes

- Baseband SoC's migrating to advanced process nodes
  - 28-nm and beyond

- Requirements for Analog Interfaces
  - Higher performance
  - Significant area and power reduction
  - Robustness for integration in advanced nodes

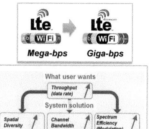

The users want high data rate, high throughput, translated into a system is that spatial diversity, channel bandwidth, spectrum efficiency, different modulation schemes and so on. And also implication of an AFE is MIMOs, sampling rate, resolutions and so on. So those are from top down requirement to the AFE.

30 · Typical Tradeoffs do Not Hold!

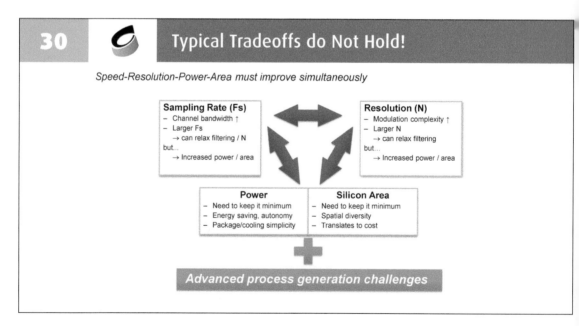

It is a challenge of AFE and always for analog design because the system requirements want to have much more function to put in, a digital content become much more complex so analog portion become small, effective and low power. But the requirement is even higher for higher speed and higher resolutions.

31 Digital Scales with Process Generation (I)

Moore's Law is the survival code in SoCs

- Every new process generation
  - Increase gate density 1.5X - 2X
  - Increase speed (f_T) 1.5X - 2X
  - Reduce consumption 1.5X - 2X

- **Digital circuits scale naturally with feature size**
  - Consist basically of interconnected switches

- SoC processing power and functionality increase in every generation
  - For the same function -> area & cost reduces
  - For the same cost -> area and function increases

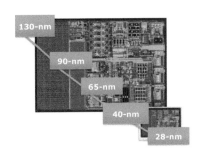

Digital scales is actually helping digital design, the gate densities, the speed and the reduced consumptions.

32 Analog Scales with Process Generation? (II)

- Every new process generation
  - Reduced analog gain
  - Increase output conductance
  - Reduce voltage headroom
  - Increased layout proximity effects

- Traditionally analog performance is controlled by:
  - Closed loop operation with negative feedback
  - Cascoding of devices to improve gain
  - High output impedance

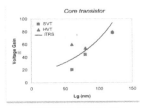

Core transistor analog gain as a function of minimum gate length *

These solutions are no longer possible!

* *source: P. Dautriche, "Analog Design trends and challenges in 28 and 20nm CMOS technology", Proceedings of ESSDERC. Sep. 2011*

However it is not really favorable too much to analog scaler because of the reduced gain, the increased output conductance, the reduced voltage headroom, the very complex layout, and the silicon effects and so on.

33 **Analog Scales with Process Generation? (III)**

Analog must exploit higher speed, matching and processing power made available by process scaling

- **Digitally-assisted analog techniques,** push complexity to digital
  - Take advantage of abundant digital gate availability in advanced nodes
  - Calibration
    - Compensation for mismatches in passive devices
    - Compensation for poor control on key parameters of active components
  - Dithering to smear-out spurs
  - Redundancy
- **Simpler, open loop, analog circuitry**
  - Less sensitive to poor analog characteristics of advanced processes
  - Do not require high supply voltage headroom
- **Parallelization**

Therefore, using digitally-assisted analog techniques are necessary.

34 **Conservative Architectural Evolution**

From enormous to gigantic!

1954: ADC – 11-bit, 50 kSPS
- 19" x 15" x 26"
- 500 W
- 150 lbs

Equivalent to:

Today: ADC – 12-bit, 80 MSPS
- **120 m3**
- **500 kW**

As an example for the architectural evolution, the pipelined ADC was previously very popular ADC, 12bit, 80MSPS, 0.86mm2, and 75mW. And now 2013, the performance of SAR ADC is just 0.04mm2 and 2.7mW. It is actually driven by technology and much more functions are made in systems.

35   SAR ADC Increasing Predominance

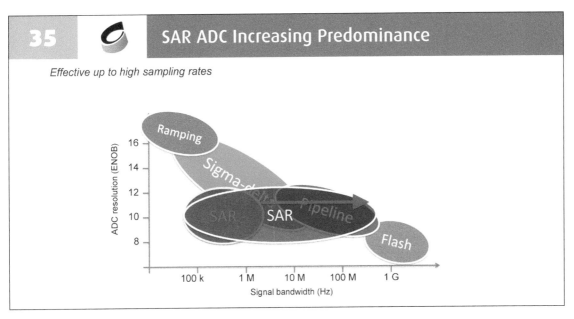

Effective up to high sampling rates

There is the traditional ADC scheme. The ramping ADC and sigma-delta ADC is normally at lower bandwidth with higher resolution, pipeline ADC with a higher frequency band and flash ADC is very high frequency. The SAR ADC is normally in this range, but becomes a very wide band.

36   Overview of ADC architectures

- ADC architectures will be described in terms of the fundamental **operations** realized inside them
- Focus on the **operations:**
  - Highlights fundamental limitations and trade-offs of different architectures
  - Provides a unified treatment that relates them all
  - Avoids the confusing **architecture shortcomings** and **limitations on the circuit solutions**

This is an overview of ADC architecture first.

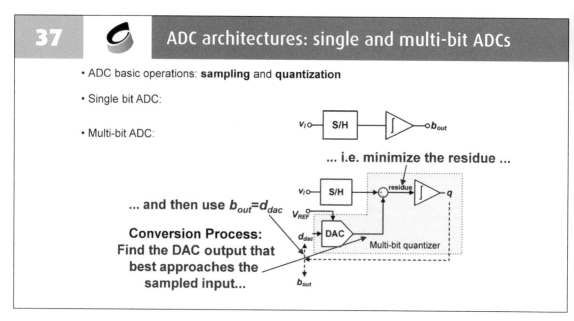

A DC architectures can be categorized into a way like the single or the multi-bit ADC. The ADC basic operations are just sampling and quantization.

The multi-bit ADC determines multi-bit at the same time. The conversion process is to find the DAC output that best approach to the sampled input.

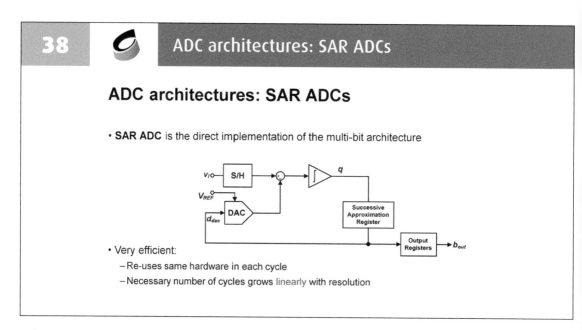

The SAR ADC is actually direct implementation of multi-bit architecture. Firstly, sample and hold. In every cycle, using the same hardware, get the residues by DAC and compare again. The necessary number of cycle grows linearly with the resolution, so that 12bit or 14bit ADC need more cycles.

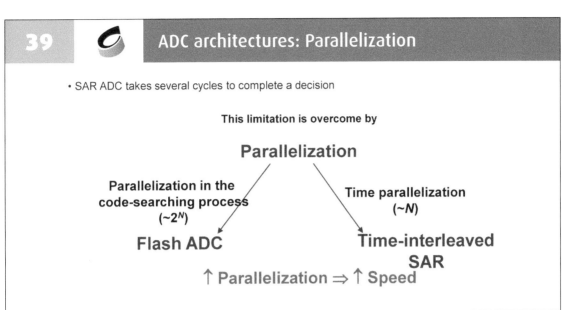

39   **ADC architectures: Parallelization**

• SAR ADC takes several cycles to complete a decision

This limitation is overcome by

Parallelization

Parallelization in the code-searching process ($\sim 2^N$)

Time parallelization ($\sim N$)

Flash ADC

Time-interleaved SAR

$\uparrow$ **Parallelization** $\Rightarrow$ $\uparrow$ **Speed**

The limitation is that multiple cycle result in reducing the speed. Parallelization can get it faster. Parallelization includes time parallelization, like time-interleaved SAR, and code searching process parallelization, like the Flash ADC.

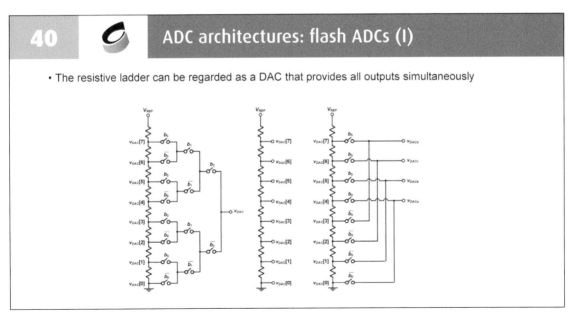

40   **ADC architectures: flash ADCs (I)**

• The resistive ladder can be regarded as a DAC that provides all outputs simultaneously

The Flash ADCs make it parallel generally by multiple references and compared at the same time. A resistive ladder can get the output reference to compare directly.

41   ADC architectures: flash ADCs (II)

• **Flash ADC**: brute-force implementation of the multi-bit architecture

Comparators determine which of the DAC outputs is nearer the sampled input

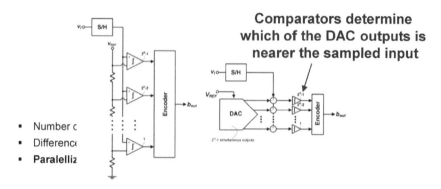

- Number c
- Difference
- Paralelliz

The Flash ADC is a brute-force implementation of the multi-bit architecture. The number of elements grows exponentially with resolutions. Multiple parallel elements will always have mismatch so that it is important to manage mismatch to make a parallelization become perfect. That's why Flash ADC is needed to take care of the offset, mismatch among all those comparators and so on. But generally the idea is just to do parallel comparisons to determine which of the DAC outputs is nearer the sampled input.

42   ADC architectures: subranging ADCs (I)

• Constituted by lower resolution quantizers that successively refine the conversion
   – DAC in the fine quantizer still needs to provide all the 2^N outputs
• Less comparators than in Flash ADC, but slower: ↓ **Parallelization** ⇒ ↓ **Speed**

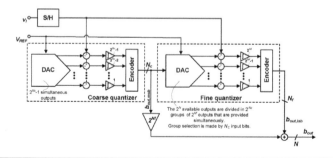

The subranging ADCs is a kind of two-step scheme. The MSB is compared firstly and the fine comparison afterwards. So this is what the coarse quantizer and fine quantizer to successfully refine the conversion. With two steps, it can save the comparator but lose the speed.

43 ADC architectures: subranging ADCs (II)

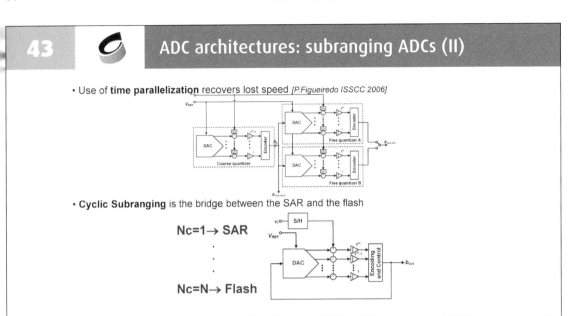

- Use of **time parallelization** recovers lost speed *[P.Figueiredo ISSCC 2006]*

- **Cyclic Subranging** is the bridge between the SAR and the flash

Nc=1→ SAR

Nc=N→ Flash

How to deal with that? One simple scheme is that the first coarse quantizer worked in full speed every clock cycle and the next fine conversions using two interleaved ADC. It will be faster than previously two steps. There is another way called cycle-subranging. It is a kind of bridge between SAR and flash. Basically it is a kind of the multi-bit SAR, doing the multi-bit parallelization on this. If Nc goes to one then it comes to SAR and Nc equal to N it becomes the flash.

44 ADC architectures: pipelining

- **Parallelization** ("in time" or "in code-searching process") allows sampling the input and provide a conversion result in every clock cycle
- Another possibility is **pipelining**

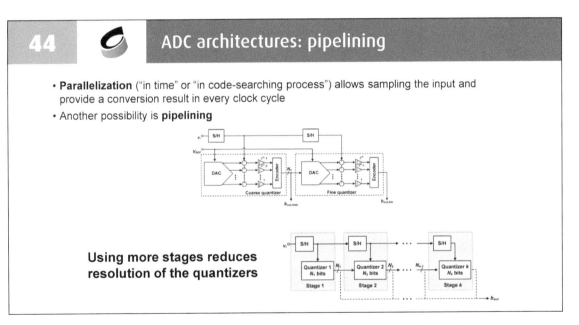

Using more stages reduces resolution of the quantizers

Another possibility is pipelining. Using subranging or two-step, it is two times conversions. By pipelining more quantilizers, the ADC uses more stages to reduce the resolution of the quantizer.

45   ADC architectures: pipelining and residue calculation + amplification (I)

- Use of several stages reduces quantizer resolution, but not complexity of their DACs:
  - e.g. last quantizer's DAC still needs to generate all the 2^N voltages.
- This limitation is overcome if each stage:
  1. performs **quantization**
  2. **calculates the residue** (i.e. the error signal corresponding to what is left to quantize)

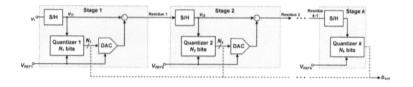

It is the very hot architecture which is called the pipelined ADC. The pipelined ADC has one advantage that it can reduce each stage resolution.

In meantime with an additional stage, it can do some digital error correction.

46   ADC architectures: pipelining and residue calculation + amplification (II)

- **Amplification** of the residue done to:
  - Ensure all stages have the same input range
  - Ease specifications of later stages
- **Pipeline ADC** - each stage constituted by:
  - a quantizer (typically, but not necessarily, a flash)
  - the residue calculator/amplifier block - MDAC

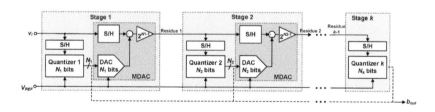

The residue becomes smaller every time. However, by adding a gain amplification of the residue to each stage, it can keep using the same stage for each stage. Each stage has the sample&hold, the quantizer, the DAC and the amplifier, and the stage can be reused. So this is the general pipeline ADC scheme.

ENERGY-EFFICIENT ADC TRENDS

47   Outline

❑ **Overview of ADC in ISSCC & VLSI**
  ❖ State-of-the-Art ADC Survey
  ❖ Performance Trend of ADC in every 5 years
  ❖ Technology Trend
❑ **Benchmarking ADCs in ISSCC**
  ❖ Speed, FoM and Resolution
❑ **Circuit-Innovation ADCs in ISSCC 2017**
❑ **Summary**

This will talk about the energy efficient ADC trends. The new technology process is not favorable to analog, but favorable to digital. Therefore the architecture that basically can take advantage of digital process will help the design. It is the idea why energy-efficient ADC is actually mostly using the dynamic scheme now.

The outline include the overview of ADC in ISSCC and VLSI, benchmarking ADCs in ISSCC and circuit-innovation ADCs in ISSCC2017.

OVERVIEW

48   Figure of Merit and Energy

❑ Widely Used Figure of Merit (FOM)

Walden FOM: $\text{FOM}_\text{W} = \dfrac{Power}{2^{ENOB} \cdot f_s}$

Energy increases 2 x per Bit (ENOB)
Empirical Perspective

Schreier FOM: $\text{FOM}_\text{s} = \text{DR}\,(dB) + 10\log\left(\dfrac{BW}{Power}\right)$

Energy increases 4 x per Bit (DR)
Thermal Noise Perspective

There are two parameters, which are the figure of merit and energy of the ADC. Walden FOM (figure of merit) focus on the practical perspective, which will make the energy increase by two times when ENOB increases one bit. Schreier FOM focus on the thermal noise perspective, which will make the energy increase by four times when ENOB increase 6dB. Schreier FOM is widely used for describing the performance of ADCs.

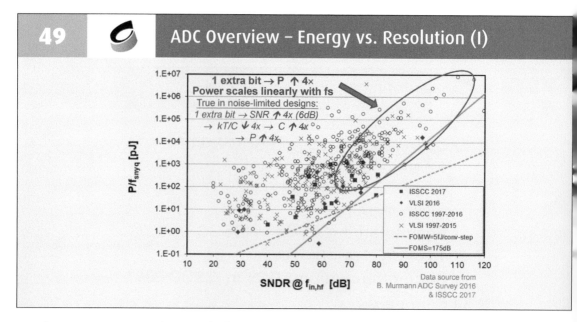

Professor Boris Murmann from Stanford University has made a survey of the ADCs. Linearity happens in the red ellipse because it's kind of noise limited design. If the SNDR increases by 6dB (one extra bit), the KT/C noise will need to be reduced by 4 times.

It's means that the C capacitors is needed to increase 4 times and the power is need to increase 4 times as well. Therefore, the right side of the intersection is limited by the noise and Schreier FOM is normally used to describe.

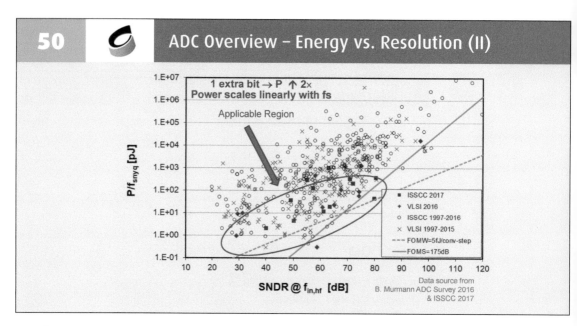

While on the other side, one extra bit of resolution only makes the power increase by two times, because it's not the noise limited design. Therefore, many techniques can be used to make it more power efficient without the limitation of thermal noise. In this region, the Walden Fom is normally used.

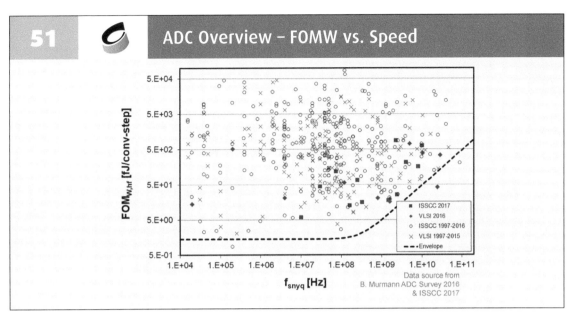

This figure shows that when the speed increase, the FOM becomes very hard to push at low FOM.

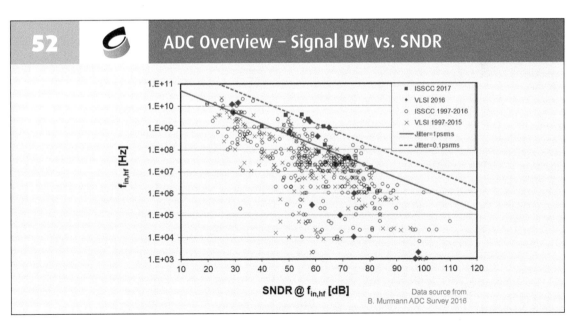

When the frequency becomes higher and higher, jitter will also be limited nowadays. In the past 10 years, the limit of jitter has been pushed from 1ps rms to 0.1 ps rms.

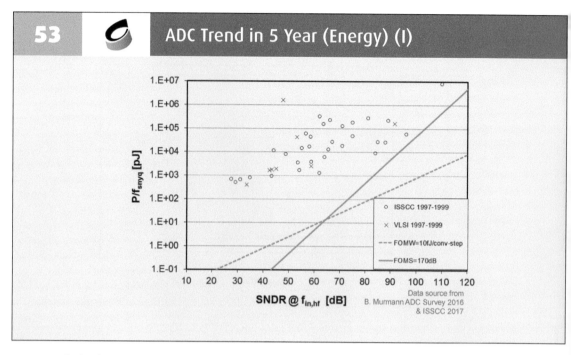

In general, the lower/medium resolutions, energy efficiency can improves by 2 times every 1~6 years and also the high resolution become much more difficult because of the thermal noise limitation.

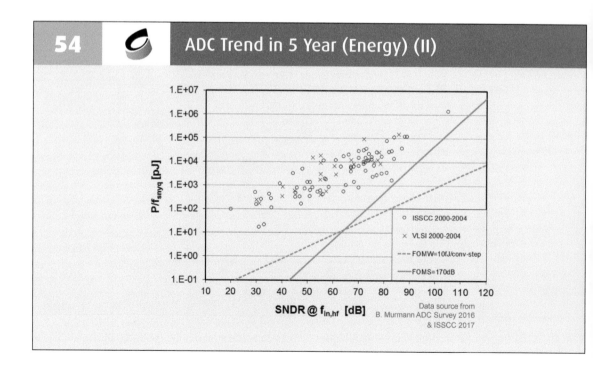

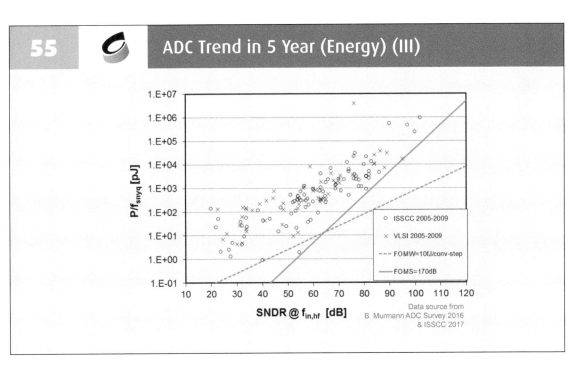

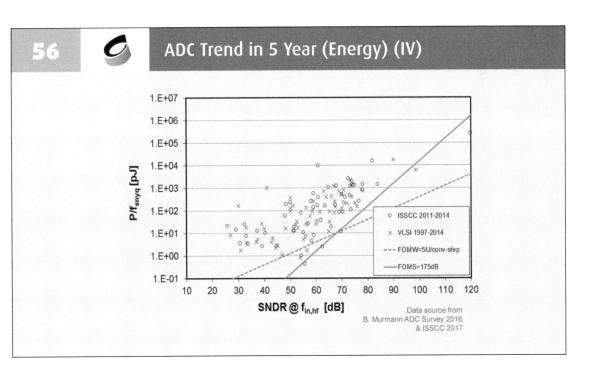

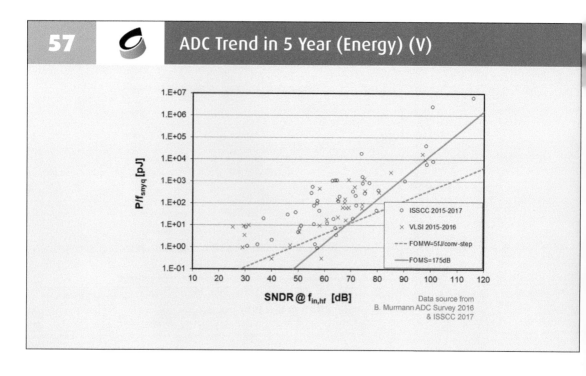

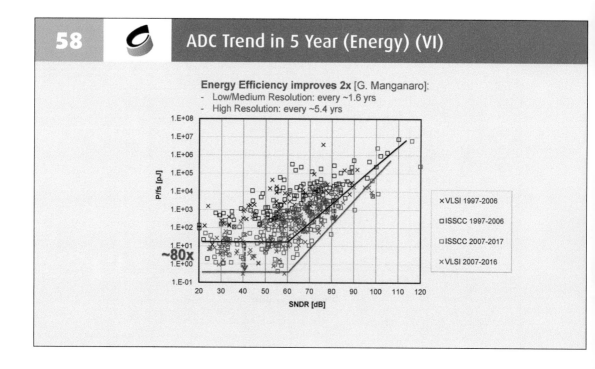

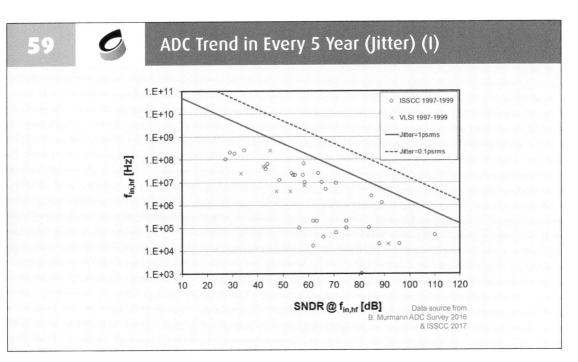

The jitter also moves very fast and reaches 0.1 ps rms region in recent years.

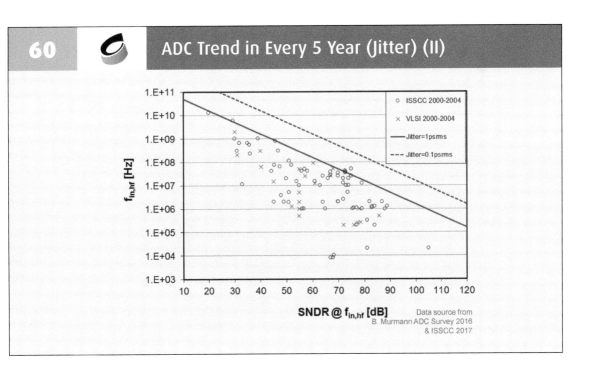

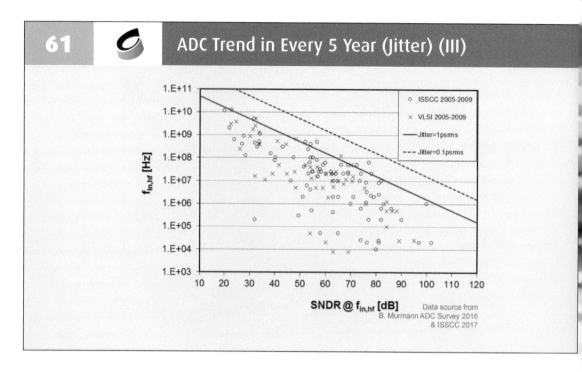

61 ADC Trend in Every 5 Year (Jitter) (III)

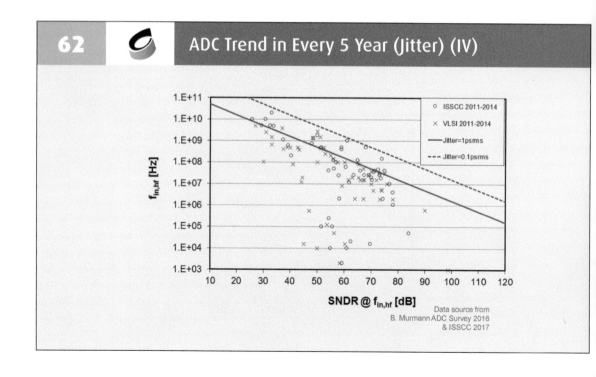

62 ADC Trend in Every 5 Year (Jitter) (IV)

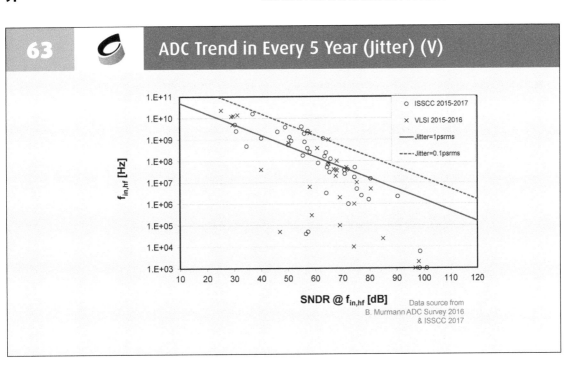

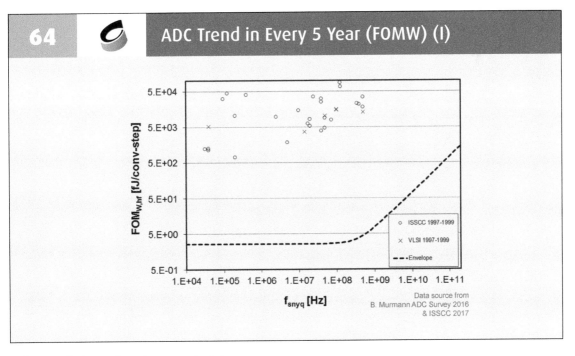

The speed of ADCs has also moved a lot.

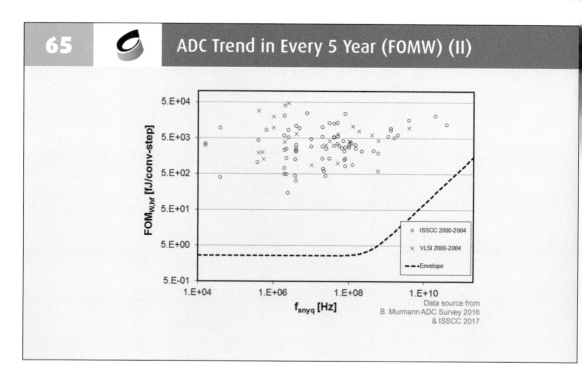

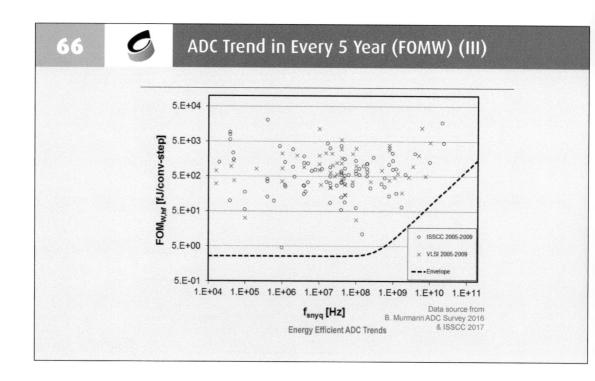

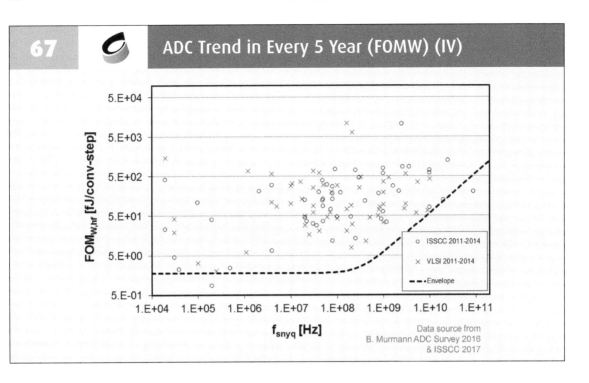

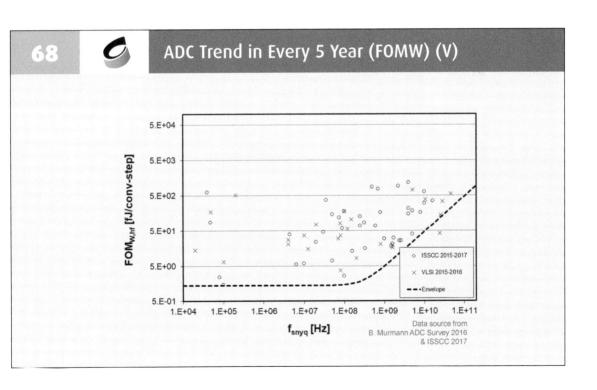

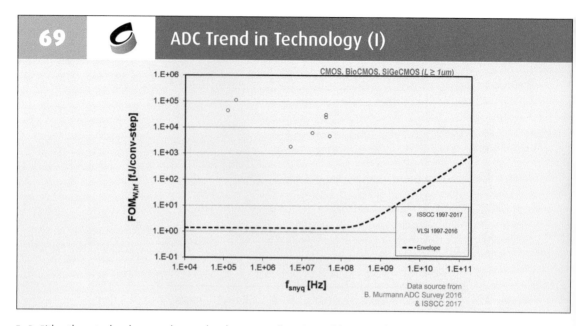

With the technology advanced, the overall performance of ADCs improved significantly. Not only the efficiency gets better, the speed pushes much higher. It is implicated that technology is really helping the ADC's speed and efficiencies. Although in previously discussion, digital processing is not favorable to analog design, the architecture of ADCs have made it much more relative to the dynamic operations to fit the technologies. Due to the purely digital type design, SAR ADCs and also so-called energy efficient type ADC have become very popular.

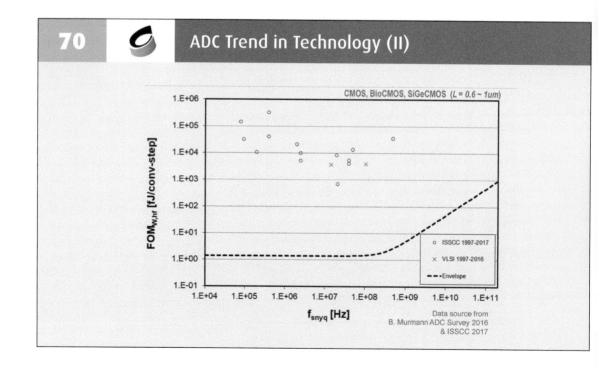

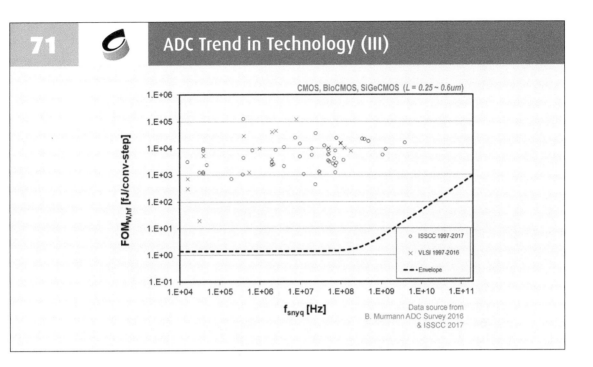

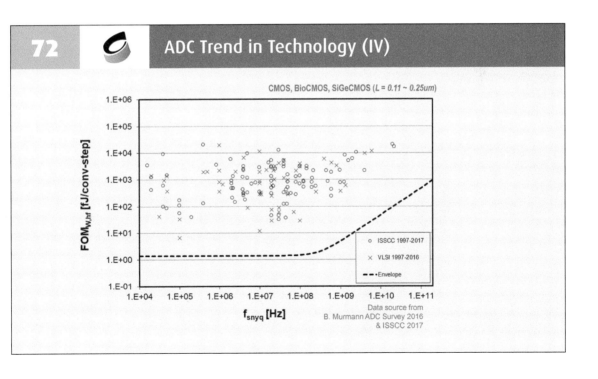

73

ADC Trend in Technology (V)

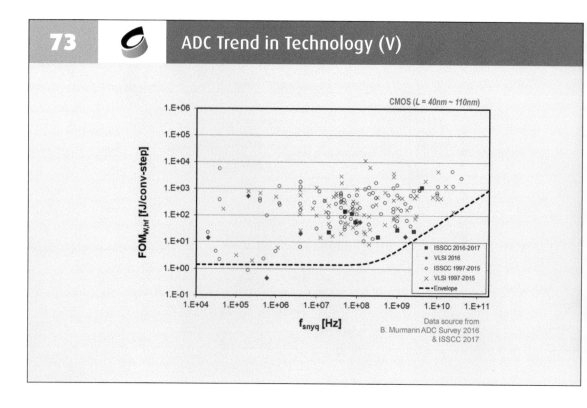

74

ADC Trend in Technology (VI)

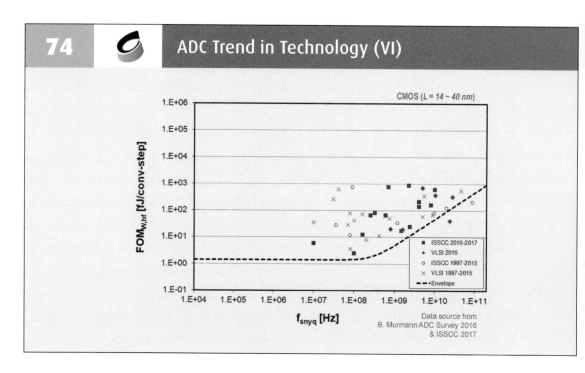

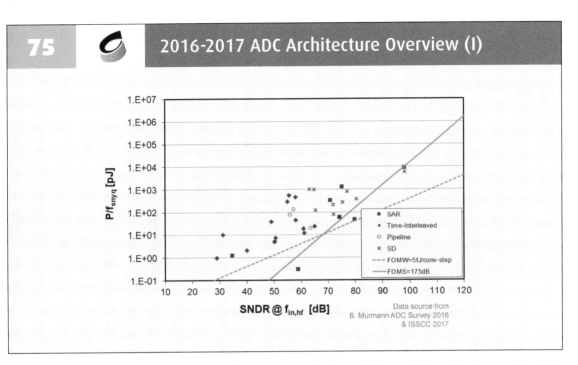

This shows the 2016~2017 ADC architecture overview. The time-interleaving ADCs are normally in the region with relatively lower resolution and SAR ADCs become very widely spread from very low resolution to high resolution. The sigma-delta ADCs and pipelined ADCs occupy in the area with medium resolutions.

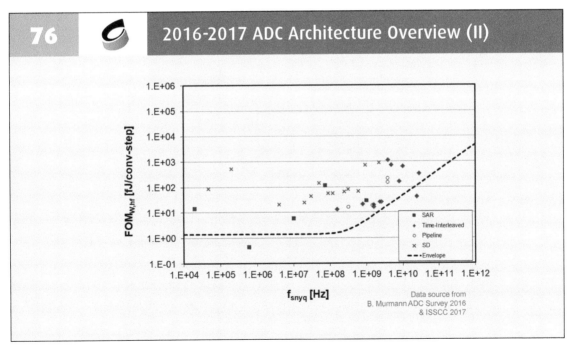

Concerned for speed, time-interleaved ADCs and pipelined ADCs have much higher speed. SAR ADCs become widely spread again and many architectures of time-interleaved ADCs are actually using SAR as well. That's the reason why SAR ADCs have become very popular.

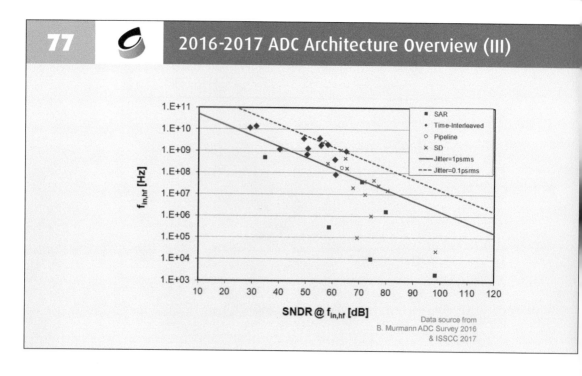

77 · 2016-2017 ADC Architecture Overview (III)

Data source from
B. Murmann ADC Survey 2016
& ISSCC 2017

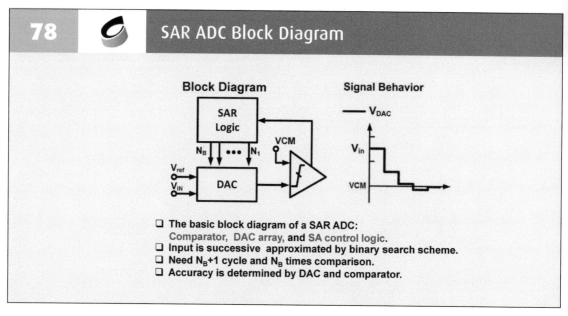

78 · SAR ADC Block Diagram

Block Diagram

Signal Behavior

❑ **The basic block diagram of a SAR ADC:**
  Comparator, DAC array, **and SA control logic.**
❑ **Input is successive approximated by binary search scheme.**
❑ **Need N_B+1 cycle and N_B times comparison.**
❑ **Accuracy is determined by DAC and comparator.**

Basically, SAR ADC only contains kind of SAR logic, successive, proximate registers. Digital design can take the advantage of the process and make a fast logic operation. The DAC is a kind of dynamic switched capacitors, resistor ladders and so on. There are Nb+1 cycle and Nb times comparison by comparator to achieve Nb bit ADC. Therefore, the accuracy is just determined by the DAC and comparator. Every time it resolve one bit then go back to get the residue then to compare next time, until the end. It's recycling, within the same hardware.

79 — Recall - Figure of Merit and Energy

❑ Compare ADC with conversion energy : FOM

Walden FOM: $\text{FOM}_W = \dfrac{Power}{2^{ENOB} \cdot f_s}$

Energy increases 2 x per Bit (ENOB)

Schreier FOM: $\text{FOM}_S = DR\,(dB) + 10\log\left(\dfrac{BW}{Power}\right)$

Energy increases 4 x per Bit (DR)

❑ If the power of ADC is mostly dynamic

Power: $P = CV_{DD}^{2} f_s$ $P \propto f_s$ or BW

❑ Ideally, both FOMs do not raise with speed, but...

Recall the FOM discussed before, if an ADC is mostly dynamic, which uses a dynamic comparator, switched capacitor DAC and digital logic (SAR logic), the power is equal to C*VDD2*Fs.

80 — Walden Figure of Merit vs. Speed

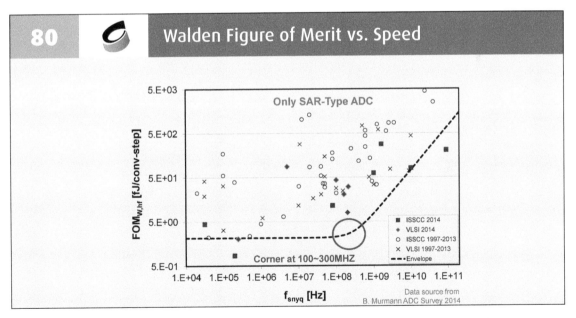

Data source from B. Murmann ADC Survey 2014

When the frequency increase, the FOM power will not be proportional so that a high frequency can cause a very large power consumption. While in the region of Walden FOM, Fs and power are in a proportional scheme so that the banding effect is not appeared.

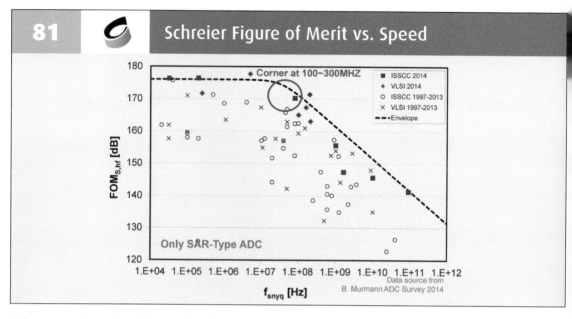

The same scheme is in the FOM for Schreier. A better efficiency can't be achieved at a higher frequency.

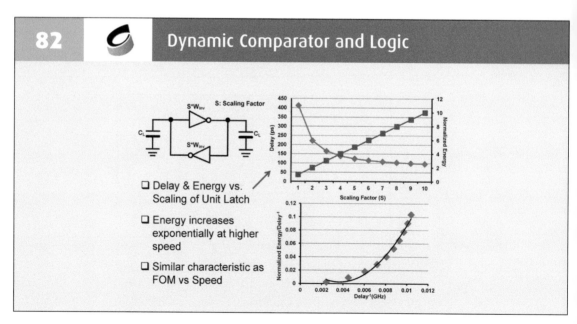

In the SAR ADC, the main building block is comparator, which can be power controlled. The delay time of a latch comparator is not linearly reduced when scaled W. The speed can not be improved so much even increasing W. It is similar to the effect in FOM vs speed.

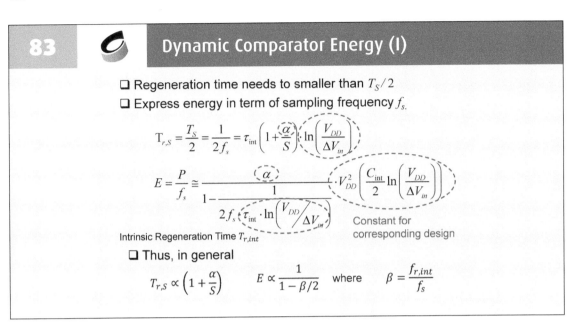

83 **Dynamic Comparator Energy (I)**

❑ Regeneration time needs to smaller than $T_S/2$
❑ Express energy in term of sampling frequency f_s.

$$T_{r,S} = \frac{T_S}{2} = \frac{1}{2f_s} = \tau_{int}\left(1 + \frac{\alpha}{S}\right) \cdot \ln\left(\frac{V_{DD}}{\Delta V_{in}}\right)$$

$$E = \frac{P}{f_s} \cong \frac{\alpha}{1 - \frac{1}{2f_s\left(\tau_{int} \cdot \ln\left(\frac{V_{DD}}{\Delta V_{in}}\right)\right)}} \cdot V_{DD}^2\left(\frac{C_{int}}{2}\ln\left(\frac{V_{DD}}{\Delta V_{in}}\right)\right)$$

Intrinsic Regeneration Time $T_{r,int}$ Constant for
                                                 corresponding design

❑ Thus, in general

$$T_{r,S} \propto \left(1 + \frac{\alpha}{S}\right) \qquad E \propto \frac{1}{1 - \beta/2} \qquad \text{where} \qquad \beta = \frac{f_{r,int}}{f_s}$$

In details, the speed of dynamic comparator is determined by its regeneration time. When the delta Vin get smaller, the comparator requires longer time to do the comparison. ? is equal to CL divided by gm, while CL contains intrinsic and external capacitance, giving the factor of alpha. The regeneration time needs to be smaller than Ts/2, and for a certain design, regeneration time of comparators is proportional to 1+?/S. The energy is actually proportionally to 1/(1-1/(2*fs)).

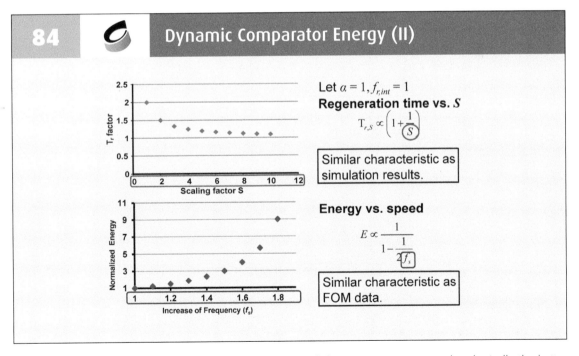

84 **Dynamic Comparator Energy (II)**

Let $\alpha = 1, f_{r,int} = 1$
Regeneration time vs. S

$$T_{r,S} \propto \left(1 + \frac{1}{S}\right)$$

Similar characteristic as simulation results.

Energy vs. speed

$$E \propto \frac{1}{1 - \frac{1}{2f_s}}$$

Similar characteristic as FOM data.

When scaling S, ? has a very small improvement but power is raised proportional to S. On the other sides, if L is reduced, intrinsic capacitances become smaller and a faster speed can be realized with lower energy. It proves that, basically the better technology, the faster speed and lower energy will have.

For some comparison purposes, SAR ADCs need one comparator so that the comparator accuracy is N bit resolution. Another architectures, like binary search ADC which is similar to Flash ADC with 2N-1 comparators, can follow binary search quantization algorithm. Different to the SAR ADC using original comparator every time, binary search ADC continuously uses different comparators and doesn't need to wait the DAC saturate. Each comparator's accuracy is the same and it doesn't relax the accuracy due to the Nb resolution accuracy. Same as the SAR ADC, total comparison count is Nb and the regeneration time is the same as well.

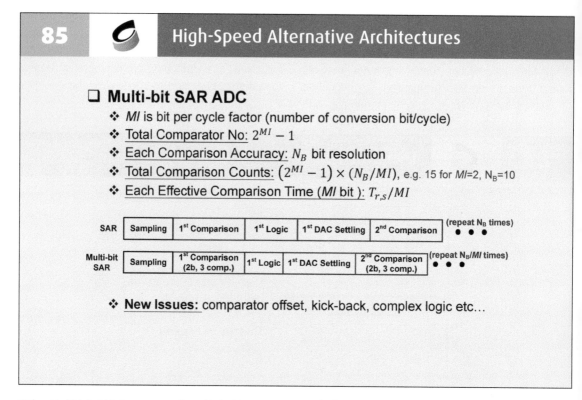

85 High-Speed Alternative Architectures

❑ **Multi-bit SAR ADC**
   ❖ *MI* is bit per cycle factor (number of conversion bit/cycle)
   ❖ Total Comparator No: $2^{MI} - 1$
   ❖ Each Comparison Accuracy: N_B bit resolution
   ❖ Total Comparison Counts: $(2^{MI} - 1) \times (N_B/MI)$, e.g. 15 for *MI*=2, N$_B$=10
   ❖ Each Effective Comparison Time (*MI* bit): $T_{r,s}/MI$

| SAR | Sampling | 1st Comparison | 1st Logic | 1st DAC Settling | 2nd Comparison | (repeat N$_B$ times) • • • |

| Multi-bit SAR | Sampling | 1st Comparison (2b, 3 comp.) | 1st Logic | 1st DAC Settling | 2nd Comparison (2b, 3 comp.) | (repeat N$_B$/MI times) • • • |

   ❖ **New Issues:** comparator offset, kick-back, complex logic etc...

The Multi-bit SAR is compared multiple bit at one time and multiple comparators and DAC arrays are needed. For example, if the ADC has a 2 bit per cycle and a 10 bit resolution, there will be 15 comparisons. Different to the SAR ADC with 10 times comparisons, multi-bit SAR havs a smaller time for regeneration due to the 2 bit comparison every time. It is the reason that so many cycles are don't needed and the time is reduced.

86 Flash + SAR ADC (I)

9b 90MS/s Design [U Fat Chio, *TSCAS-II* 2010]

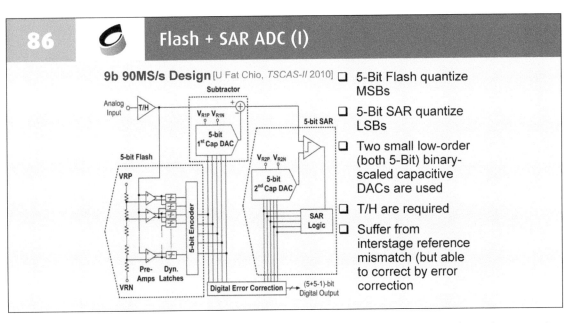

- ❏ 5-Bit Flash quantize MSBs
- ❏ 5-Bit SAR quantize LSBs
- ❏ Two small low-order (both 5-Bit) binary-scaled capacitive DACs are used
- ❏ T/H are required
- ❏ Suffer from interstage reference mismatch (but able to correct by error correction)

Nowadays, many high efficient ADCs in ISSCC or other papers, are using kind of hybrid architecture. Hybrid architecture means that the design is combining different architecture, taking the advantage of them, then combine them together for a new architecture. For example, Flash SAR is a hybrid architecture which use the Flash to quantize the MSB in the beginning and SAR quantize the LSB.

87 Flash + SAR ADC (Ii)

9b 150MS/s Design [Ying-Zu Lin, *VLSI 2010*]

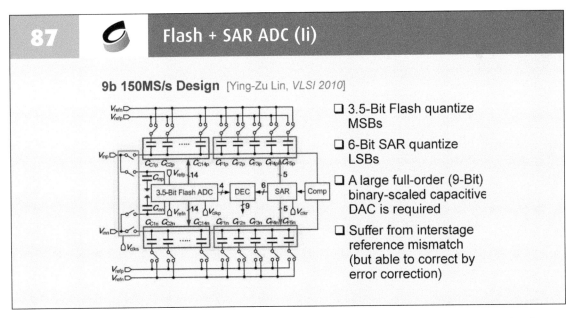

- ❏ 3.5-Bit Flash quantize MSBs
- ❏ 6-Bit SAR quantize LSBs
- ❏ A large full-order (9-Bit) binary-scaled capacitive DAC is required
- ❏ Suffer from interstage reference mismatch (but able to correct by error correction)

The 3.5 bit Flash ADC starts the comparison first. After getting the Flash ADC output, it immediately puts the output to the SAR ADC and decodes the rest to quantize the LSB.

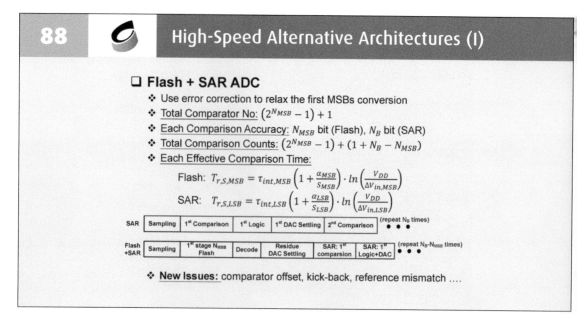

88 High-Speed Alternative Architectures (I)

❑ **Flash + SAR ADC**
- ❖ Use error correction to relax the first MSBs conversion
- ❖ Total Comparator No: $(2^{N_{MSB}} - 1) + 1$
- ❖ Each Comparison Accuracy: N_{MSB} bit (Flash), N_B bit (SAR)
- ❖ Total Comparison Counts: $(2^{N_{MSB}} - 1) + (1 + N_B - N_{MSB})$
- ❖ Each Effective Comparison Time:

$$\text{Flash: } T_{r,S,MSB} = \tau_{int,MSB}\left(1 + \frac{\alpha_{MSB}}{S_{MSB}}\right) \cdot ln\left(\frac{V_{DD}}{\Delta V_{in,MSB}}\right)$$

$$\text{SAR: } T_{r,S,LSB} = \tau_{int,LSB}\left(1 + \frac{\alpha_{LSB}}{S_{LSB}}\right) \cdot ln\left(\frac{V_{DD}}{\Delta V_{in,LSB}}\right)$$

SAR	Sampling	1st Comparison	1st Logic	1st DAC Settling	2nd Comparison	(repeat N_B times) •••

Flash +SAR	Sampling	1st stage N_{MSB} Flash	Decode	Residue DAC Settling	SAR: 1st comparsion	SAR: 1st Logic+DAC	(repeat N_B-N_{MSB} times) •••

- ❖ **New Issues:** comparator offset, kick-back, reference mismatch ....

In the Flash SAR ADC, the total comparator number is the sum of Flash ADC comparators, which is 2NMSB-1 and SAR ADC comparator which is only one. The resolution has no change where the Flash ADC convert MSB resolution and SAR ADC convert the rest. MSB can relaxed for Flash ADC, but the Nb resolution is still needed for SAR ADC.

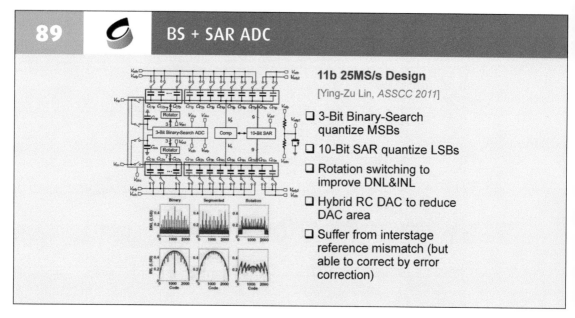

89 BS + SAR ADC

11b 25MS/s Design
[Ying-Zu Lin, *ASSCC 2011*]

- ❑ 3-Bit Binary-Search quantize MSBs
- ❑ 10-Bit SAR quantize LSBs
- ❑ Rotation switching to improve DNL&INL
- ❑ Hybrid RC DAC to reduce DAC area
- ❑ Suffer from interstage reference mismatch (but able to correct by error correction)

Similar like Flash SAR ADC, BS SAR ADCs use the binary search in the front and SAR ADC afterwards. For multi stage, also called pipelining, the error correction can be used to relax the first MSB conversion. It is the advantage of multi bit. For example, if a 10 bit resolution is needed, an extra 1 bit can be designed to relax the first MSB conversion.

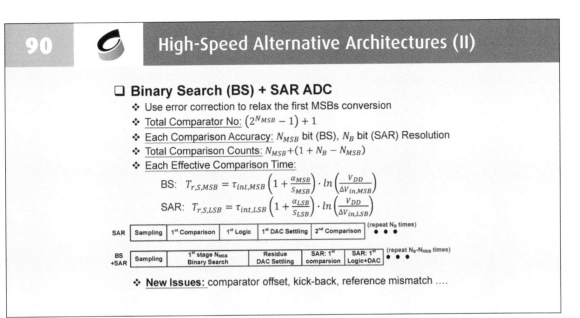

90 High-Speed Alternative Architectures (II)

❑ **Binary Search (BS) + SAR ADC**
  ❖ Use error correction to relax the first MSBs conversion
  ❖ Total Comparator No: $(2^{N_{MSB}} - 1) + 1$
  ❖ Each Comparison Accuracy: N_{MSB} bit (BS), N_B bit (SAR) Resolution
  ❖ Total Comparison Counts: $N_{MSB} + (1 + N_B - N_{MSB})$
  ❖ Each Effective Comparison Time:

  $$\text{BS:} \quad T_{r,S,MSB} = \tau_{int,MSB}\left(1 + \frac{\alpha_{MSB}}{S_{MSB}}\right) \cdot \ln\left(\frac{V_{DD}}{\Delta V_{in,MSB}}\right)$$

  $$\text{SAR:} \quad T_{r,S,LSB} = \tau_{int,LSB}\left(1 + \frac{\alpha_{LSB}}{S_{LSB}}\right) \cdot \ln\left(\frac{V_{DD}}{\Delta V_{in,LSB}}\right)$$

SAR	Sampling	1st Comparison	1st Logic	1st DAC Settling	2nd Comparison	• • • (repeat N_B times)

BS +SAR	Sampling	1st stage N_{MSB} Binary Search	Residue DAC Settling	SAR: 1st comparsion	SAR: 1st Logic+DAC	• • • (repeat N_B-N_{MSB} times)

  ❖ **New Issues:** comparator offset, kick-back, reference mismatch ....

The BS SAR ADC has the same total comparators, accuracy, counts with the Flash SAR ADC.

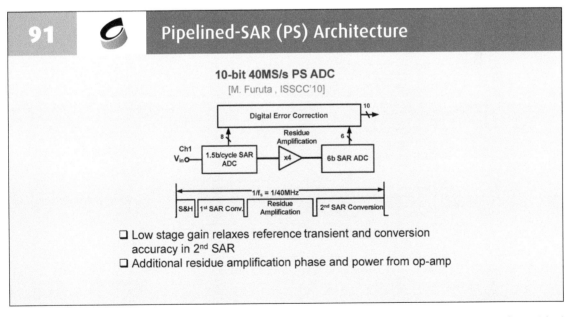

91 Pipelined-SAR (PS) Architecture

10-bit 40MS/s PS ADC
[M. Furuta , ISSCC'10]

❑ Low stage gain relaxes reference transient and conversion accuracy in 2nd SAR
❑ Additional residue amplification phase and power from op-amp

The pipelined SAR can use the error correction to relax the first MSB conversion, meantime use the residual amplification gain to relax the LSB conversion.

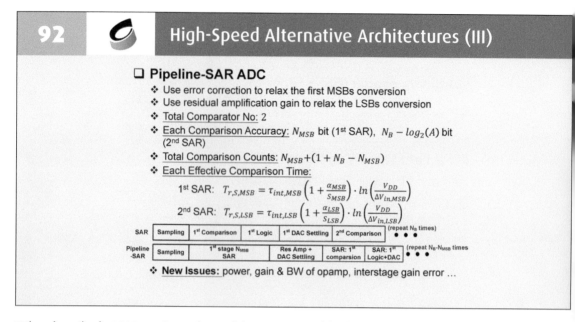

92 High-Speed Alternative Architectures (III)

❏ **Pipeline-SAR ADC**
- ❖ Use error correction to relax the first MSBs conversion
- ❖ Use residual amplification gain to relax the LSBs conversion
- ❖ Total Comparator No: 2
- ❖ Each Comparison Accuracy: N_{MSB} bit (1st SAR), $N_B - log_2(A)$ bit (2nd SAR)
- ❖ Total Comparison Counts: $N_{MSB}+(1 + N_B - N_{MSB})$
- ❖ Each Effective Comparison Time:

1st SAR: $T_{r,S,MSB} = \tau_{int,MSB} \left(1 + \frac{\alpha_{MSB}}{S_{MSB}}\right) \cdot ln\left(\frac{V_{DD}}{\Delta V_{in,MSB}}\right)$

2nd SAR: $T_{r,S,LSB} = \tau_{int,LSB} \left(1 + \frac{\alpha_{LSB}}{S_{LSB}}\right) \cdot ln\left(\frac{V_{DD}}{\Delta V_{in,LSB}}\right)$

SAR	Sampling	1st Comparison	1st Logic	1st DAC Settling	2nd Comparison	(repeat N_B times) •••

Pipeline -SAR	Sampling	1st stage N_{MSB} SAR	Res Amp + DAC Settling	SAR: 1st comparsion	SAR: 1st Logic+DAC	(repeat N_B-N_{MSB} times) •••

- ❖ **New Issues:** power, gain & BW of opamp, interstage gain error ...

Therefore, the first SAR requires only Nmsb bit accuracy and for the second stage requires Nb-log2(A) where A is the gain of the residue amplifier.

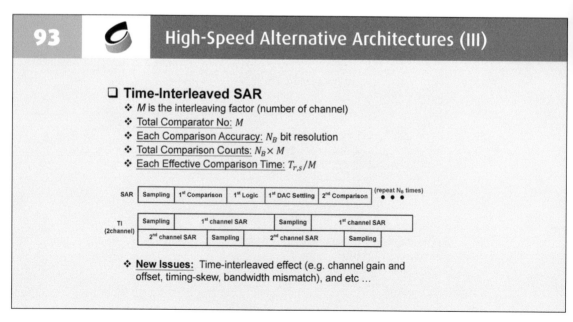

93 High-Speed Alternative Architectures (III)

❏ **Time-Interleaved SAR**
- ❖ M is the interleaving factor (number of channel)
- ❖ Total Comparator No: M
- ❖ Each Comparison Accuracy: N_B bit resolution
- ❖ Total Comparison Counts: $N_B \times M$
- ❖ Each Effective Comparison Time: $T_{r,s}/M$

SAR	Sampling	1st Comparison	1st Logic	1st DAC Settling	2nd Comparison	(repeat N_B times) •••

TI (2channel)	Sampling	1st channel SAR		Sampling	1st channel SAR	
	2nd channel SAR		Sampling	2nd channel SAR		Sampling

- ❖ **New Issues:** Time-interleaved effect (e.g. channel gain and offset, timing-skew, bandwidth mismatch), and etc ...

In time-interleaved ADC, the speed can be higher due to the smaller and more effective comparison time. But the resolution of the comparison doesn't change because of the Nb.

This slide shows many parameters, include the comparator number, the comparison accuracy, the number of comparisons and comparator effective time.

94   Comparator Energy, Resolution and Speed

Convert to different resolution comparator (e.g. for MSB & LSB):

$$\frac{t_{comp,10b,S}}{t_{comp,5b,S}} = \frac{\ln\left(V_{DD}\big/\Delta V_{in,10b}\right)}{\ln\left(V_{DD}\big/\Delta V_{in,5b}\right)} \cong 1.83$$

Energy ratio when convert to different resolution:

$$\frac{P_{comp,10b,S}}{P_{comp,5b,S}} = \frac{f_{r,10b,S}}{f_{r,5b,S}} \cdot \frac{\ln\left(V_{DD}\big/\Delta V_{in,10b}\right)}{\ln\left(V_{DD}\big/\Delta V_{in,5b}\right)} \Rightarrow \frac{E_{comp,10b,S}}{E_{comp,5b,S}} = \frac{1.833}{1.833} = 1$$

❑ For the same comparator design, when the resolution requirement is relaxed from 10 to 5b, and under the same energy, the speed can enhance 1.833 times..

For the same comparator design, when the resolution requirement is relaxed from 10 bit to 5 bit, and under the same energy, the speed can enhance 1.833 times.

95   Architecture Trend (Energy vs Speed) (I)

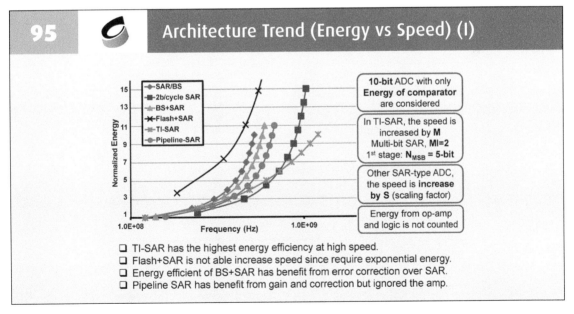

❑ TI-SAR has the highest energy efficiency at high speed.
❑ Flash+SAR is not able increase speed since require exponential energy.
❑ Energy efficient of BS+SAR has benefit from error correction over SAR.
❑ Pipeline SAR has benefit from gain and correction but ignored the amp.

As shown in this slide, TI-SARs have the highest energy efficiency at high speed. They cannot increase linearly and still have to be exponential. Flash SARs are not able increase speed since the energy is exponentially increased. Therefore, the Flash SAR is not so good in terms of the energy efficiency. The multi-bit SAR with 2 bit per cycle is very close to the time-interleaved ADC. Pipelined SAR ADCs also have the better efficiency but the power of the amplifier is not counted.

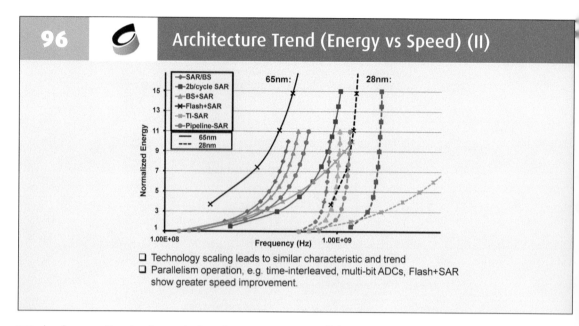

96   **Architecture Trend (Energy vs Speed) (II)**

❑ Technology scaling leads to similar characteristic and trend
❑ Parallelism operation, e.g. time-interleaved, multi-bit ADCs, Flash+SAR
show greater speed improvement.

Technology scaling leads to similar characteristic and trend. In 28nm, Flash SAR become better than the normal SAR ADC and the binary SAR ADC. A great speed improvement can be gained by using parallelism operations, like time-interleaved, multi-bit ADC and Flash. Therefore, in a very advanced technology, the parallelism should be used as possible.

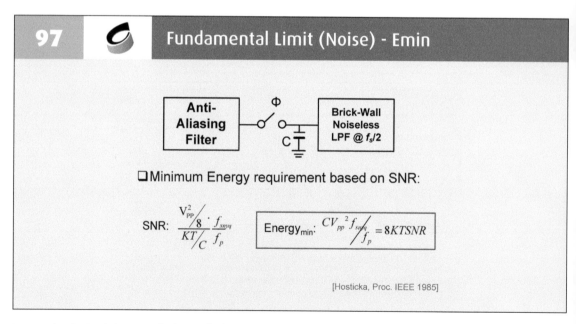

97   **Fundamental Limit (Noise) - Emin**

❑ Minimum Energy requirement based on SNR:

SNR: $\dfrac{V_{pp}^2/8}{KT/C} \cdot \dfrac{f_{snyq}}{f_p}$ Energy$_{min}$: $\dfrac{CV_{pp}^2 f_{snyq}}{f_p} = 8KTSNR$

[Hosticka, Proc. IEEE 1985]

In a noise limited design, which is called Emin, the KT/C noise is the minimum noise requirement of the design. The KT/C noise cannot be avoided because the sampling need the capacitor. Therefore, the minimum energy can be calculated.

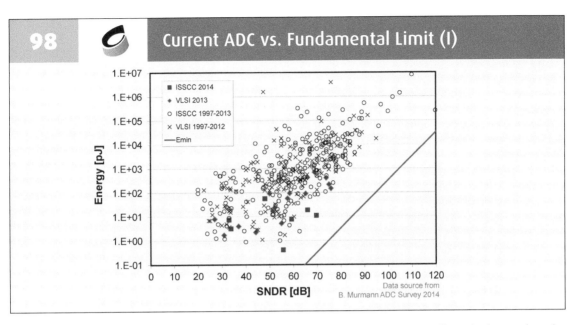

In this slide, the red line is the Emin, and theoretically, the ADC's performance will not be better than the Emin.

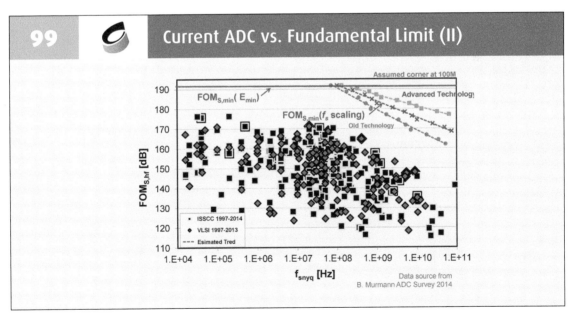

The Emin cannot be achieved basically and with the scaling factor, there is a pending effect because of intrinsic parasitic capacitance and so on. However with an advanced technology, the line will come much closer to the Emin, which means the effect of scaling factors becomes smaller.

ENERGY OF THE SAR-TYPE ADCS VS SNDR

The energy versus SNDR will be discussed.

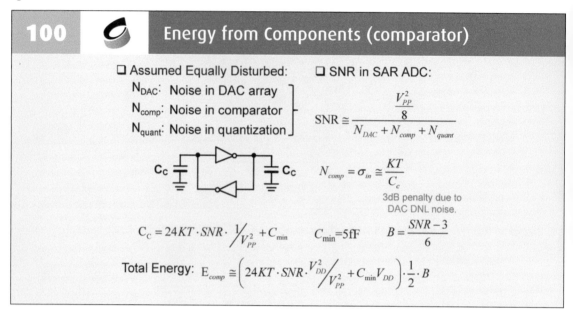

In a SAR ADC, the noises are from the DAC noise, comparator noise, quantizers and so on. The total energy from comparator can be obtained with respect to the SNR.

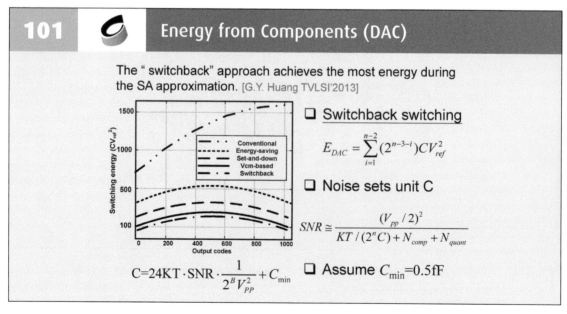

The energy from DAC is the switching energy. How to reduce the switching energy has been discussed frequently.

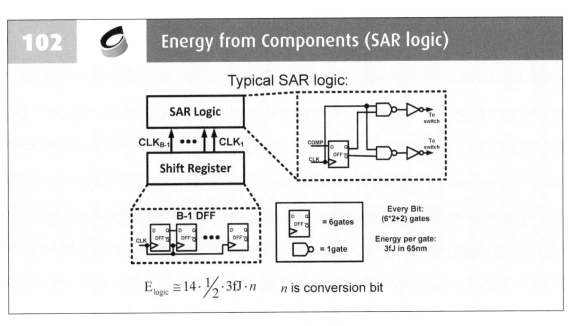

102 — **Energy from Components (SAR logic)**

Typical SAR logic:

$$E_{logic} \cong 14 \cdot \frac{1}{2} \cdot 3fJ \cdot n \qquad n \text{ is conversion bit}$$

Another energy is the logic power which is the typical digital logic power.

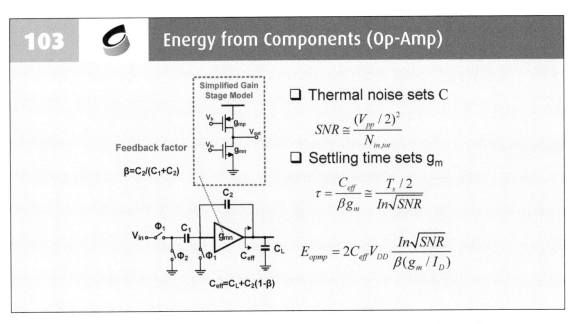

103 — **Energy from Components (Op-Amp)**

Simplified Gain Stage Model

Feedback factor
$$\beta = C_2/(C_1+C_2)$$

$$C_{eff} = C_L + C_2(1-\beta)$$

❏ Thermal noise sets C

$$SNR \cong \frac{(V_{pp}/2)^2}{N_{in,tot}}$$

❏ Settling time sets g_m

$$\tau = \frac{C_{eff}}{\beta g_m} \cong \frac{T_s/2}{In\sqrt{SNR}}$$

$$E_{opmp} = 2C_{eff}V_{DD}\frac{In\sqrt{SNR}}{\beta(g_m/I_D)}$$

The energy of Op-Amp, which is needed for residue amplification, will be included in pipeline SAR.

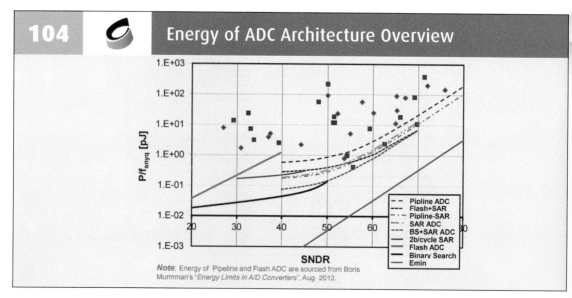

104 — **Energy of ADC Architecture Overview**

Note: Energy of Pipeline and Flash ADC are sourced from Boris Murmman's *"Energy Limits in A/D Converters"*, Aug. 2012.

The binary search ADCs actually can only going medium/low resolutions as a better efficiency. But the binary search SAR can extend the efficiency up to a certain resolution due to the digital corrections and so on. It can be found that the pipelined SAR ADC has higher efficiency compared to traditional pipeline ADC. Besides, they can achieve higher resolutions by residue amplifier to relax the resolution of the second stage.

BOUNDARY-EXTENSION / BENCHMARKING HIGH-PERFORMANCE ADCS

The next section is some boundary design/benchmarking high performance ADC.

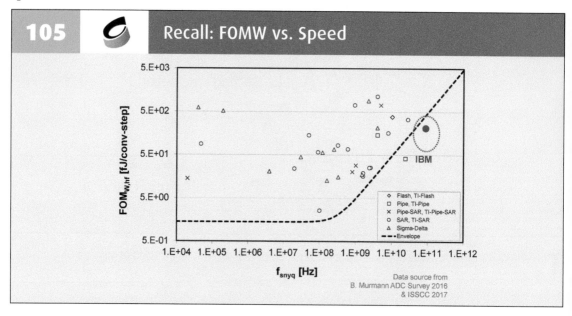

105 — **Recall: FOMW vs. Speed**

Data source from
B. Murmann ADC Survey 2016
& ISSCC 2017

This is the ADC from IBM developed, which is achieving the highest speed with 90 Giga samle per second.

90GS/s 8b TI SAR ADC

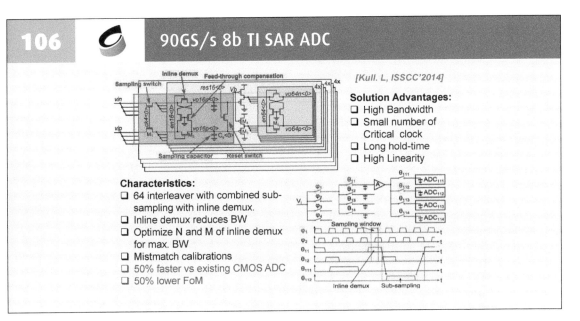

106

It is a time-interleaved ADC and achieves very high bandwidth. The main idea of it is actually using a kink of hierarchical interleaving scheme, which is not a new thing. In this design, 64 interleaver with combined sub-sampling with inline demux are used, and a buffer is given until a certain leaver. It achieves very good performance with a fifty percent faster vs existing CMOS ADC and fifty percent lower FOM. It is the way to optimise M and N inline demux for maximum bandwidth.

Recall : Energy vs. Resolution

107

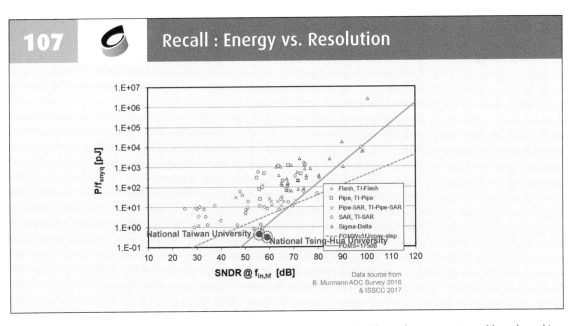

An ADC with the fastest performance or the lowest energy is called boundary extension of benchmarking ADC. These two ADCs achieve the lowest FOM nowadays.

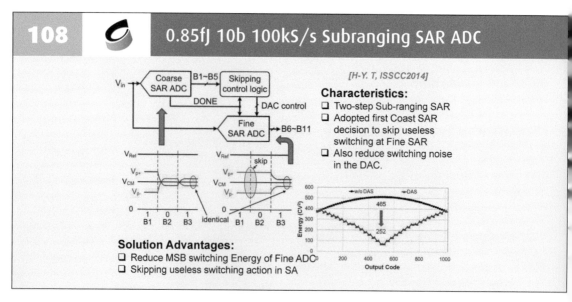

108   0.85fJ 10b 100kS/s Subranging SAR ADC

[H-Y. T, ISSCC2014]

Characteristics:
- ❑ Two-step Sub-ranging SAR
- ❑ Adopted first Coast SAR decision to skip useless switching at Fine SAR
- ❑ Also reduce switching noise in the DAC.

Solution Advantages:
- ❑ Reduce MSB switching Energy of Fine ADC
- ❑ Skipping useless switching action in SA

The first ADC is using a two-step approach which can get a better efficiency but a reduced speed. However, the sample rate in this application is only 100kHz. A coarse SAR ADC is used first and a traditional SAR ADC is realized to do some conversions converge to the residues. Therefore, some unnecessary switching can be skipped and the energy can be reduced much more. The ADC achieves at zero point with 0.85pJ for conversion step.

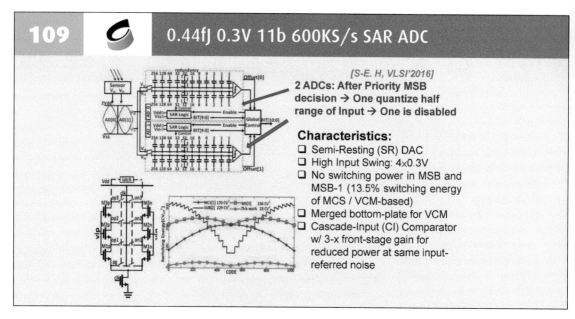

109   0.44fJ 0.3V 11b 600KS/s SAR ADC

[S-E. H, VLSI'2016]

2 ADCs: After Priority MSB decision → One quantize half range of Input → One is disabled

Characteristics:
- ❑ Semi-Resting (SR) DAC
- ❑ High Input Swing: 4×0.3V
- ❑ No switching power in MSB and MSB-1 (13.5% switching energy of MCS / VCM-based)
- ❑ Merged bottom-plate for VCM
- ❑ Cascade-Input (CI) Comparator w/ 3-x front-stage gain for reduced power at same input-referred noise

This ADC which is called semi-resting scheme, achieve smaller than 0.5pJ in conversion step at 0.3V supply. There are two ADCs and after the priority MSB, one quantize only has the half range of input and the other one is disabled. Therefore, the MSB capacitor, which is the biggest one, can be skipped and the power can be saved. In addition, the merging scheme is used so that the VCM buffer is unnecessary, because the positive and negative are sampling together and the nature form will be still get the similar level as the VCM. A cascade input comparator with 3-x front stage gain is used to reduce power at the same input referred noise. Therefore, it achieves very good FOM in 0.44pJ.

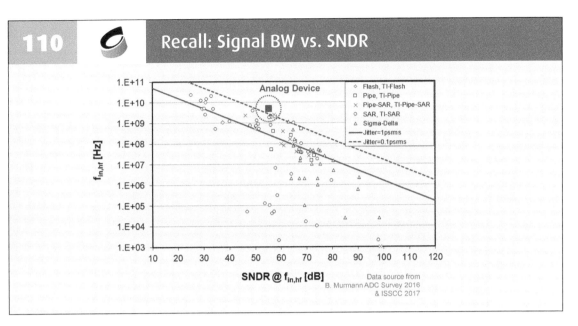

The next one is the lowest jitter design for analog device.

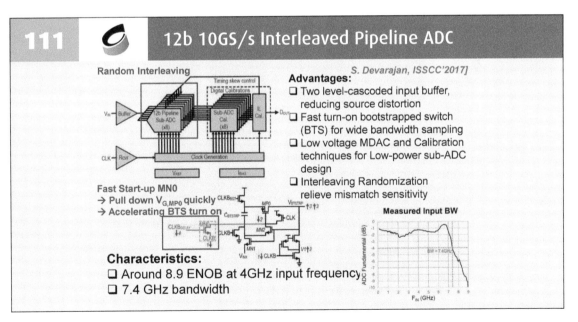

The architecture is an interleaving pipelined ADC, and it achieves a very wide band with smaller than 0.1ps-rms jitter. A two-level cascode input buffer is used in the scheme to reduce the source distortion. The design adds one more pull down stage to prepare in advance so that it makes the overall switch sampling faster than the traditional bootstrap switch. It achieves 7.4GHz input bandwidth with 12 bit resolution and 10GS/s, which is the best FOM so far.

CIRCUIT TECHNIQUES AND INNOVATIONS FOR ADCS IN ISSCC 2017

The next section will talk about the circuit techniques and innovations for ADCs in ISSCC 2017.

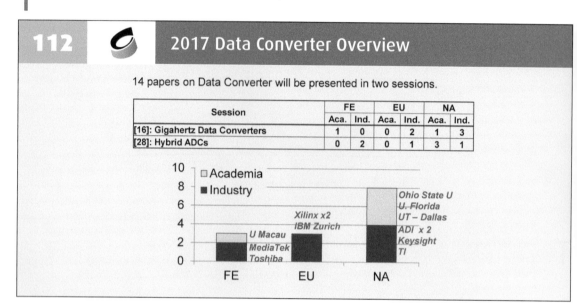

112

2017 Data Converter Overview

14 papers on Data Converter will be presented in two sessions.

Session	FE		EU		NA	
	Aca.	Ind.	Aca.	Ind.	Aca.	Ind.
[16]: Gigahertz Data Converters	1	0	0	2	1	3
[28]: Hybrid ADCs	0	2	0	1	3	1

There are 14 papers in the data converter session. Part of them are from industry, such as Media Tek, Toshiba, IBM, Xilinx, ADI, Keysight and TI, and the rest from the universities, such as Ohio State University, University of Florida, UT-Datlas and University of Macau. All of those speed become higher.

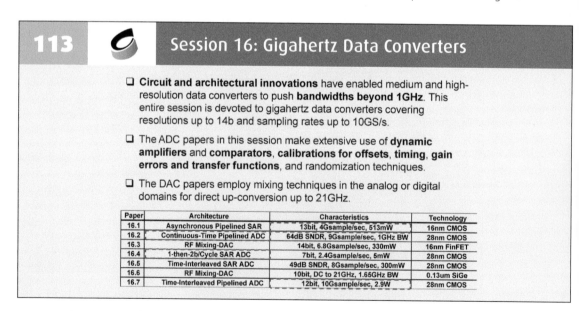

113

Session 16: Gigahertz Data Converters

- ❏ **Circuit and architectural innovations** have enabled medium and high-resolution data converters to push **bandwidths beyond 1GHz**. This entire session is devoted to gigahertz data converters covering resolutions up to 14b and sampling rates up to 10GS/s.

- ❏ The ADC papers in this session make extensive use of **dynamic amplifiers** and **comparators, calibrations for offsets, timing, gain errors and transfer functions**, and randomization techniques.

- ❏ The DAC papers employ mixing techniques in the analog or digital domains for direct up-conversion up to 21GHz.

Paper	Architecture	Characteristics	Technology
16.1	Asynchronous Pipelined SAR	13bit, 4Gsample/sec, 513mW	16nm CMOS
16.2	Continuous-Time Pipelined ADC	64dB SNDR, 9Gsample/sec, 1GHz BW	28nm CMOS
16.3	RF Mixing-DAC	14bit, 6.8Gsample/sec, 330mW	16nm FinFET
16.4	1-then-2b/Cycle SAR ADC	7bit, 2.4Gsample/sec, 5mW	28nm CMOS
16.5	Time-Interleaved SAR ADC	49dB SNDR, 8Gsample/sec, 300mW	28nm CMOS
16.6	RF Mixing-DAC	10bit, DC to 21GHz, 1.65GHz BW	0.13um SiGe
16.7	Time-Interleaved Pipelined ADC	12bit, 10Gsample/sec, 2.9W	28nm CMOS

For Gigahertz ADC, the circuit and architectural innovation push the bandwidth beyond 1GHz. The ADC papers in this session make extensive use of dynamic amplifiers and comarators, calibrations for offsets, timing, gain erroes and transfer functions, and randomization techniques. They are mostly in 28nm and 16nm process.

114   Session 28: Hybrid ADCs

❏ Analog-to-digital converters (ADCs) continue to evolve from classical architectures into **hybrid converters** that combine the strengths of various ADC types.

❏ This session demonstrates multiple hybrid ADCs ranging from MHz to GHz bandwidths, employing combinations of **SAR**, **pipeline** and **oversampling** architectures.

❏ Pipelined-SAR ADCs utilizing **PVT-stabilized** dynamic amplifiers, separate comparators per decision and digital amplifiers are disclosed. Noise-shaping phase-domain excess-loop-delay compensation as well as a double noise-shaping quantizer are employed in the ADCs to improve the conversion efficiency.

Paper	Architecture	Characteristics	Technology
28.1	Noise-Shaping SAR	80dB SNDR, 5MHz BW, 0.46mW	28nm CMOS
28.2	Continuous-Time ADC	80dB SNDR, 15MHz BW, 11.4mW	0.13um CMOS
28.3	VCO-Based Continuous-Time ADC	72dB SNDR, 125MHz BW, 54mW	16nm CMOS
28.4	Pipelined SAR ADC	12bit, 330Msample/sec, 6.2mW	65nm CMOS
28.5	Pipelined SAR ADC	10bit, 1.5Gsample/sec, 7.0mW	14nm FinFET
28.6	Two-Step Split SAR ADC	78.5dB SNDR, 75Msample/sec, 25mW	65nm CMOS
28.7	Pipelined SAR ADC	12bit, 160Msample/sec, 1.9mW	28nm CMOS

The hybrid ADCs basically combine the SAR, pipeline and oversampling ADC. The scheme includes PVT-stabilized scheme, noise shaping SAR, continuous time ADC, VCO based continuous time ADC, pipeline SAR and two-step split scheme. The 0.13um and 65nm can still be used by hybrid ADC and in the previous session there are some hybrid schemes. Therefore, the hybrid schemes have become a very hot topic now.

115   Data Converter Trend – Power Efficiency

● This figure plots power dissipated relative to the effective Nyquist sampling rate (P/fsnyq), as a function of signal-to-noise and distortion ratio (SNDR), to give a measure of ADC power efficiency. Contributions at ISSCC 2017 are depicted by the colored legends representing various converter architectures.

● ISSCC 2017 heralds the first converter achieving the benchmark of 5fJ/conversion-step while simultaneously demonstrating nearly 80 dB SNDR through the use of noise shaping and dynamic amplification techniques.

This is the data convert trend in the part of power efficiency, speed and so on. This figure plots power dissipated relative to the effective Nyquist sampling rate, as a function of SNDR, to give a measure of ADC power efficiency. ISSCC 2017 heralds the first converter achieving the benchmark of 5fJ/conversion-step while simultaneously demonstrating nearly 80dB SNDR through the use of noise shaping and dynamic amplification techniques.

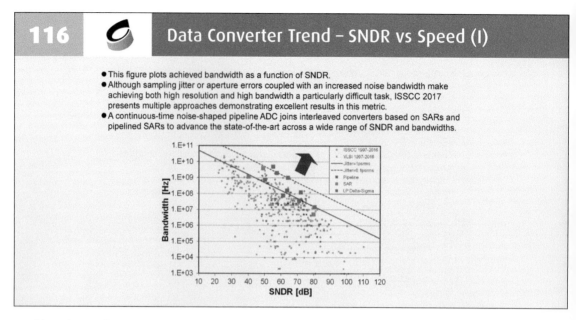

Although sampling jitter of aperture errors coupled with an increased noise bandwidth make achieving both high resolution and high bandwidth a particularly difficult task, ISSCC 2017 presents multiple approaches demonstrating excellent results in this matric.

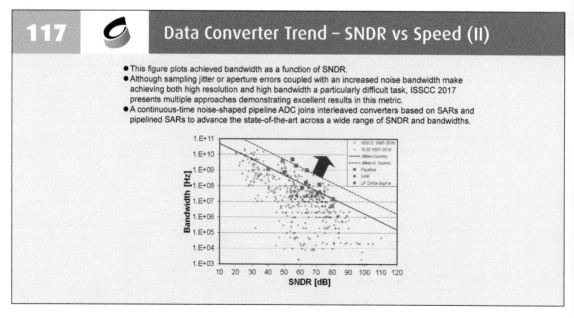

A continuous-time noise-shaped pipeline ADC joins inerleaved converters based on SARs and pipelined SARs to advance the state-of-the-art across a wide range of SNDR and bandwidths.

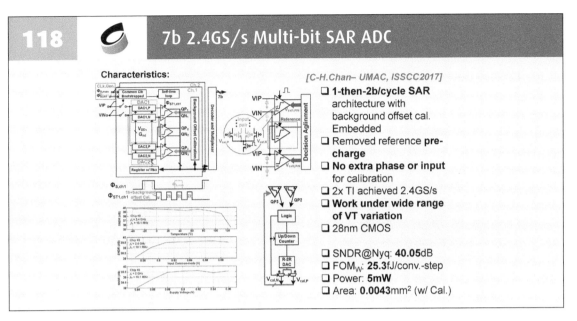

118

7b 2.4GS/s Multi-bit SAR ADC

[C-H.Chan– UMAC, ISSCC2017]

- ❏ **1-then-2b/cycle SAR** architecture with background offset cal. Embedded
- ❏ Removed reference **pre-charge**
- ❏ **No extra phase or input** for calibration
- ❏ 2x TI achieved 2.4GS/s
- ❏ **Work under wide range of VT variation**
- ❏ 28nm CMOS

- ❏ SNDR@Nyq: **40.05dB**
- ❏ FOM$_W$: **25.3fJ/conv.-step**
- ❏ Power: **5mW**
- ❏ Area: **0.0043mm²** (w/ Cal.)

This one is the multi-bit SAR from university of Macau. The multi-bit SAR is quite competitive with the time-interleaving, because it doesn't need to calibrate the gain and timing. It's the first time to propose a kind of 1-then-2b/cycle SAR for 7 bit and 2.4GS/s. There are three comparators, one is for comparison and the other two are for offset calibrations. The first bit comparison determines the power priorities of the signal so that a large precharge scheme can be avoided because the MSB charging is the biggest step. It achieves a good FOM, with 25 fJ/conversion step and 5mW, 0.0043mm2 with calibration on chip.

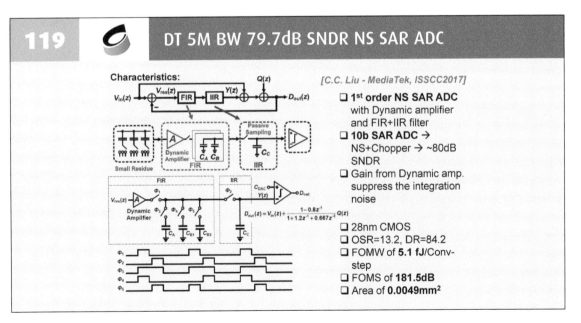

119

DT 5M BW 79.7dB SNDR NS SAR ADC

[C.C. Liu - MediaTek, ISSCC2017]

- ❏ **1st order NS SAR ADC** with Dynamic amplifier and FIR+IIR filter
- ❏ **10b SAR ADC →** NS+Chopper → ~80dB SNDR
- ❏ Gain from Dynamic amp. suppress the integration noise

- ❏ 28nm CMOS
- ❏ OSR=13.2, DR=84.2
- ❏ FOMW of **5.1 fJ/Conv-step**
- ❏ FOMS of **181.5dB**
- ❏ Area of **0.0049mm²**

$$D_{out}(z) = V_{in}(z) + \frac{1-0.8z^{-1}}{1+1.2z^{-1}+0.667z^{-2}} Q(z)$$

This one is the noise shaping SAR ADC from Media Tek, which uses FIR and IIR scheme to noise-shape the residue and improve the SNR. It uses a passive sampling for power saving and a proposed dynamic amplifier to keep the gain. All the integration noise including comparator noise and input noise can be shaped and it achieves a very good FOM at 181.5dB with 5.1 fJ/conversion step.

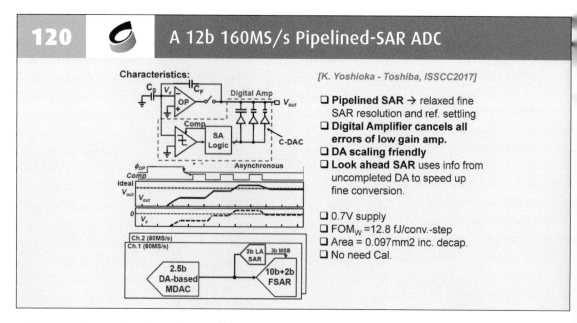

120 **A 12b 160MS/s Pipelined-SAR ADC**

Characteristics:

[K. Yoshioka - Toshiba, ISSCC2017]

❑ **Pipelined SAR** → relaxed fine SAR resolution and ref. settling
❑ **Digital Amplifier cancels all errors of low gain amp.**
❑ **DA scaling friendly**
❑ **Look ahead SAR** uses info from uncompleted DA to speed up fine conversion.

❑ 0.7V supply
❑ FOM_W =12.8 fJ/conv.-step
❑ Area = 0.097mm2 inc. decap.
❑ No need Cal.

This one is the pipelined SAR from Toshiba. It uses digital amplifiers with a SAR logic to compensate the virtual ground node and save the power. It also realizes the first part with a digital MDAC. The smart way to save some times is using the information from incomplete digital amplifier to speed up the fine conversion. Therefore, when the digital amplifiers process complete, the fine SAR ADC already have three bits known.

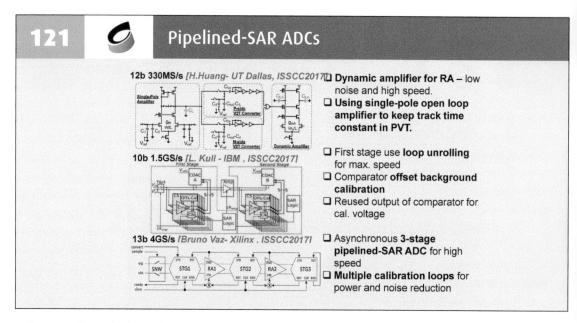

121 **Pipelined-SAR ADCs**

12b 330MS/s *[H.Huang- UT Dallas, ISSCC2017]*

10b 1.5GS/s *[L. Kull - IBM , ISSCC2017]*

13b 4GS/s *[Bruno Vaz- Xilinx , ISSCC2017]*

❑ **Dynamic amplifier for RA** – low noise and high speed.
❑ **Using single-pole open loop amplifier to keep track time constant in PVT.**

❑ First stage use **loop unrolling** for max. speed
❑ Comparator **offset background calibration**
❑ Reused output of comparator for cal. voltage

❑ Asynchronous **3-stage pipelined-SAR ADC** for high speed
❑ **Multiple calibration loops** for power and noise reduction

This one is the multiple pipelined SAR ADC from IBM. It uses a single pole open loop dynamic amplifier to keep track time constant of PVT. It uses unrolling comparator at first stage to enhance the speed, do the calibrations and so on. It achieves three stage pipelined SAR ADC at 13 bit 4GS/s with multiple calibration loop for power and noise reduction.

122 **Summary**

❑ ISSCC presents State-of-the-Art Data Converters with either **Boundary-Extension / Benchmarking High Performance** or **High-Potential Technologies with Innovations** at

❖ Architectural Design – Hybrid

❖ Circuit Techniques – Hybrid

❖ Digital-Assisted Calibration Techniques

❖ Technology-Driven Optimization

❖ Application-Driven Optimization

❖ … and more

In conclusion, parallelism, digital assist calibrations are actually widely used, and hybrid scheme which combines all these together becomes very popular. This is the summary of ADC trend.

123 **ADC implementations in advanced technologies**

• Technology Scaling – the bad
  – Reduction of g_m/g_{ds}
  – Headroom limitations caused by VDD reduction
  – Bad CMOS switches
  – Higher variability
  – Transistor properties and matching more and more dependent on surroundings
  – Higher interconnect delays

• … and the good
  – Faster devices
  – Digital processing is increasingly powerful ... able to overcome limitations in the anal...

Although the technology scaling leads to the gm and headroom reduction, bad CMOS switch, higher variability, higher interconnect delays and transistor properties and matching more and more dependent on surroundings, there are still some advantages which makes the devices much faster and the digital processing much more powerful. That's the reason why kinds of digital assistance schemes are needed.

124 ADC architectures: digital gain calibration in pipeline ADCs (I)

- Quantizer specifications are relaxed.   MDAC non-idealities limit performance
- Gain error of S/H, DAC and residue amplifier cause G_{Ei} and $G_{Eo} \neq 1$
- **Digital gain calibration**: Multiply by $1/G_{Eo}$

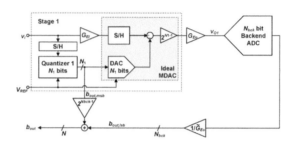

The pipelined ADCs are still showing the good performance and achieving highest bandwidth in the industry nowadays. This slide shows the digital gain calibration in pipelined ADC. There are some certain errors, such as a gain error of sample and hold, the DAC error and also the GEi and GEo error caused by the residue amplifier. The digital calibration can track the gain error and multiply the gains in digital, which is a normally way for gain calibration.

125 ADC architectures: digital gain calibration in pipeline ADCs (II)

- Determination of digital coefficients

Foreground
(Fast Startup)

Background
(Adapt coef. as VDD/Temp varies)

The foreground calibration, which is fast startup, will be done at the beginning. It makes the reference at a certain level so that the compensation coefficients can be known roughly through the gain measurement. With the rough coefficients to compensate the gain of the residue amplifier, the coefficients which drift at the temperature and different supply voltages can also be known at the background. Therefore, with a pseudo-random sequence, the gain can be measured and adjusted continuously.

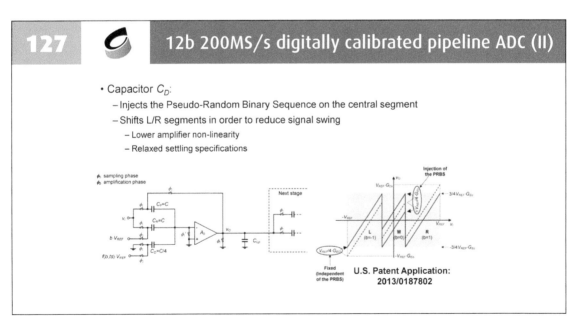

126   **12b 200MS/s digitally calibrated pipeline ADC (I)**

- Calibration of amplifier finite gain and capacitor mismatches
  - Fast startup time and robustness against VDD/Temp variations
- Stages with reduced output swing
- Opamp switching technique with no speed or signal swing limitations

This is an example that actually showing 12 bit 200MS/s digital calibrated pipeline ADC. There are ten stages which is 1.5 bit calibrated stage, three uncalibrated stages due to the small error, and a 3 bit Flash ADC at last. As mentioned, the foreground calibration will start one by one, so that the background calibration will continuously be done.

127   **12b 200MS/s digitally calibrated pipeline ADC (II)**

- Capacitor C_D:
  - Injects the Pseudo-Random Binary Sequence on the central segment
  - Shifts L/R segments in order to reduce signal swing
    - Lower amplifier non-linearity
    - Relaxed settling specifications

U.S. Patent Application:
2013/0187802

A pseudo-random binary sequence is injected on the central segment when a reference is injected to the MDAC. Actually, different sequences will affect the output swing of the pipeline stage. Therefore, it should be made sure that the output is not over-ranged of the next stage so that the digital error correction can be controlled.

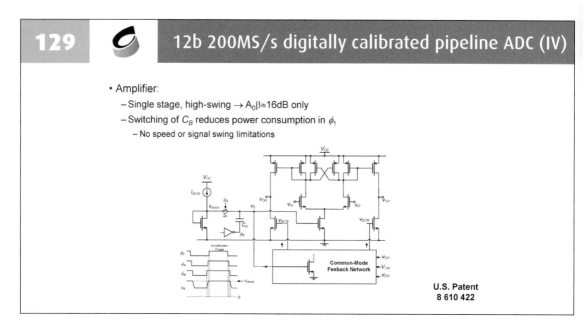

| 128 | 12b 200MS/s digitally calibrated pipeline ADC (III) |

- Foreground measurements:
  - Gain error
  - C_D/C_R ratio, so that the removal of the PRBS in digital domain is done by the right amount

U.S. Patent Application:
2013/0187801

Actually, the gain of amplifier is not so linear over a full range. Therefore, during the foreground measurements, it is necessary to measure multiple times to obtain the gain errors. Besides, the capacitance mismatches can also be averaged during multiple times measurements.

| 129 | 12b 200MS/s digitally calibrated pipeline ADC (IV) |

- Amplifier:
  - Single stage, high-swing → $A_0\beta \approx 16dB$ only
  - Switching of C_B reduces power consumption in ϕ_1
    - No speed or signal swing limitations

U.S. Patent
8 610 422

Due to the digital calibration, a single stage amplifier with roughly 16dB gain is used. CB is used to reduce the power consumption at the different phases.

130   12b 200MS/s digitally calibrated pipeline ADC (V)

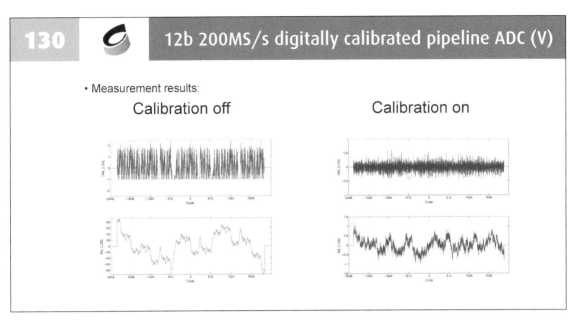

The measure shows different results between calibration off and on. When calibration off, the DNL reaches -1 to 2 and the INL is also very big. But after the calibration, it achieves a very good performance. That's the traditional way to calibrate gain errors with amplifier in pipelined ADC.

131   SAR ADC Basic Block Diagram

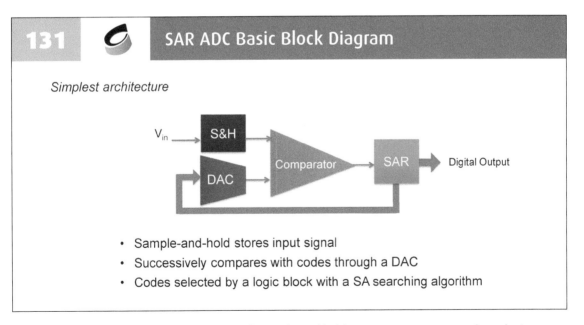

The SAR ADCs have a simple architecture with sample and hold, a DAC, a comparator and SAR logic.

132   Advantages of SAR ADC Architecture

- Compact area:
  - Sample-and-hold and DAC merged into a single capacitor array
- Low power consumption:
  - Zero static consumption possible on all blocks
- Suitable for analog "unfriendly" processes
  - No power-hungry precision amplifier

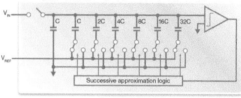

Example: 6-bit SAR ADC

This is a traditional SAR ADC with the capacitor array. The sample and hold and the DAC are normally merged to get a single capacitor array. There is no static power obviously because of the capacitor array and dynamic comparator. That's why it is so suitable for analog processes.

133   Limitations of SAR ADC Architecture

Traditional implementation has drawbacks

- Inherently slow due to the need of oversampled clock
  - Test all bits within one conversion cycle

- Large area due to the size of the capacitor array
  - Increases exponentially with resolution

- Can be improved with a multitude of techniques
  - For low power, low area and higher speed

It is inherently slow due to the need of oversampled clock and a large area due to the size of capacitor array. And the capacitor array increases exponentially for a high resolution.

134 SAR ADC Optimization Techniques

Remove need for high frequency clock

- Bit decision operations sequenced by comparator output
- Comparator outputs (differential) are both reset to VDD before each comparison
- After comparison the two outputs become complementary
- A NAND gate detects the comparison ready
- Together with a delay block this signal can sequence the operations

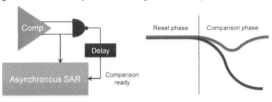

Using asynchronous SAR scheme, the comparator output can trigger for the next operation. It is different to the synchronous SAR ADC, where an assigned clock and a fixed timing are needed for each comparison.

135 SAR ADC Optimization (I)

Reducing size of capacitor array

- Size of capacitor array should be determined by thermal noise only
- Segmentation into a main array and a secondary array
- Calibration of matching errors
- Simplifies driving of ADC input

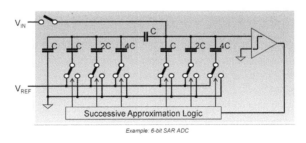

Example: 6-bit SAR ADC

A split capacitance scheme can be used for the high resolution to avoid the very large capacitance. The floating capacitor should be designed carefully because the parasitics will affect the performance. And the calibrations of matching errors are necessary as well.

136   SAR ADC Optimization (II)

Reducing size of capacitor array

- Calibration allows to reduce the value of the unit capacitor (C)
  - Limited by thermal noise only, not matching
- Measurement of matching errors makes use of binary nature of capacitor array
  - Nominal value of each capacitor is equal to the sum of the remaining lower order capacitors (plus a unity capacitor)
  - 32 = 16 + 8 + 4 + 2 + 1 +1
- Comparing capacitors by charge redistribution:

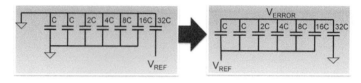

Calibration allows to reduce the value of the unit capacitor and measurement of matching errors makes use of binary nature of capacitor array.

137   SAR ADC Optimization (III)

Other techniques can be applied

- Bottom-plate sampling
  - Better linearity due to no signal dependent charge injection

- Pre-charging to $V_{REF}/2$ instead of V_{REF}
  - Lower consumption due to lower voltage steps in capacitors

- Introducing redundant capacitors for error correction
  - Higher speed because of reduced time allowed for settling

- Time-interleaving
  - Higher speed by parallelization

The bottom-plate sampling can avoid signal dependent charge injection and the Vcm-base switching scheme basically can make the precharging of capacitor from Vref to Vref/2. Besides, the redundant capacitors can be introduced for error corrections and increase the tolerant range. The time interleaving can also be used for high speed architecture.

138   Synopsys High Speed SAR ADC

Architectural advantages

- Purely Dynamic biasing
    - Perfect scaling of power consumption with sampling frequency

- Self timed operation
    - Input clock at same frequency as sampling rate

- Parallelization to achieve highest sampling rates

- Digitally calibrated capacitor array
    - Sizing by the noise constraint only, and not matching

- Calibration run only at start-up
    - Only capacitor ratios need to be corrected: stable with supply/temp variations
    - Fast power up: calibration constants reused after power-down cycles
    - In interleaving, calibration corrects also offset and gain mismatches

The biasing can be purely dynamic so that the bias current increases when the sampling frequency increases. The self timed operation is used in the digital asynchronous calibration scheme and parallelization such as time-interleaved scheme is used for high frequency. Digital calibrated capacitor array can make the capacitor array small and calibrations only run at start-up.

139   Design Example: 12b 80MS/s digitally calibrated SAR ADC (I)

- Asynchronous architecture. Operation independent of clk duty cycle
- Low noise fully dynamic comparator
- Use of time-interleaving: 12b 160MS/s and 320MS/s ADCs

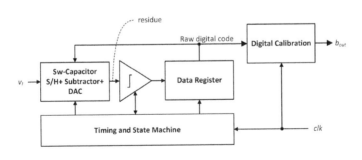

This is an example of a 12 bit 80MS/s digitally calibrated SAR ADC with asynchronous architecture. The operations are independent of clock cycle. Therefore, it can be used in the interleaving scheme for the 12 bit 160MS/s or 320MS/s ADC.

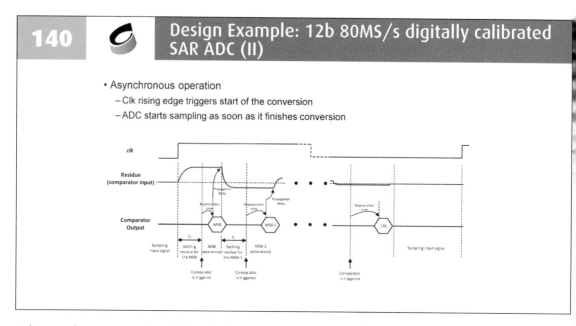

140

Design Example: 12b 80MS/s digitally calibrated SAR ADC (II)

- Asynchronous operation
  - Clk rising edge triggers start of the conversion
  - ADC starts sampling as soon as it finishes conversion

The asynchronous operation is that the last comparison can immediately trigger for the next steps until the end of comparison.

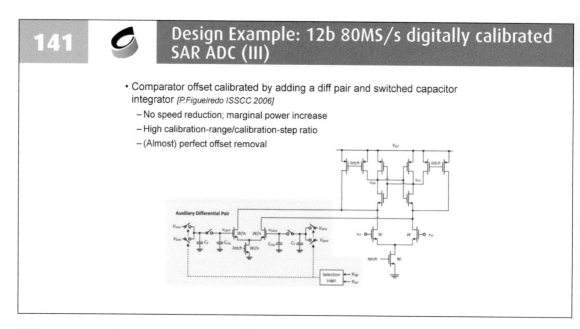

141

Design Example: 12b 80MS/s digitally calibrated SAR ADC (III)

- Comparator offset calibrated by adding a diff pair and switched capacitor integrator [P.Figueiredo ISSCC 2006]
  - No speed reduction; marginal power increase
  - High calibration-range/calibration-step ratio
  - (Almost) perfect offset removal

The comparator in SAR ADC is using an additional differential pair with a switch capacitor to charge the voltage level for offset calibration.

142 Design Example: 12b 80MS/s digitally calibrated SAR ADC (IV)

- DAC with capacitive dividers avoids the exponential increase on the number of (small) unit capacitors

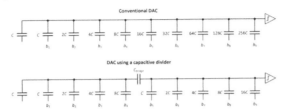

- **Digital calibration**: addresses random capacitor mismatches and sensitivity to parasitics in the capacitive divider nodes.
  - Measures capacitor ratios at startup
  - Corrects the raw code provided by the SAR

This is the DAC array. With the digital calibration, capacitor array mismatches and sensitivity to parasitics in the capacitive divider node can be measured and calibrated.

143 Pipeline vs SAR comparison

- SAR ADC implementation:
  - Smaller power and area
  - Better adapted to technology scaling

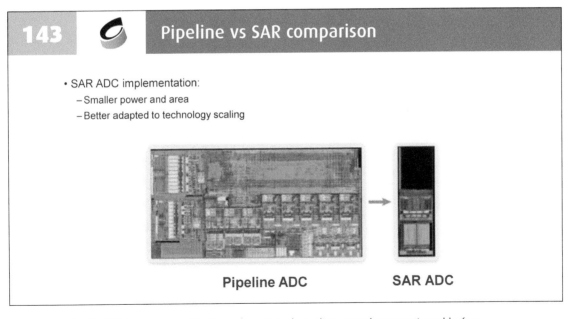

Pipeline ADC   **SAR ADC**

The area of SAR ADC is much smaller than the original pipeline ADC that mentioned before.

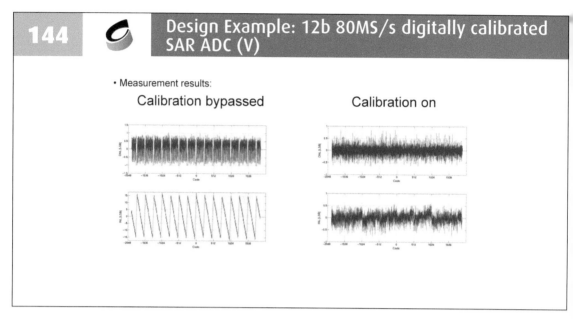

The performance is much better when calibration on than bypassed.

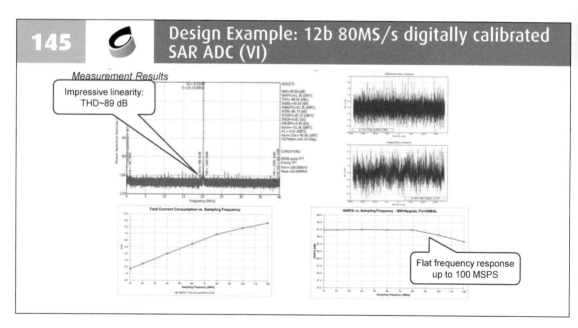

This is the measurement at 320MS/s with the interleaving scheme. The flat frequency response is up the 340MS/s and the highest spur is at -72dBFS. It is much more effective to design a SAR ADC compared to the pipelined ADC. Because in SAR ADC, it can be put in parallel to speed up but in pipelined ADC, it should be spent effort to design many of stages. That's the reason why SAR ADC is much more popular than pipelined ADC.

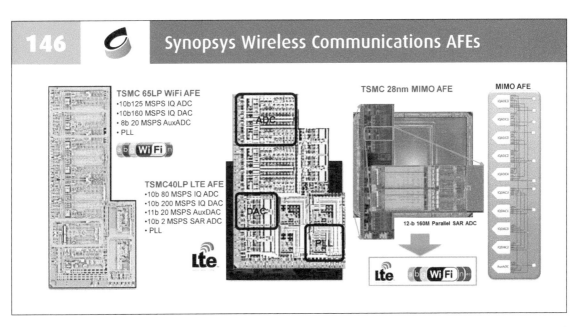

In the analog front end, there are multiple ADCs, e.g. IQ ADC for reception, IQ DAC for transmission and GPDAC for power amplifier and VGA control. The chip also has the PLL to generate clock and the general purposed ADC to receive the RSSI.

PART 3. THE NEXT TOPIC IS FOR THE MULTIMEDIA SOC AFE

The next topic is for the multimedia SoC AFE.

Basically, all consumer electronic products, e.g. the TVs, mobile phones, pads, tablets, sets of box, cameras, have lots of audio functions.

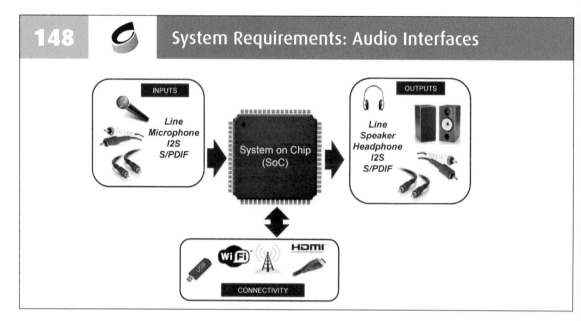

The audio system requires the audio inputs, which are microphones, line in and S/PDIF (digital input), and outputs, which are headset, speaker, line out and so on. Besides, it has the connectivity like the USB and WIFI and these audio interface is actually prevalent in many applications.

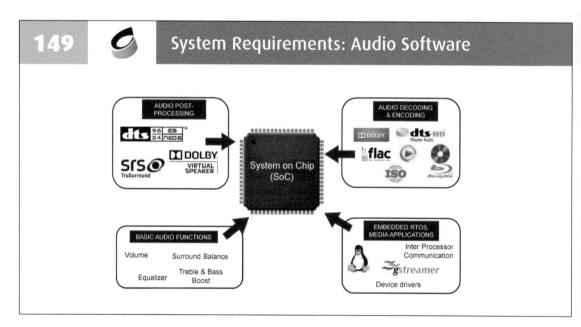

There are lots of audio functions in SoCs nowadays. Software is also needed to process, like the equalizers, surrounding sound and all those dts, dolby. Now, the voice recognition is very popular and it needs very compact function of a voice recognition, wake up the system and tell the system what to do. It doesn't need a HiFi quality audio but a very precise voice recognition which is also rely on the audio.

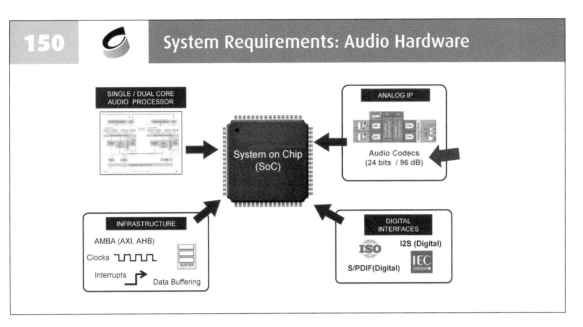

n the hardware perspective, the audio codec is typically 24 bit and 96dB dynamic range.

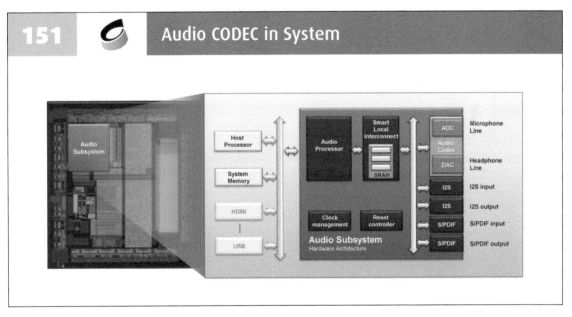

This is the function of the audio subsystems, which contain the audio processor, the audio codec and interface, clock managements and the other smart local interconnections with the other stuff. There are also some stand alone chip, that connects with the audio function and clock management. But it is now quite popular for audio with clock management. Audio starts to integrate with the power management unit because the audio actually serves for the voice and it has large signal range and correlated matrix. Therefore, a kind of IO voltage designs have to be used, which also fit the needs for the power management. It is the reason why the audio is not always integrated in the application processor with digital.

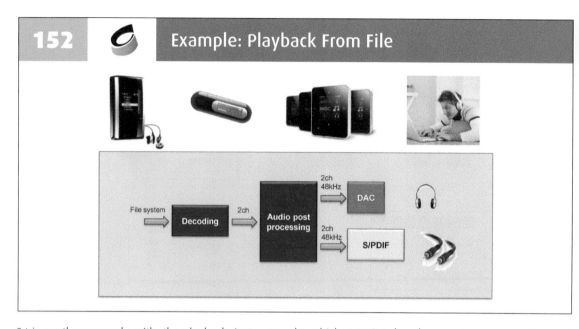

t is another example with the playback, just a recorde, which contains decoding, post processing, DAC and so on.

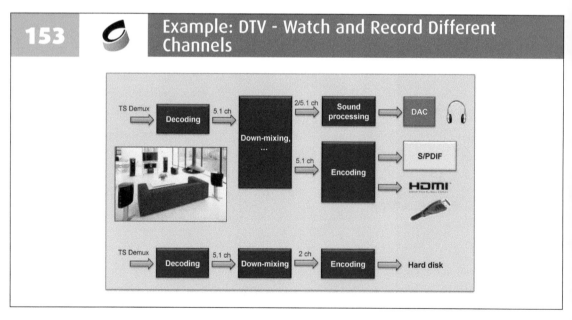

t is the digital TV, which also contains the sound processing, encoders, DAC and so on.

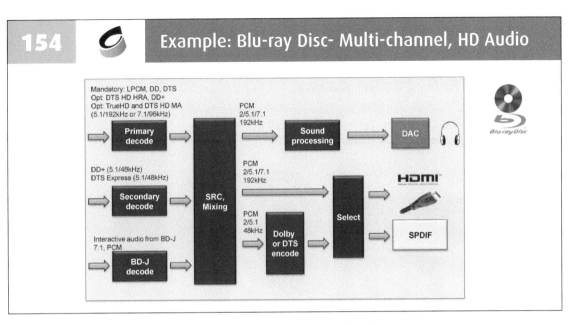

The Blu-ray disc and multi-channel HD audio, are also similar to the above.

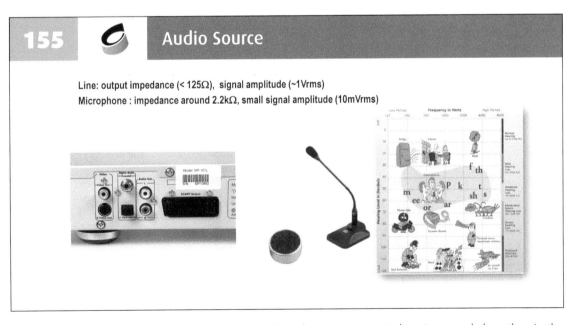

Basically, audio source has two typical types of analog input. One is line input and the other is the microphone, both of which are mostly used nowadays.

156 Audio Load

Analog output loading:
Line Load: High impedance > 3kΩ (Typical 10kΩ)
Headphone : low impedance around 16 - 500Ω
Loudspeaker: low impedance from 4 -16Ω

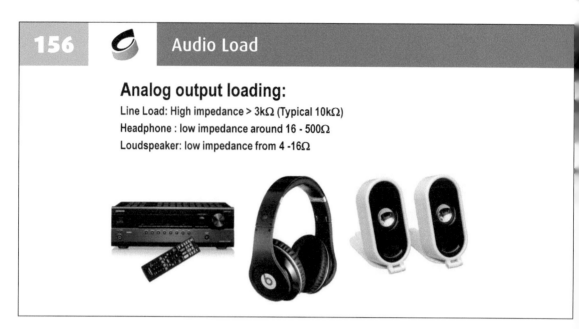

The typical loadings are line load, headphone and speaker. The output loading of line load is typically bigger than 3kΩ and 10kΩ is the standard. The headphone is around 16~500Ω and the most typical is 32Ω. Loudspeaker is 4 to 16Ω and most typical is 8Ω.

157 Digital Audio Interface

S/PDIF interface
OPTICAL
USB
Wireless

The digital audio interface includes S/PDIF interface, optical, USB, wireless connections. But the analog portion is the mainly focused on.

158 Electret Condenser Microphone

The Electret Condenser Microphones which is nowadays one of most commonly used microphone for consumer electronics.

Based on the theorem that using charging and discharging motion between electric capacity conductors where electret is a stable dielectric material with a permanently embedded static electric charge

Due to its high output impedance, a typical electret microphone requires embedded preamp which uses an FET in a common source configuration.

The two-terminal electret capsule contains an FET which must be externally powered by supply voltage V+. The resistor sets the gain and output impedance. The audio signal appears at the output, after a DC-blocking capacitor

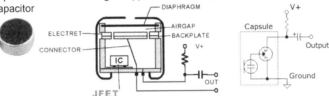

The electret condenser microphone is actually the most popular microphone nowadays. The equivalent circuit is similar to the common source configuration.

159 Headphone / Speaker

A **loudspeaker** is an electroacoustic transducer that produces sound in response to an electrical audio signal input. The most common form of loudspeaker uses a paper cone supporting a voice coil electromagnet acting on a permanent magnet. **Headphones** are a pair of small loudspeakers that are designed to be held in place close to a user's ears.

Typically, the loudspeaker has the impedance $4 - 16\Omega$. headphone has the impedance $16 - 64\Omega$, and some professional headphone can be up to 500Ω

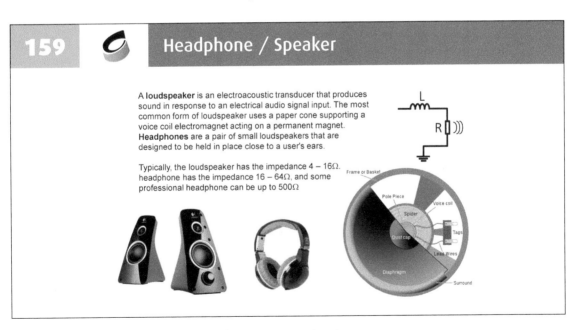

The headphone and speaker is just like the resistance with inductors.

160 Headphone jack

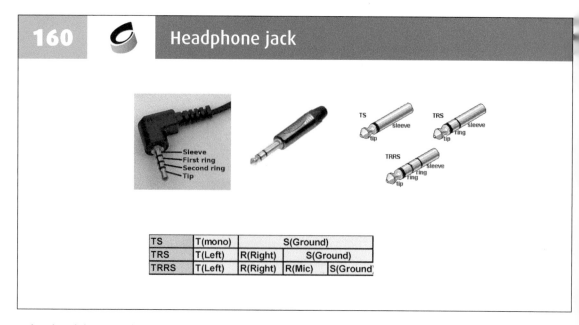

TS	T(mono)	S(Ground)		
TRS	T(Left)	R(Right)	S(Ground)	
TRRS	T(Left)	R(Right)	R(Mic)	S(Ground)

he headphone jack often has different rings. Actually it contains one tip, two rings and one sleeve. The last one is the ground and the tip is left channel, two rings are the right channel and the microphone input.

161 Audio Codec Specification Parameter

- A-weighting
- Dynamic Range
- THD+N
- SNR
- Channel crosstalk
- Output full scale
- Output Power

For the audio codec, specification parameters are well-known, including A-weighting, Dynamic Range, THD+N, SNR, channel crosstalk, output full scale and output power and so on.

162   A-weighting

- A-weighting is the most commonly used of a family of curves defined in the International standard IEC 61672:2003 and various national standards relating to the measurement of sound pressure level. A-weighting is applied to instrument-measured sound levels in effort to account for the relative loudness perceived by the human ear, as the ear is less sensitive to low audio frequencies.

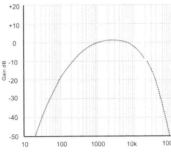

A-weighting is actually the most commonly used of a family of curves defined in the international standard, that basically according to the human ears, that loudness perceived by the human ear. In other words, the ear is actually a filter, which is not so sensitive to the low frequency and the high frequency up to 20kHz. It means that the output of the DAC will be filtered out like the A-weighting curves.

163   Dynamic Range (I)

- Dynamic range – is a measure of the difference between the highest and lowest portions of a signal. Normally a THD+N measurement at 60dB below full scale. The measured signal is then corrected by adding the 60dB to it.

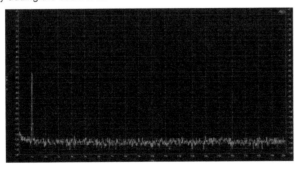

At the audio standard, dynamic range can be measured according to the following method. Measure the SNR at 60dB below full scale and add back the 60dB. It is totally different to the definition in other communication system, because in audio there are no harmonics obviously in -60dB full scale input.

164 Dynamic Range (II)

- Dynamic range – is a measure of the difference between the highest and lowest portions of a signal. Normally a THD+N measurement at 60dB below full scale. The measured signal is then corrected by adding the 60dB to it.

Audio Source	Dynamic Range
Analog TV	60 dB
Phonograph Record	65 dB
FM Radio	70 dB
Chamber Orchestra	70 dB
Full Symphony Orchestra	90 dB
Typical Rock Band	120 dB
Compact Disc	92-96 dB

This slide shows the different kinds of definitions of dynamic range in different applications.

165 Signal-to-Noise Ratio

- SNR is a measure of the difference in level between the full scale output and the output with no signal applied.

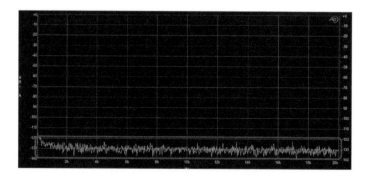

SNR is a measure of the difference in level between the full scale ouput and the output with no signal applied.

166 THD

- The total harmonic distortion (THD) if a signal is a measurement of the harmonic distortion present and is defined as the ratio of the sum of the powers of all harmonic components to the power of the fundamental frequency. THD is used to characterize the linearity of audio systems.

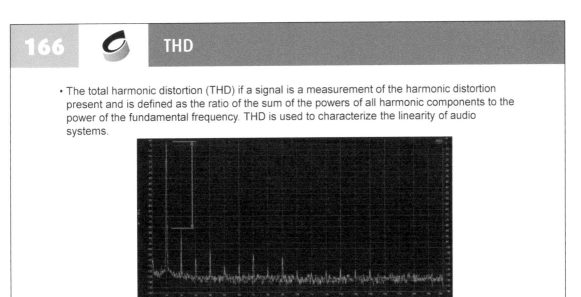

THD is similar to other definition, which is the ratio of the sum of the powers of all harmonic components to the power of the fundamental frequency.

167 THD+N

- THD+N (Total Harmonic Distortion plus Noise) is a ratio, of the rms value of (Noise + Distortion)/Signal. Using the THD+N plot, can find out all the important performance parameter of the audio codec

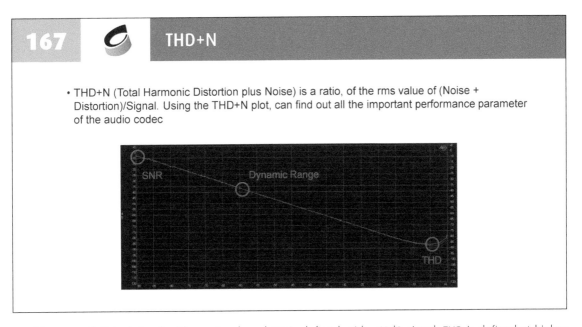

In this curve, SNR is defined with no signal, and DR is defined with -60dB signal. THD is defined at higher input level so that it becomes dominant for the performance.

168 Output Full Scale

- Full scale is in term of Vrms:
- For single-end output, output full scale = $\frac{Vpp}{VDD} = 1Vrms$
- For differential-end output, output full scale = $\frac{Vpp}{VDD} = 2Vrms$

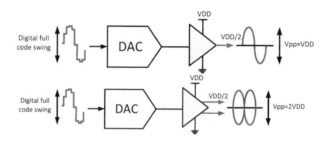

Output full scale is normally 1V rms in single end output and 2V rms in differential output. The output scale is already higher than the supply voltage of advanced technology, like 28nm. Therefore, the audio codec is always designed by IO devices or in a higher, old technology process.

169 Output Power

- In the market, typical impedance of headphone is 16 or 32 Ω
- Audio headphone drivers are specified as Maximum Power of mW (Range from 10mW ~ 40mW).
- Typical output swing of 1Vrms @ 3.3V supply
  - For 32Ω headphone, maximum output power @ 3.3V supply:
  - $P = \frac{V_{RMS}^2}{R} = \frac{1V_{RMS}}{32\Omega} = 31.25mW$

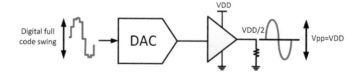

The typical impedance of headphone is 16~32Ω in the market. The maximum output power at 3.3V supply is 31mW for 32Ω headphone. The power is much higher for lots of loudspeakers.

AFES FOR MULTIMEDIA SOCS

AUDIO CODEC FEATURES AND BUILDING BLOCKS

The next part is about the audio codec function and building blocks.

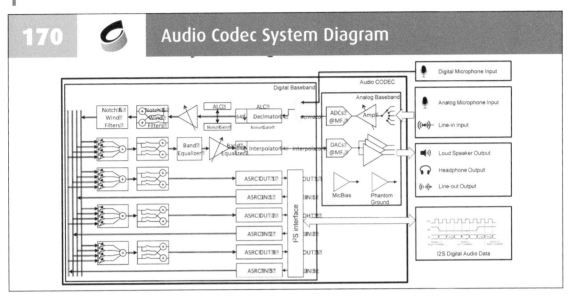

170 Audio Codec System Diagram

In the modern audio codec, the input path includes analog microphone input, line input, a mux, a PGA with a volume control and the ADC. The audio ADC is over sampling ADCs with noise shaping, which is basically the sigma-delta ADC and the digital decimation filters are also required.

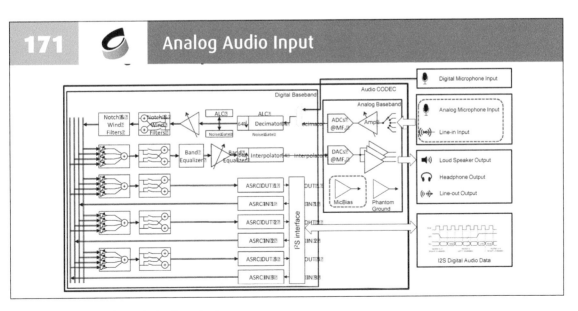

171 Analog Audio Input

The audio inputs also include the MicBias.

172 Analog Line-in Input Level

- Consumer electronic devices concerned with audio (for example sound cards) often have a connector labeled line-in and/or line-out

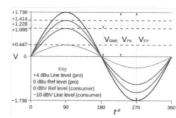

Use	Nominal level	Nominal level, V_{RMS}	Peak amplitude, V_{PK}	Peak-to-peak amplitude, V_{PP}
Professional audio	+4 dBu	1.228	1.736	3.472
Consumer audio	−10 dBV	0.316	0.447	0.894

In the professional audio, the nominal value is +4dBu, where 0dBu denotes the signal level driven into 600Ω at 1mW. The professional audio defines +4dBu, which is also equal to 1.228 Vrms and 3.472 peak to peak voltage. Its voltage level is higher than that of the consumer audio, which is 0.894 peak to peak voltage.

173 Analog Microphone Input Level

- Microphone sensitivity is typically measured with a 1 kHz sine wave at a 94 dB sound pressure level (SPL), or 1 pascal (Pa) pressure.
- For a microphone with a sensitivity rating of 20 mV/Pa
  - Conversation: 60dB SPL => 0.02 Pa => 0.4mV
  - Normal Max :80dB SPL => 0.2 Pa => 4mV
  - Threshold of Pain:120 dB SPL => 20 Pa => 400mV

Sound pressure and Sound pressure level

Description	Sound Level	Example
barely audible	0 dB	Threshold of hearing
	10 dB	Rustling leaf
very quiet	20 dB	Quiet room
	30 dB	Soft whisper
quiet	40 dB	Quiet library
	50 dB	Average home
moderately loud	60 dB	Ordinary conversation, Light traffic
	70 dB	Vacuum cleaner, Heavy traffic
very loud	80 dB	Garbage disposal
	90 dB	Diesel truck (10 m away)
uncomfortably loud	100 dB	Newspaper press
	110 dB	Jet flyover at 300 m
painful	120 dB	Threshold of pain, Thunderclap

Microphone sensitivity is typically measured with a 1kHz sine wave at a 94dB sound pressure level (SPL). The conversation is showing that 60dB SPL is equal to 0.4mV and the normal maximum of microphone is up to 4mV. Therefore, the microphone's input level is very small and the amplification is required.

174 · Input level Specifications

Parameter	Conditions	TYP	Unit
Audio input @0dB gain	Input level for full-scale digital output (0dBFS level) Measured at -10dBFS @ 1kHz output, extrapolated to full-scale	1 x AVDD/3.3	Vrms
Audio input @20dB gain		0.1 x AVDD/3.3	
Audio input @40dB gain		0.01 x AVDD/3.3	

Full-scale signal changes in proportion with AVDD and Gain to avoid Clipping occurs

A PGA amplifier is required to amplify the small signal to the full scale of the ADC. The definition of the gain is as follows: 0dB gain means 1V rms and 20dB gain means 0.1V rms.

175 · Analog Line and Microphone Input Configuration

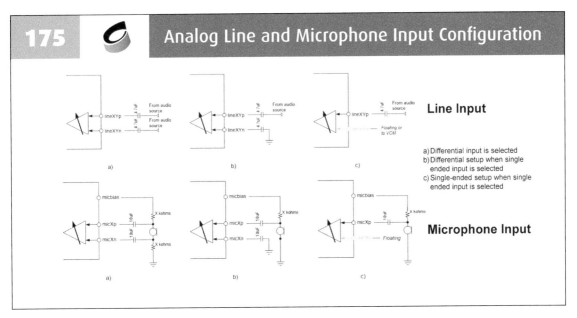

Line Input

a) Differential input is selected
b) Differential setup when single ended input is selected
c) Single-ended setup when single ended input is selected

Microphone Input

The line input is always like a single ended. The equivalent circuit of microphone is a micbias with ac compensate capacitor, which can be single ended or differential input. Therefore, the PGA for audio is required to make the single end to differential conversion and accept differential as well.

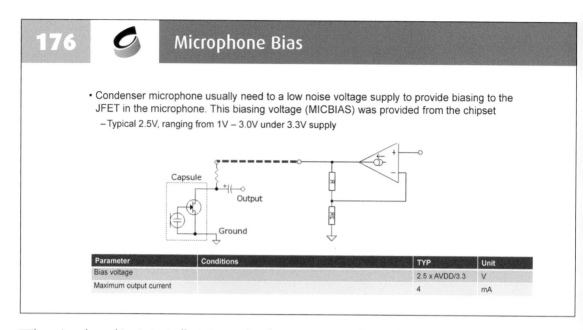

176 — Microphone Bias

- Condenser microphone usually need to a low noise voltage supply to provide biasing to the JFET in the microphone. This biasing voltage (MICBIAS) was provided from the chipset
  - Typical 2.5V, ranging from 1V – 3.0V under 3.3V supply

Parameter	Conditions	TYP	Unit
Bias voltage		2.5 x AVDD/3.3	V
Maximum output current		4	mA

The microphone bias is typically 2.5V, ranging from 1V to 3.3V. The micbias is like a simple regulator with normally 4mA output current driven.

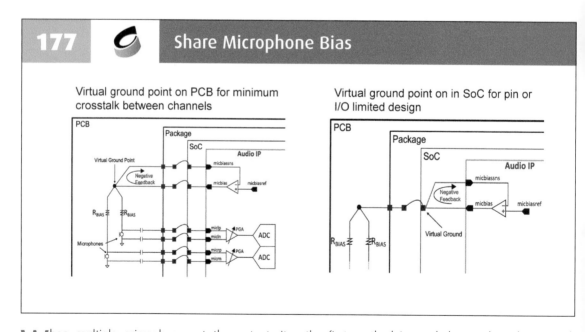

177 — Share Microphone Bias

Virtual ground point on PCB for minimum crosstalk between channels

Virtual ground point on in SoC for pin or I/O limited design

When multiple microphones at the output, it should be careful to connect. Basically, the microphone sense and bias pins are connected to two pins in the SoC, two bonding wires in the package and they are connected together on the PCB. However, they can also be connected together in the SoC so that one pin or I/O can be saved. Obviously, the first method is much better than the second, due to the crosstalk among different microphones. The dynamic range can be reduced from 90dB to 60 or 70dB because of the crosstalk. Therefore it is important to share the microphone bias carefully and the first method is recommended.

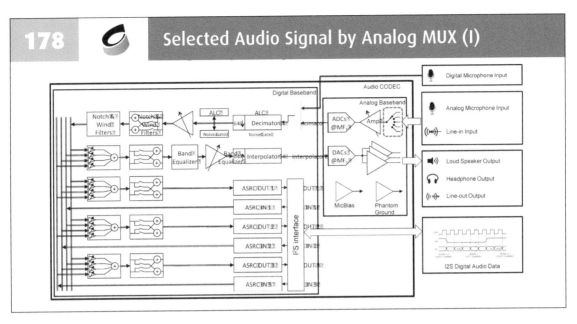

A mux is required to select different inputs.

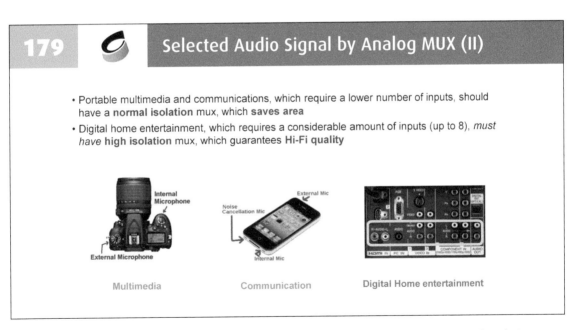

There will be different kinds of microphones, such as external microphone, internal microphone, noise cancellation microphone and so on. The mux can switch between them. Portable multimedia and communications, which are not always used at the same time and the crosstalk of which are not so serious, should have a normal isolation mux. However, digital home entertainment, which requires a considerable amount of inputs and continuously connected, must have high isolation mux to avoid the crosstalk.

180 — Selected Audio Signal by Analog MUX (III)

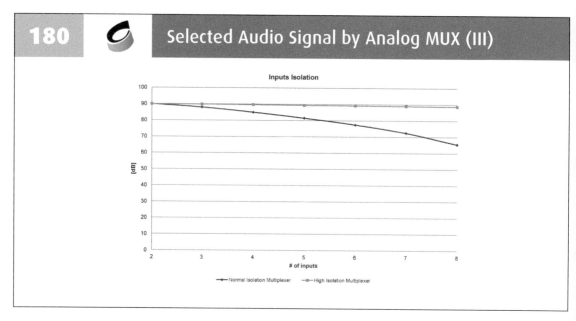

t shows the isolations around 90dB for the digital home entertainment required. But for the multimedia, the isolation can be much more relaxed.

181 — Scaled Audio Signal

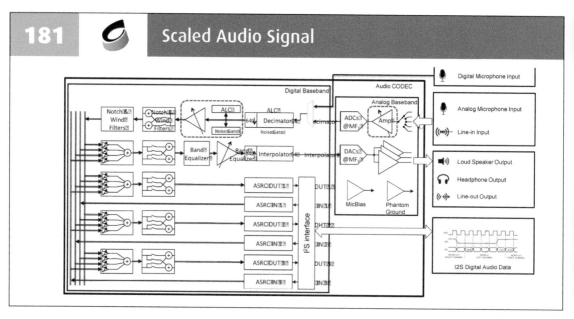

The audio signal can be scaled both in analog and in digital.

182 PGA Options - Which Gain Range?

- The volume control can be using the PGA for the input and Gain control driver for the output.

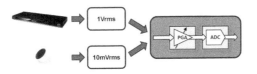

- **Different application will have different gain range requirement**
  - Electret microphones require amplification up to 40dB (or up to 60dB) – **matching application requirements**
  - Line level sources (1Vrms) typically require lower amplification and attenuation – **area savings**

The PGA gain range need to cover full range between the large signal from line input and the small signal from the microphone. At least, 40dB or 60dB gain is requirement for microphone input and lower amplification or even some attenuation is required for line input. Therefore, it needs to define a 60dB gain PGA.

183 Input Impedance Specifications

Parameter	Conditions	TYP	Unit
Input impedance	0dB gain	154	kOhm
	22dB gain	22	kOhm

Most of PGA implementation is based on an inverting amplifier topology.
A gain adjustment obtained by changing Ri value, will directly affect the input impedance value.

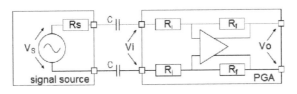

Most of PGA implementation is based on an inverting amplifier topology. A gain adjustment obtained by changing Ri value, will directly affect the input impedance value. Normally, the impedance should be larger than 10 or 20kΩ at all gain range.

184 ⬤ Input Impedance vs PGA Gain

- Low-end Microphone has output impedance around 2Kohms

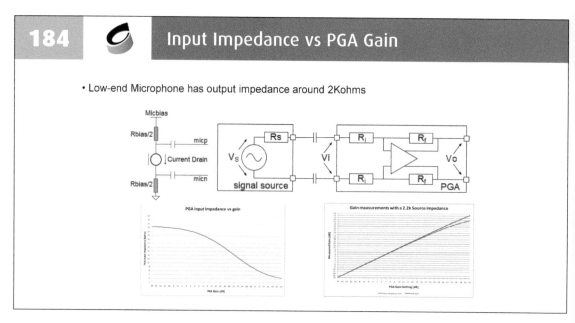

The impedance input changes to small when the gain is large. Due to the output impedance of microphone is around 2kΩ, there will be a division of the output impedance of microphone and the input impedance of the PGA. Therefore, the PGA input impedance should be carefully handled, avoiding the actual effective gain dropped.

185 ⬤ Microphone Boost

- The signal from the microphone usually is smaller than from line-in. Microphone input need a Gain boost amplifier plus Programmable Gain Amplifier (PGA) before to the ADC, the amplifier need to be high impedance and low capacitance, otherwise the input impedance will attenuate the overall gain.

Microphone boost is a selection for microphone input and an extra 20dB gain will be added.

186　PGA Options - Which Step?

- Volume control steps should be as **small** as possible to minimize ZIP noise
- Zip noise is further suppressed with digital signal processing (*soft ramping* with 0.1dB step)

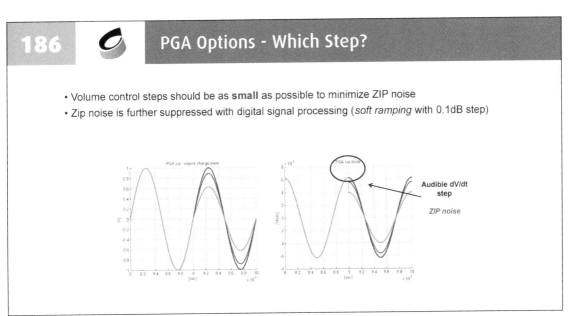

Volume control steps should be as small as possible to minimize zip noise. PGA can be adjust 1dB/step, but it still causes the zip noise when changing the gain. To reduce the zip noise, the soft ramping with 0.1dB/step is required.

187　Soft Gain Ramping

- Soft-ramping is applied to both the digital and analog volume controls. Basically, when a new volume setting is applied, the volume will change one step at a time from the original setting to the new setting.

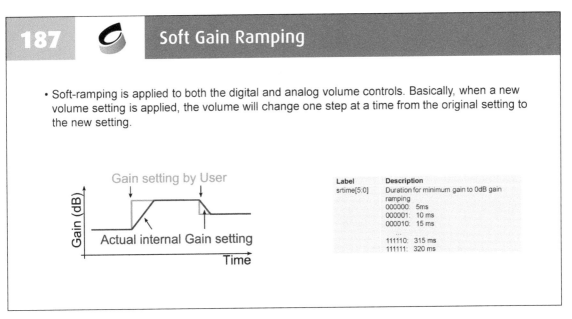

Soft ramping is applied to both the ditial and analog volume controls. Basically when a new volume setting is applied, the volume will change one step at time from original setting to the new setting. But with the soft gain ramping, there will be a tunable time defined by the user for the step growing.

188 — What is Zipper Noise

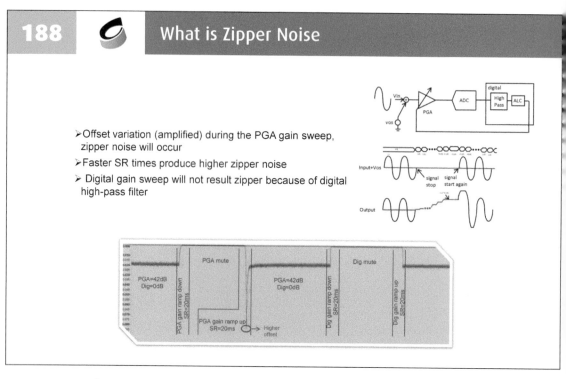

> Offset variation (amplified) during the PGA gain sweep, zipper noise will occur
> Faster SR times produce higher zipper noise
> Digital gain sweep will not result zipper because of digital high-pass filter

Zipper noise will come out because of the offset variation during the PGA gain sweep. Digital gain sweep will not result zipper noise because of the digital high-pass filter. The ramp up process is a feedback loop.

189 — PGA Options - Automatic Level Control

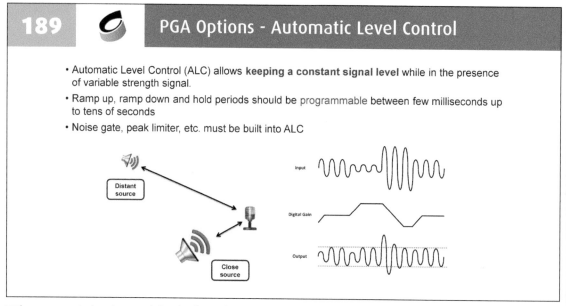

- Automatic Level Control (ALC) allows **keeping a constant signal level** while in the presence of variable strength signal.
- Ramp up, ramp down and hold periods should be programmable between few milliseconds up to tens of seconds
- Noise gate, peak limiter, etc. must be built into ALC

The automatic level control (ALC) allows keeping a constant signal level while in the presence of variable strength signal, which is very useful now foer every consumer electronic applications.

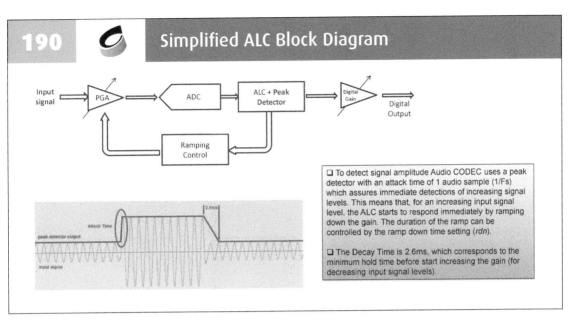

It is the simplified ALC block diagram including PGA, ADC and the ALC with peak detector. To detect signal amplitude, audio codec uses a peak detector with an attack time of 1/Fs which assures immediate detections of increasing signal level. This means that, for an increasing input signal level, the ALC starts to respond immediately by ramping down the gain. The duration of the ramp can be controlled by the ramp down time setting. The decay time corresponds to the minimal hold time before starting increasing the gain, for decreasing the input level.

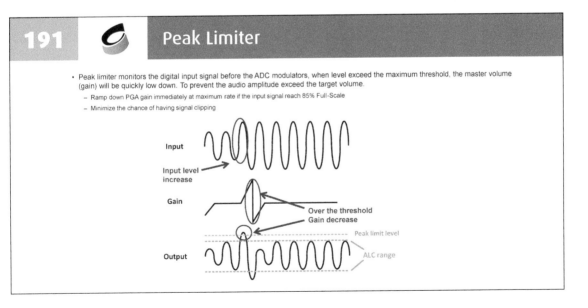

The peak detector will monitor the digital input signal before the ADC modulator where the level exceed the maximum threshold, the master volume gain will be quickly low down to prevent the audio amplitude exceed the target volume. It is well known that the signal clipping causes the harmonics. To avoid the signal clipping, if the input level reaches to the 85% of the full scale, the gain will immediately attenuate so that the output can be kept in the ALC range.

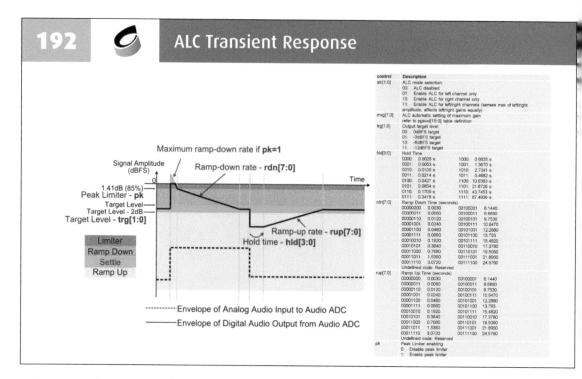

The ALC transient response shows how the ALC is doint. The ramp-down rate, ramp-up rate, the target level and peak limiter, all of these information can be controlled by users.

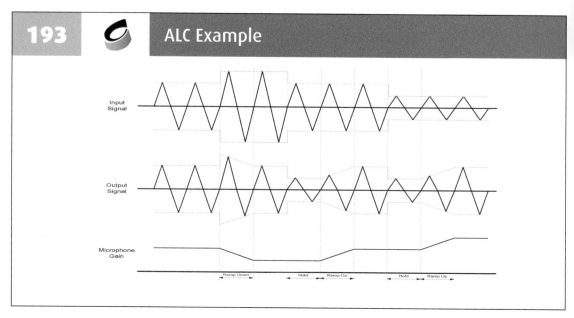

The gain is changing to make the output signal in the similar level.

194   ALC Steady State Transfer Function

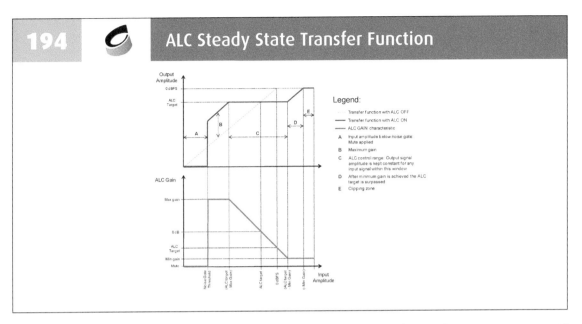

This is another ALC gain transfer functions. If the people don't say anything, the system will consider the noise as weak signal and amplify it.

Therefore it required to set the noise gating to avoid amplifying noise.

195   Noise Gate (I)

- The signal level is monitored before the digital volume.
- To detect a silent state, the signal level at the peak detector must be below a threshold. When the signal raises again above the silent threshold plus the hysteresys the control of the digital volume is given back to ALC.

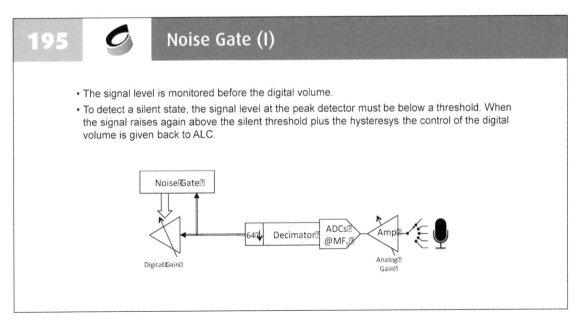

The noise gating is used to detect the silent state and the signal level at the peak detector must be below a threshold. When the signal raises again above the silent threshold plus the hysteresys, the control of the digital volum is given back to the ALC.

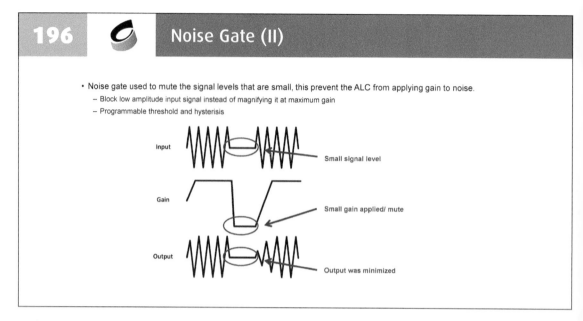

t shows that if the signal is quiet and the gain is actually muted, the output is minimized. The programmable threshold and hysterisis is required to define the level of noise gating threshold by user.

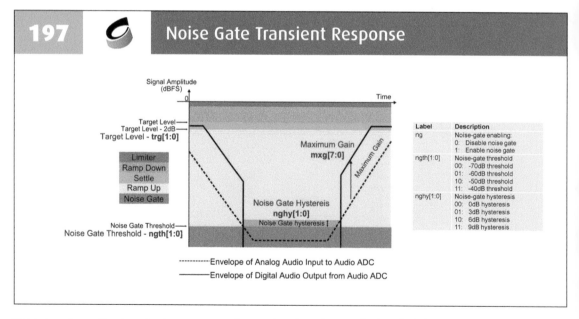

This is noise gating transient responses. When a signal reaches to the level of the noise gating threshold, it will be completely muted.

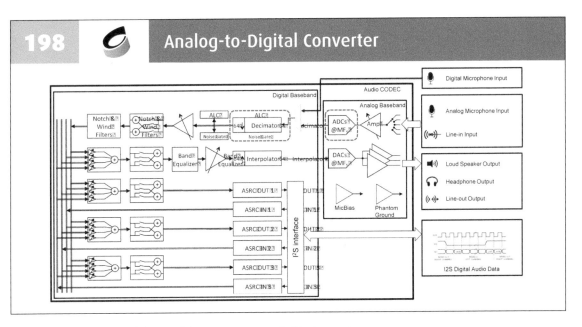

198   Analog-to-Digital Converter

In audio codec, sigma-delta modulators are most suitable to used.

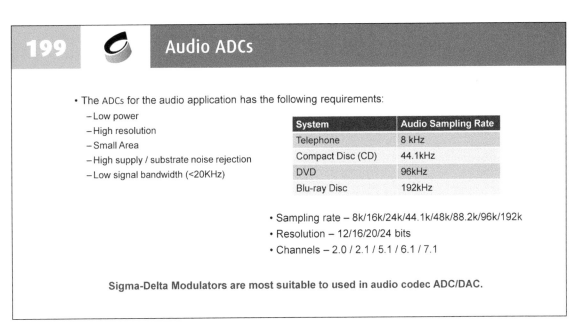

199   Audio ADCs

- The ADCs for the audio application has the following requirements:
  - Low power
  - High resolution
  - Small Area
  - High supply / substrate noise rejection
  - Low signal bandwidth (<20KHz)

System	Audio Sampling Rate
Telephone	8 kHz
Compact Disc (CD)	44.1kHz
DVD	96kHz
Blu-ray Disc	192kHz

- Sampling rate – 8k/16k/24k/44.1k/48k/88.2k/96k/192k
- Resolution – 12/16/20/24 bits
- Channels – 2.0 / 2.1 / 5.1 / 6.1 / 7.1

Sigma-Delta Modulators are most suitable to used in audio codec ADC/DAC.

The ADCs for audio application have the following requirements, such as the low noise, low power, high resolution, small area, high supply rejection ratio and also low signal bandwidth. The sampling rate should be supported from 8kHz, which is used for the telephone, up to 192kHz, which is used for the blueray disc. The resolution is from 12 to 24 bit and the channel need to support 2.0/2.1/5.1/6.1/7.1.

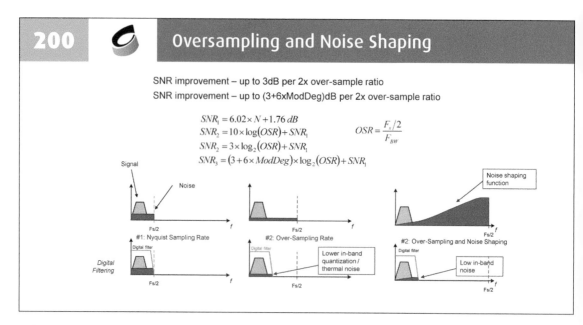

The SNR can be improved by 3*log2(OSR) because of the oversampling. And the SNR can be improved further more due to the noise shaping. Therefore, a better SNR will be got not only with the OSR, but also with the noise shaping effect.

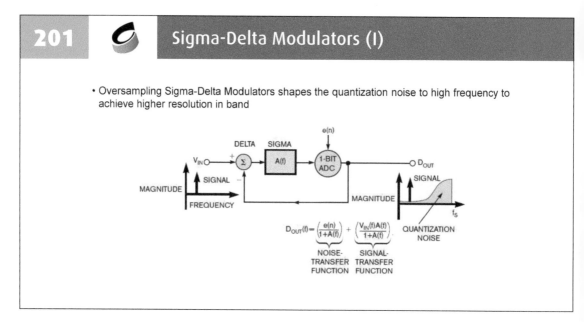

Basically, sigma-delta ADC can shape the quantization noise to push it to the high frequency range. In the sigma-delta ADC, there is a feedback loop and the delta will be integrated. The noise transfer function, realized by the feedback loop, is a high pass filter because of the integrator. Therefore, the high pass filter will push the noise into a high frequency and shape the noise.

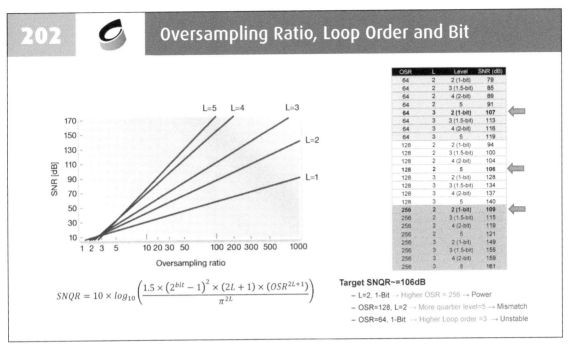

202 Oversampling Ratio, Loop Order and Bit

OSR	L	Level	SNR (dB)
64	2	2 (1-bit)	79
64	2	3 (1.5-bit)	85
64	2	4 (2-bit)	89
64	2	5	91
64	3	2 (1-bit)	107
64	3	3 (1.5-bit)	113
64	3	4 (2-bit)	116
64	3	5	119
128	2	2 (1-bit)	94
128	2	3 (1.5-bit)	100
128	2	4 (2-bit)	104
128	2	5	106
128	3	2 (1-bit)	128
128	3	3 (1.5-bit)	134
128	3	4 (2-bit)	137
128	3	5	140
256	2	2 (1-bit)	109
256	2	3 (1.5-bit)	115
256	2	4 (2-bit)	119
256	2	5	121
256	3	2 (1-bit)	149
256	3	3 (1.5-bit)	155
256	3	4 (2-bit)	159
256	3	5	161

$$SNQR = 10 \times log_{10}\left(\frac{1.5 \times \left(2^{bit} - 1\right)^2 \times (2L + 1) \times (OSR^{2L+1})}{\pi^{2L}}\right)$$

Target SNQR~=106dB

- L=2, 1-Bit → Higher OSR = 256 → Power
- OSR=128, L=2 → More quartier level=5 → Mismatch
- OSR=64, 1-Bit → Higher Loop order =3 → Unstable

The SNR is determined by over sampling ration, loop order and the quartier level. For example, a 64 OSR with third order one bit can reach 107dB SNR. In the case of 128 OSR, it can also reach 106dB with 5 level and second order. It can achieve 109dB if 256 OSR with one bit second order is used. Therefore, there will be a multiple choice for different kinds of things. Basically, the higher OSR means the higher sampling rates, which leads to more power. And the second one requires 5 levels so that a multi-bit DAC is needed and the mismatch should be concerned. The higher order also leads to a stability issues. There are some trade-off between them.

203 More Bit level (Bit >1)

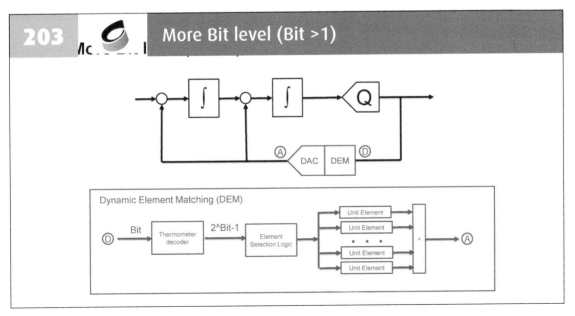

The mismatch is required to consider in multi-bit level and dynamic element matching is mostly used. Basically, the elements can be rotated according to a sequence so that the noise will be average.

204 Data Weight Average

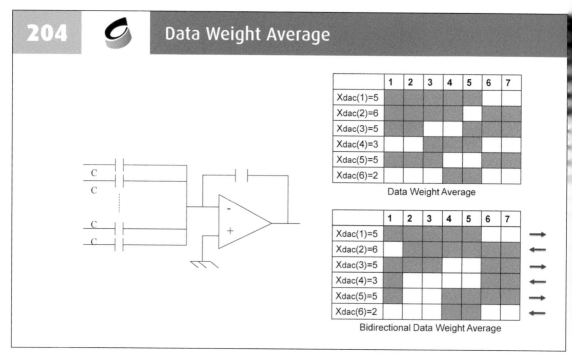

	1	2	3	4	5	6	7
Xdac(1)=5							
Xdac(2)=6							
Xdac(3)=5							
Xdac(4)=3							
Xdac(5)=5							
Xdac(6)=2							

Data Weight Average

	1	2	3	4	5	6	7
Xdac(1)=5							
Xdac(2)=6							
Xdac(3)=5							
Xdac(4)=3							
Xdac(5)=5							
Xdac(6)=2							

Bidirectional Data Weight Average

With the digital weight average, the capacitors can be rotated in order to have a different kind of additional shaping of the mismatch. With the bidirectional data weight average, two rotating directions are alternate.

205 Recent Sigma Delta Modulator Examples (I)

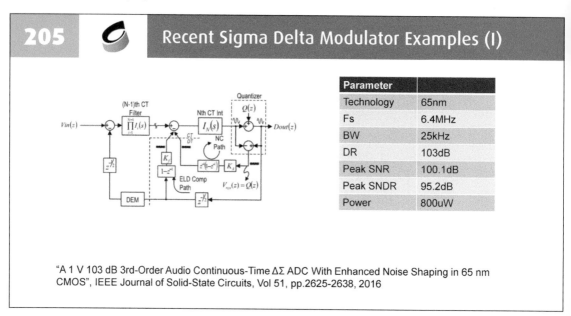

Parameter	
Technology	65nm
Fs	6.4MHz
BW	25kHz
DR	103dB
Peak SNR	100.1dB
Peak SNDR	95.2dB
Power	800uW

"A 1 V 103 dB 3rd-Order Audio Continuous-Time ΔΣ ADC With Enhanced Noise Shaping in 65 nm CMOS", IEEE Journal of Solid-State Circuits, Vol 51, pp.2625-2638, 2016

In this literature, it achieves 103dB dynamic range and 100dB peak SNR with 800uW power consumption, 1V power supply and 6.4MHz Fs in 65nm. A third order continuous time sigma-delta ADC with enhanced noise shaping is used.

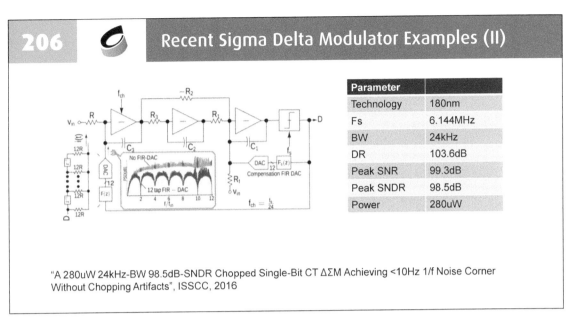

206 Recent Sigma Delta Modulator Examples (II)

Parameter	
Technology	180nm
Fs	6.144MHz
BW	24kHz
DR	103.6dB
Peak SNR	99.3dB
Peak SNDR	98.5dB
Power	280uW

"A 280uW 24kHz-BW 98.5dB-SNDR Chopped Single-Bit CT ΔΣM Achieving <10Hz 1/f Noise Corner Without Chopping Artifacts", ISSCC, 2016

Another example is very popular now, which uses a kind of FIR filter inside the loop. It is a single bit continuous time sigma-delta modulator achieving a 98dB SNDR with 280uW power consumption.

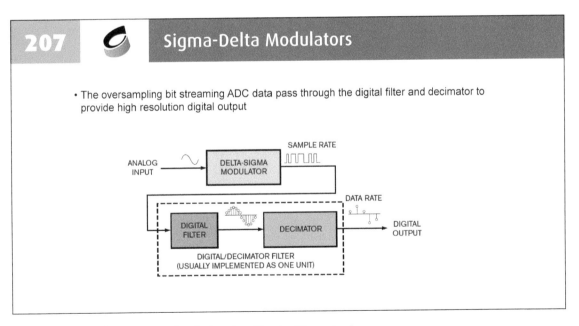

207 Sigma-Delta Modulators

- The oversampling bit streaming ADC data pass through the digital filter and decimator to provide high resolution digital output

In the actual implementation, the decimation filter is still required.

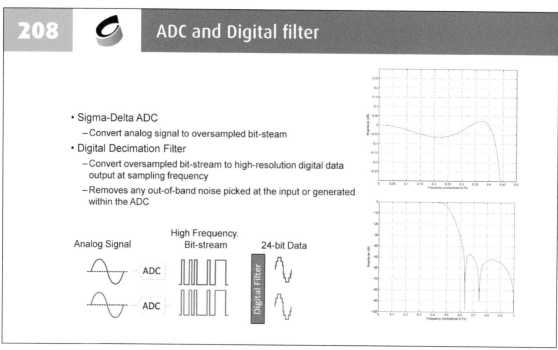

208 — **ADC and Digital filter**

- Sigma-Delta ADC
  - Convert analog signal to oversampled bit-steam
- Digital Decimation Filter
  - Convert oversampled bit-stream to high-resolution digital data output at sampling frequency
  - Removes any out-of-band noise picked at the input or generated within the ADC

Analog Signal — High Frequency. Bit-stream — 24-bit Data

ADC — Digital Filter

Because noise is actually shifted to the high frequency, the decimation filter can filter it out.

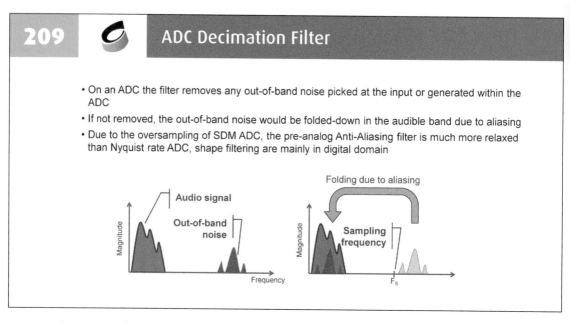

209 — **ADC Decimation Filter**

- On an ADC the filter removes any out-of-band noise picked at the input or generated within the ADC
- If not removed, the out-of-band noise would be folded-down in the audible band due to aliasing
- Due to the oversampling of SDM ADC, the pre-analog Anti-Aliasing filter is much more relaxed than Nyquist rate ADC, shape filtering are mainly in digital domain

Folding due to aliasing

Audio signal

Out-of-band noise

Sampling frequency

F_s

Magnitude — Frequency

Due to the oversampling of sigma-delta ADC, the anti-aliasing filter is much more relaxed than Nyquist rate ADC.

210 SINC Filter

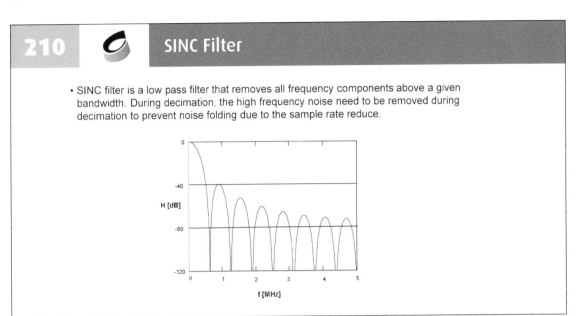

- SINC filter is a low pass filter that removes all frequency components above a given bandwidth. During decimation, the high frequency noise need to be removed during decimation to prevent noise folding due to the sample rate reduce.

The sinc filter can be used to filter all at each Fs and just a delay line can achieve a sinc filter. Each notch happens at Fs and so on.

211 Decimator (I)

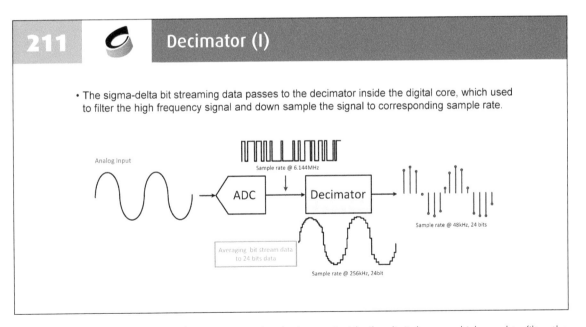

- The sigma-delta bit streaming data passes to the decimator inside the digital core, which used to filter the high frequency signal and down sample the signal to corresponding sample rate.

The sigma-delta bit streaming data passes to the decimator inside the digital core, which used to filter the high frequency signal and down sample the signal to corresponding sample rate.

212 Decimator (II)

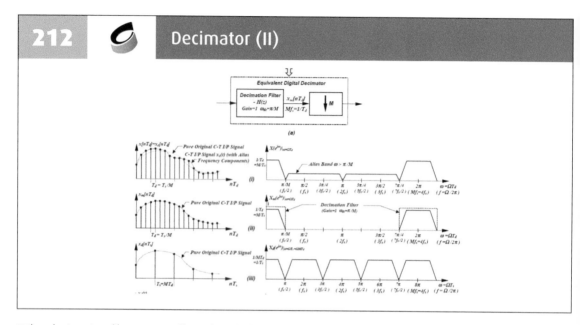

The decimation filter is actually a digital filter and down samplers. The digital filter can filter the high frequency noise out to avoid it aliasing in band.

213 Input Path Specifications

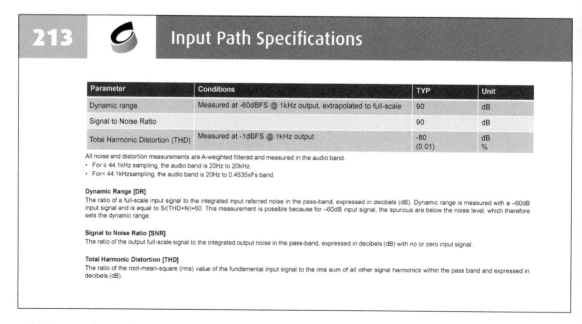

Parameter	Conditions	TYP	Unit
Dynamic range	Measured at -60dBFS @ 1kHz output, extrapolated to full-scale	90	dB
Signal to Noise Ratio		90	dB
Total Harmonic Distortion (THD)	Measured at -1dBFS @ 1kHz output	-80 (0.01)	dB %

All noise and distortion measurements are A-weighted filtered and measured in the audio band.
- For ≥ 44.1kHz sampling, the audio band is 20Hz to 20kHz.
- For< 44.1kHzsampling, the audio band is 20Hz to 0.4535xFs band.

Dynamic Range [DR]
The ratio of a full-scale input signal to the integrated input referred noise in the pass-band, expressed in decibels (dB). Dynamic range is measured with a –60dB input signal and is equal to S/(THD+N)+60. This measurement is possible because for –60dB input signal, the spurious are below the noise level, which therefore sets the dynamic range.

Signal to Noise Ratio [SNR]
The ratio of the output full-scale signal to the integrated output noise in the pass-band, expressed in decibels (dB) with no or zero input signal.

Total Harmonic Distortion [THD]
The ratio of the root-mean-square (rms) value of the fundamental input signal to the rms sum of all other signal harmonics within the pass band and expressed in decibels (dB).

The input path specifications are as followed, 90dB dynamic range, 90dB SNR and -80dB THD at 1kHz.

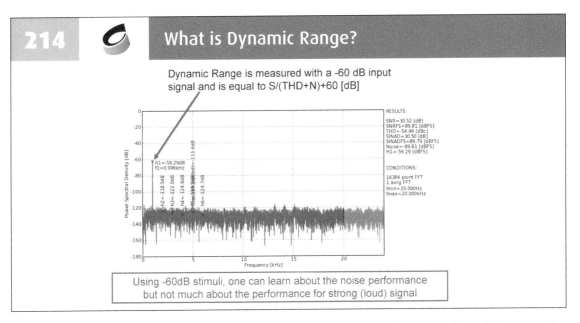

ynamic range is measured with a -60dB input signal. The SNR at -60dB is 30.52dB and the SNR at full scale will be added 60dB and become 89.81dB.

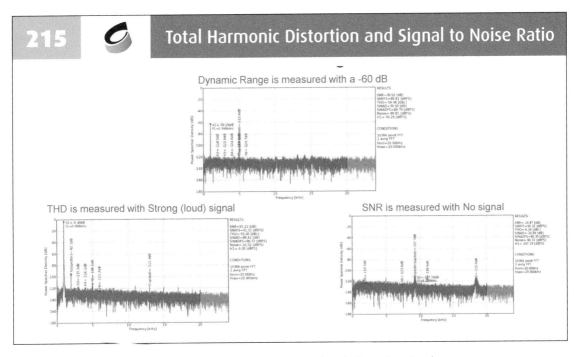

THD is measured with a strong signal and SNR is measured with the noise signal.

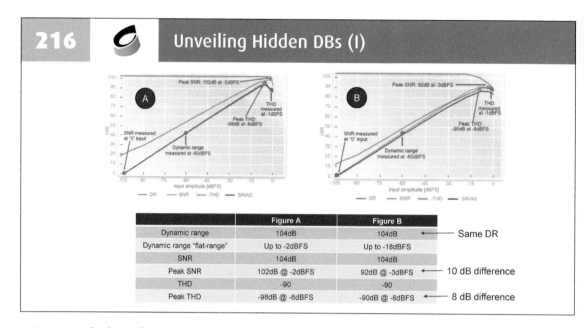

Figure A and B have the same dynamic range at 104dB. However, there are 10dB difference in peak SNR and 8dB difference in peak THD.

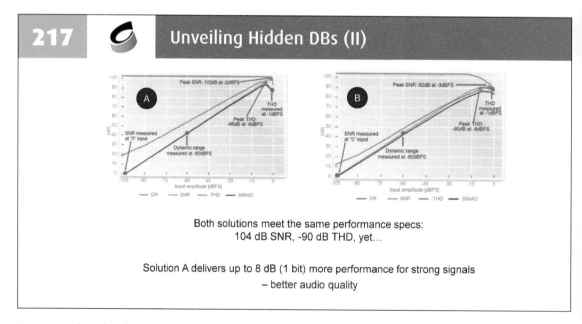

It means that the dynamic range of audio is not enough and the peak SNR and peak THD are also required. With strong signal, there is a big distortion in figure B. Therefore, it is the hidden dBs that need to be known as a system designer.

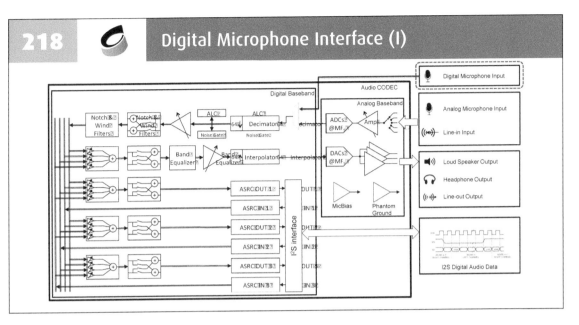

owadays, some microphone has the digital function interface.

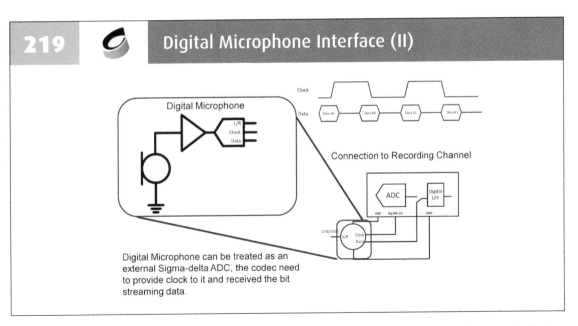

igital microphone can be treated as an external sigma-delta ADC and the codec need to provide clock to it and received the bit streaming data.

220 Shaping Filter

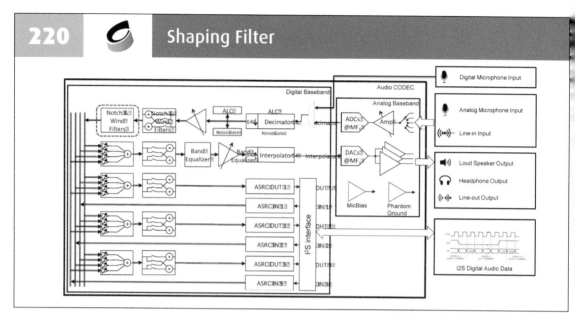

In the audio application, there are different kinds of shaping filter requirements. For example, the user takes a photo and wants to zoom and zoom out, and in the meantime he are recording. The mechanisms voice is also recorded but can not be heard, because of the notch wind filter which filters the voice or noise out. The wind noise can also be tented by the shaping filter during the windy area in mobile phone.

221 Other Special Digital Filtering

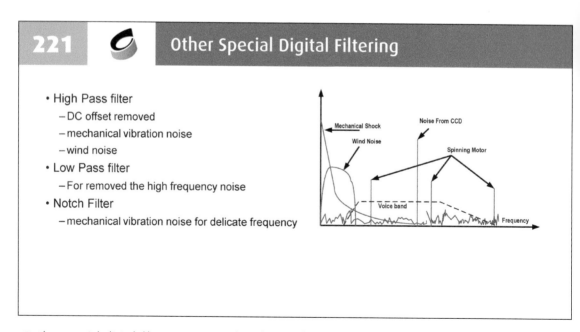

- High Pass filter
  - DC offset removed
  - mechanical vibration noise
  - wind noise
- Low Pass filter
  - For removed the high frequency noise
- Notch Filter
  - mechanical vibration noise for delicate frequency

Other special digital filters are required such as high pass filter for dc offset removed, mechanical vibration noise and wind noise, low pass filter for the high frequency noise removed and notch filter for mechanical vibration noise at delicate frequency.

222 Notch Filter

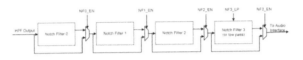

- Notch Filters includes 4 independent notch filters to remove narrow-band noise
- A Notch filter uses a combination of a zero in the unitary circle and a nearby pole with high Q-factor

$$H(z) = \frac{1 + a_1 z^{-1} + z^{-2}}{a_0 + a_1 z^{-1} + a_2 z^{-2}}$$

- These coefficients need to be represented with enough resolution to preserve the frequency response with reasonable accuracy. For large ratio of sampling frequency to notch filter the coefficient sensitivity increases rapidly.
- For example, assuming FBB=96kHz:
  - Fnotch= 2000Hz, Q=10, 1% accuracy requires 11 bit coefficients
  - Fnotch= 200Hz, Q=10, 1% accuracy requires 16 bit coefficients

Notch filter includes 4 independent notch filters to remove narrow-band noise, which uses a combination of a zero in the unitary circle and a nearby pole with high Q-factor. These coefficients need to be represented with enough resolution to preserve the frequency response with reasonable accuracy. For large ratio of sampling frequency to notch filter, the coefficient sensitivity increases rapidly.

223 Wind filter

- The wind filter is a simple second-order highpass filter with programmable cut-off frequency between 85Hz and 1000Hz.

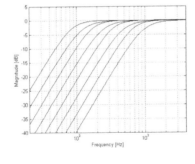

In the wind filter, it is required to make a programmable high pass cut off frequency according to the system requirement.

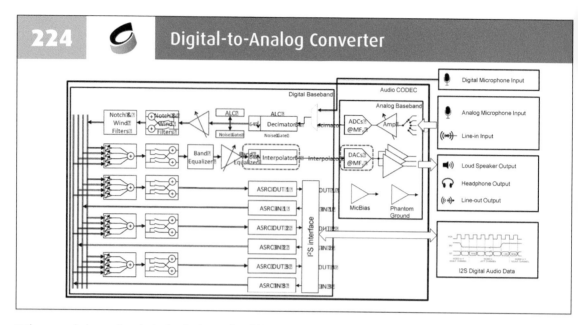

The record channel and playback channel will be introduced next.

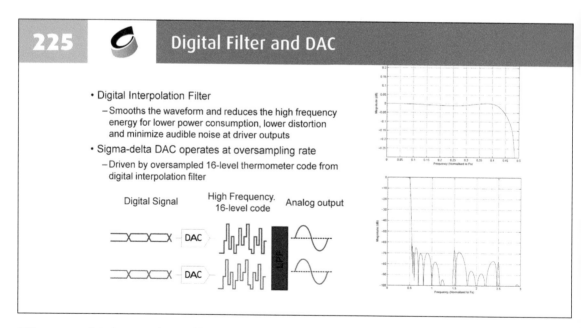

There are digital interpolation filter and sigma-delta DAC operating at the oversampling rate. The sigma-delta DAC is driven by oversampled 16-level thermometer code from digital interpolation filter. Therefore, a digital stream will get to the analog output.

226   Interpolator (I)

- Digital audio data will be converted to lower bit, higher frequency DAC input through interpolation. Interpolation can smooth the waveform shape, shift the noise to high frequency and minimize the signal distortion.

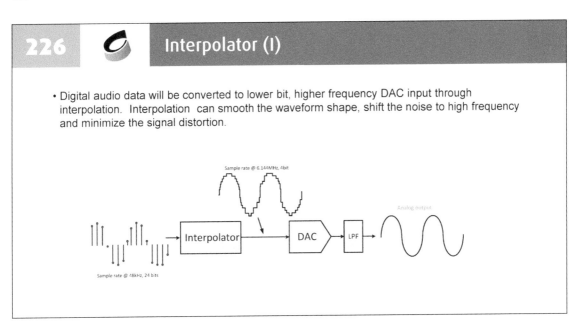

Digital audio data will be converted to lower bit, higher frequency DAC input through interpolation. Interpolation can smooth the waveform shape, shift the noise to high frequency and minimize the signal distortion.

227   Interpolator (II)

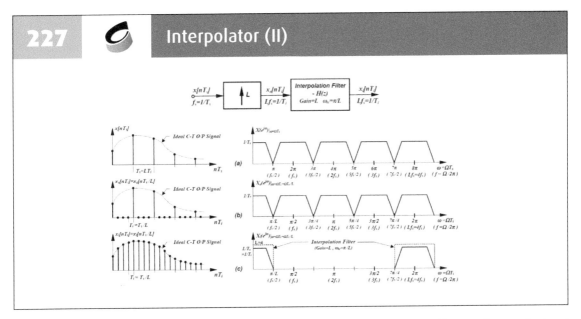

Interpolator is in opposite directions with decimator. With the zero patterning and filtering, the simple signal at the lower sampling frequency achieves the higher sampling rate.

228 Stereo Mixer (I)

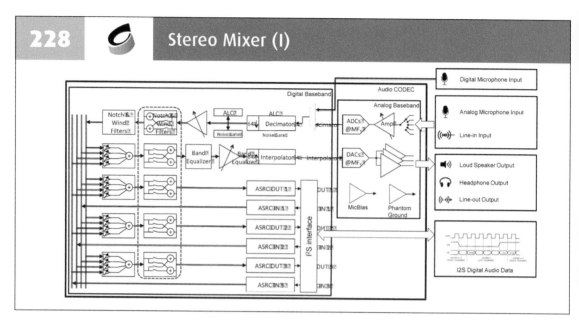

There are also lots of stereo mixers.

229 Stereo Mixer (II)

- This block allows a number of operations on the stereo channels as shown in Figure 28. These are as follows:

  a) Direct Feedthrough, the signals pass unaffected.

  b) L&R swap. The L and R channels are swaped.

  c) (L+R)/2 mix. The L and R channels are mixed to produce a Mono output on both L and R channels.

  d) Mono replication (from R or L channels). It is used in ADC, DAC and ASRC_OUT channels. Mono Replication is shown for one channel. It is supported also for the other channel.

  e) Inversion. Either channel can be independently inverted. This feature is in addition to any of the previous ones. An example application with only one channel inverted is for supporting Mono material with a pseudo-differential output in bridged amplifier configurations.

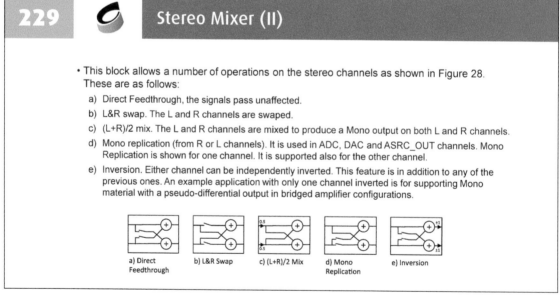

a) Direct Feedthrough   b) L&R Swap   c) (L+R)/2 Mix   d) Mono Replication   e) Inversion

Stereo mixer allows a number of operations on the stereo channels, such as direct feed through, left and right channel swap, left and right channel mixed, mono replications from left or right channel and inversion. All of them are implemented in digital because the digital mixer is much simpler than analog mixer.

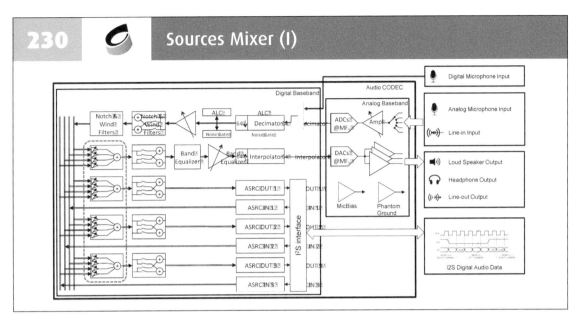

There are multiple sources of input and they will be mixed together.

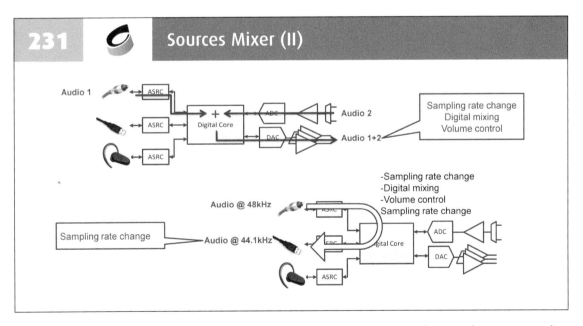

If the sources are at the different sampling rate, it is necessary to convert them at the same sampling frequency and then mix them together.

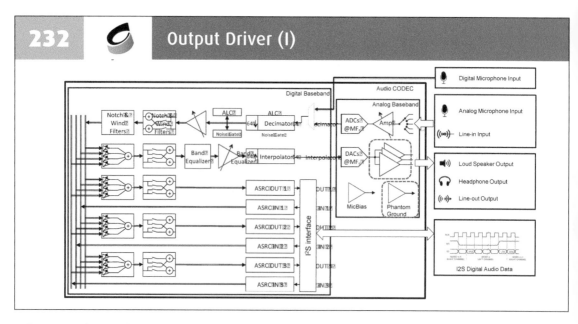

The output driver is the key of the playback channel.

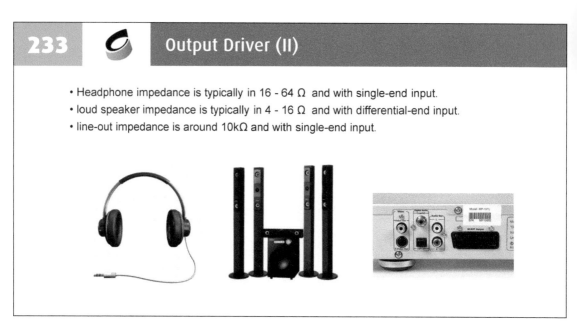

Normally, the line output have a typical loading at 10kΩ.

234 Line out Specifications

Parameter	Conditions	TYP	Unit
Output load resistance	Line output, to ground, AC coupled	10	kOhm
Output impedance	Line output, to ground, AC coupled	125	Ohm
Output level for full-scale digital input (0dBFS level)	Measured at -10dBFS @ 1kHz input, extrapolated to full-scale	1x AVDD/3.3	Vrms
Dynamic range (DR)	Measured at -60dBFS @ 1kHz input, extrapolated to full-scale	96	dB
Signal to Noise Ratio		96	dB
Total Harmonic Distortion (THD)	Measured at -3dBFS @ 1kHz input	-86 (0.005)	dB (%)

The output impedance is around 125Ω. The performance in playback channel requires 96dB SNR and 86dB THD, which is actually 6dB better than the record channel at least, because the sources in playback have a higher quality, such as MP3 or Hi-Fi bits stream.

235 Headphone Driver (I)

- Headphone is expecting a ground referenced AC signal.
- Headphone drivers are providing VCM (AVDD/2) referenced AC signal.
- Difference in DC references cause DC current flow from Audio Drivers to headphones

- The DC current causes:
- Excessive and useless power consumption
- IC and headphone overheating
- Signal clipping at the ESD protections

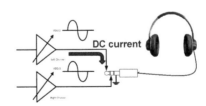

Headphone is expecting a ground referenced AC signal. Headphone drivers are providing vcm, which is VDD/2, referenced AC signal.

236 Headphone Driver (II)

- DC levels isolated by AC coupling capacitors.
- Disadvantage:
  - Affects audio quality (capacitor linearity and bass cut-off)
  - Cut-off frequency = $1/2\pi RC = 1/(2\pi*16*220\mu) \approx 45Hz$
  - (assume using AC couping Capacitor 220 Ω Fand headphone load with 16Ω resistance)
  - Needs to charge during power up (pop-noise/ slow startup)
  - Are bulky (height, footprint)
  - Smaller SMD type capacitor are expensive

Bass cut-off frequency: fc = 1 / 2πR$_{HP}$C
Charge on each capacitor: Q = C × VDD/2

VDD
3.3V typ.

VDD
VDD/2
0V

0V

Left

C= 220μF

Right

C= 220μF

0V

0V

Headphone impedance
R$_{HP}$ = 16 or 32 Ohm

Normally, the DC level is isolated by ac coupled capacitor. It is required to have a cut-off frequency at 45Hz so that a very big ac coupled capacitor is required due to the 16Ω loading. It is a kind of high pass filter to attenuate some low frequency noise.

Due to cost and area of the 220uF capacitor, another solution which is called the virtual ground solution has taken the place of the ac coupled capacitor in the headset.

237 Headset Specifications

Parameter	Conditions	TYP	Unit
Output load resistance	Headset output to ground, AC coupled	16~32	ohm
Output level for full-scale digital input (0dBFS level)	Measured at -10dBFS @ 1kHz input, extrapolated to full-scale	1x AVDD/3.3	Vrms
Maximum Output Power	Output load resistance at 16 ohm	40	mW
Dynamic range (DR)	Measured at -60dBFS @ 1kHz input, extrapolated to full-scale	94	dB
Signal to Noise Ratio		94	dB
Total Harmonic Distortion (THD)	Measured at -1.95dBFS (20mW) and 32ohm Load @ 1kHz input	-74 (0.02)	dB %
	Measured at -5dBFS (20mW) and 16ohm Load @ 1kHz input	-74 (0.02)	dB %

The output load of headset is normally 16-32Ω. The dynamic range is 94dB and the maximum output power is 40mW at the 16Ω. THD is -74dB under the measurement at -1.95dB or -5dB full scale at 20mW with 1kHz input.

238 **POP noise**

- When the output driver was enable, VCM requires fixed RC time constant to charge the capacitance. The output voltage on the headphone will caused a jump and created a "Pop" sound.
- To minimize the "Pop" noise, need to increase the pre-charging time, optimize the ramping shape

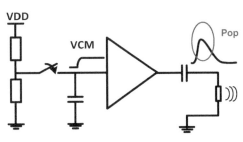

The ac coupled capacitor will generate additional problem which is called the pop noise. When the output driver was enable, vcm requires fixed RC time constant to charge the capacitance. The output voltage on the headphone will caused a jump and created a pop sound.

 239 **Conventional pre-charge scheme for low pop-noise**

- Fixed RC time constant during pre-charge
  - Single pre-charge rate (RC) in the whole pre-charge period
  - Tradeoff between pop-noise level and time

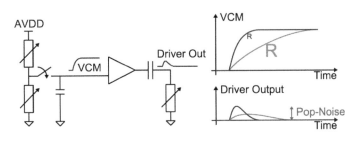

The pre-charge can be made very slowly, but it will take a long time to get the right vcm level. Therefore, it is a tradeoff between pop noise level and time.

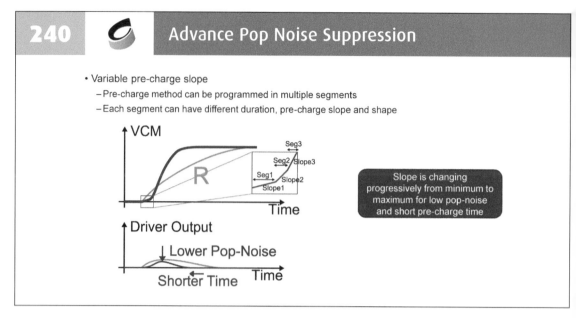

The slope can be controlled in the segment so that the pop noise can become smaller.

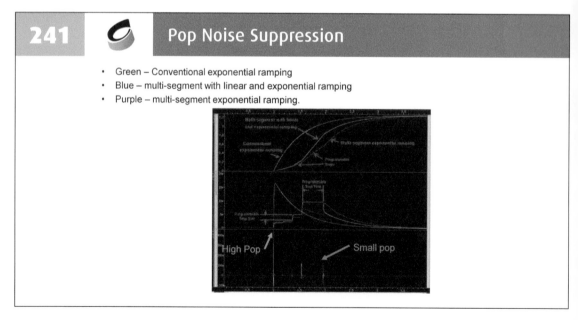

The normal ramping up takes a big pop noise. With the multi-segment exponential ramping, the pop noise can be reduced as small as possible.

242   Phantom Ground

- Using a VCM buffer for providing a "virtual ground" reference for headphone.
- Disadvantage:
  - Added silicon area and power for the on-chip buffer
  - Residual pop-noise due to buffer/drivers offsets
  - Introduces ground loops when "virtual ground" is connected to 0V ground

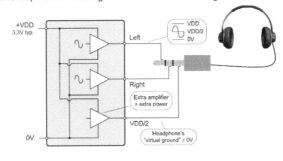

Using a vcm buffer, it provides a virtual ground reference for headphone so that the ac coupled capacitor is not required.

243   Differential Driver: Earpiece and Loudspeaker Driver

- Loud speaker driver required more driving power and its' output is in differential
  - Delivering maximum 500mW to 8Ω loudspeaker with a 3.3V power supply
  - Differential output provide 2VRMS (5.6Vdiffp-p) signal swing to Loudspeaker

    P=VRMS2/R=22/8=500mW

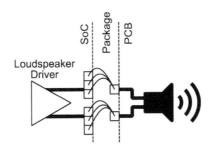

The loudspeaker requires to drive 4 or 8Ω. Delivering maximum 500mW power to 8Ω, the differential output of loudspeaker provides 2Vrms (5.6 Vpp) signal swing, which is larger than the supply voltage 3.3V.

244 Loudspeaker Specifications

Parameter	Conditions	TYP	Unit
Output load resistance	Loudspeaker Positive output to Negative output, AC coupled	8	ohm
Output level for full-scale digital input (0dBFS level)	Measured at -2dBFS @ 1kHz digital input with 1ohms total equivalent resistance from power source to IP core power pins and 0.4ohms from IP output to 8 ohms load	4.342x AVDD/3.3	Vpp
Maximum Output Power	Output load resistance at 8 ohm Measured with 1ohms total equivalent resistance from power source to IP core power pins and 0.4ohms from IP output to 8 ohms load	450	mW
Dynamic range (DR)	Measured at -60dBFS @ 1kHz input, extrapolated to full-scale	94	dB
Signal to Noise Ratio		94	dB
Total Harmonic Distortion (THD)	Measured at -4dBFS (200mW) @ 1kHz input	-60 (0.1)	dB (%)

The loudspeakers normally have the differential output due to the large swing signal. Its specifications are showed in this slide.

245 Audio Driver Type - Class AB

- Class AB is widely considered a good compromise for audio power amplifiers, since much of the time the audio is quite enough that the signal stays in the Class A region, where it is amplified with good fidelity. For the large single swing, the transistor works in class B amplifier operation with less distortion.
- Both output devices are allowed to conduct at the same time, but just a small amount near the crossover point (i.e. each device conducts for slightly more than half a cycle, but less than a whole cycle).

Class A
Both output transistors are continuously conducting with bias current flowing in the output devices. Most linear (lowest distortion) but also the least power efficient, around 20%.
Class B
No continuous bias current flowing in the output devices with one device conducting in the positive and the other in the negative region. Improved efficiency of around 50% but with crossover distortion during the transition, particularly evident at low output levels where the non-linearity at crossover is a larger proportion of the output signal.

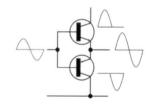

Class AB is widely used as the loudspeaker driver. Class A can achieve a best quality without any crossover distortions, but the lowest power efficiency. Class B has less crossover distortions only for the large single swing. Therefore, in the partial application, Class AB is mostly used.

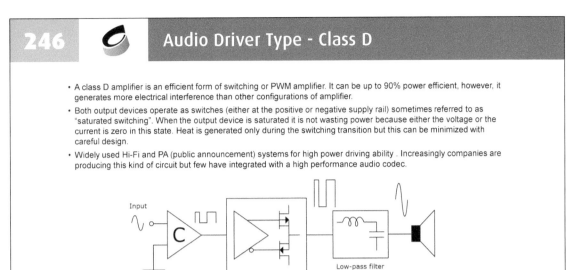

246 **Audio Driver Type - Class D**

- A class D amplifier is an efficient form of switching or PWM amplifier. It can be up to 90% power efficient, however, it generates more electrical interference than other configurations of amplifier.
- Both output devices operate as switches (either at the positive or negative supply rail) sometimes referred to as "saturated switching". When the output device is saturated it is not wasting power because either the voltage or the current is zero in this state. Heat is generated only during the switching transition but this can be minimized with careful design.
- Widely used Hi-Fi and PA (public announcement) systems for high power driving ability . Increasingly companies are producing this kind of circuit but few have integrated with a high performance audio codec.

A class D amplifier is an efficient form of switching or PWM amplifier with up to 90% power efficiency. It is just like a switch and its output is like a bit stream. Therefore, the low pass filter is required to filter out on the bit stream.

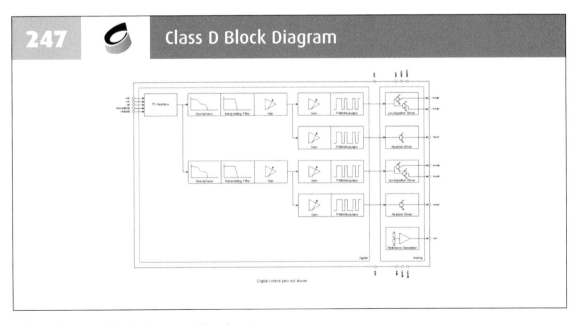

247 **Class D Block Diagram**

This is the typical block diagram of the class D.

248 Class D BTL loudspeaker driver Specifications

Parameter	Conditions	TYP	Unit
Signal Amplitude		16 X AVDD/2.5	Vrms
Output power	Measured at 0dBFS and 8ohm Load @ 1kHz input	300	mW
Power efficiency	Measured at 300mW and 8ohm Load @ 1kHz input	80	%
Dynamic range (DR)	Measured at 8ohm Load @ 1kHz input	96	dB
Signal to Noise Ratio	Measured at -1dBFS and 8ohm Load @ 1kHz input	96	dB
Total Harmonic Distortion (THD+N)	Measured at 300mW and 8ohm Load @ 1kHz input	-70 0.03	dB %
	Measured at 125mW and 8ohm Load @ 1kHz input	-74 0.02	dB %

Signal to Total Harmonic Distortion plus Noise [S/(THD+N) or SINAD]
The ratio of the root-mean-square (rms) value of the fundamental input signal to the rms sum of all other spectral components (harmonics, spurious, noise, etc.) within the pass band and expressed in decibels (dB).

This is the specifications for Class D BTL loudspeaker driver.

249 High Efficiency Driver For SoCs

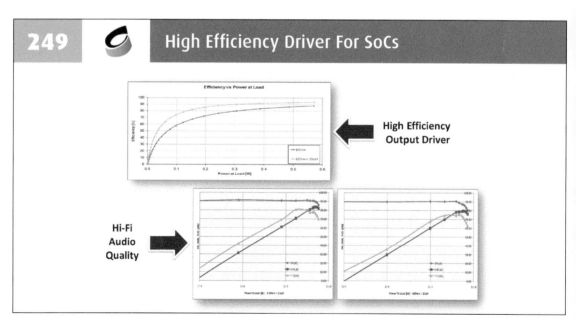

This is the efficiency of the driver for SoCs. The inductance of the loading will also affect the efficiency.

250 Audio Driver Type - Class G/H (I)

- Class G/H amplifier is an enhanced version of class AB, the supply and ground level will change according to the input signal amplitude to minimize power consumption or increased power efficiency
  - Audio signals have considerable crest factors (peak-to-average power ratio) – above 12 dB, sometimes 18 dB
  - 18 dB crest factor means that for a 2 mW average signal, peak signals can reach 1.6 Vpp
  - Reducing the supply by half would increase efficiency by 2X but would also clip peak signals

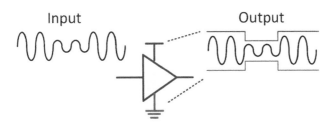

Class G and H scheme are also widely used in some applications such as some mobile phone. Due to the peak to average power ratio in audio signal is around 12dB to 18dB, in most cases the signal is very small. The idea for Class G/H is that the supply voltage is depend on the input. Because in playback channel the input is from digital, the signal level has been known. Then the supply voltage can be adjusted by the signal level.

251 Audio Driver Type - Class G/H (II)

- Larger area and with needs of components for Charge Pump for negative supply

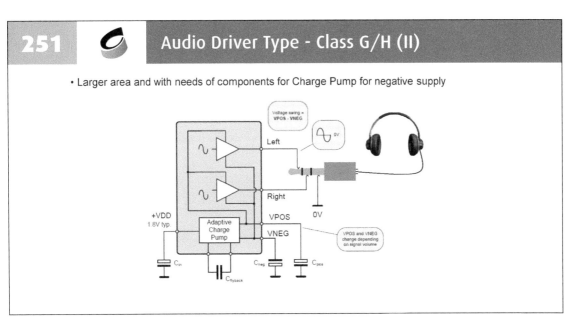

A charge pump is used to adjust the supply in the Class G/H. The charge pump also needs additional area and additional capacitance externals so that the Class G/H are not really used too much in a very deep supply process.

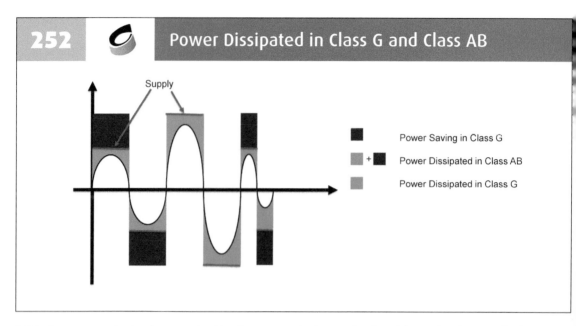

252   Power Dissipated in Class G and Class AB

Supply

■ Power Saving in Class G

■ + ■ Power Dissipated in Class AB

■ Power Dissipated in Class G

This is an example implemented with Class AB and Class G in the older technology. With the power management unit, the Class G can save a lot of power in the small signal.

253   Class G driver Specifications

Parameter	Conditions	TYP	Unit
Output load resistance	Headset output to ground, AC coupled	16	ohm
Maximum Output Power	Output load resistance at 16 ohm with THD+N=0.1%	32	mW
	Output load resistance at 16 ohm with THD+N=1%	34	mW
DC Offset		0.5	mV
Signal to Noise Ratio		104	dB
Total Harmonic Distortion (THD+N)	Measured at (20mW) and 16ohm Load	-91 0.003	dB %
	Measured at (5mW) and 16ohm Load	-87 0.004	dB %

This is the Class G driver specifications.

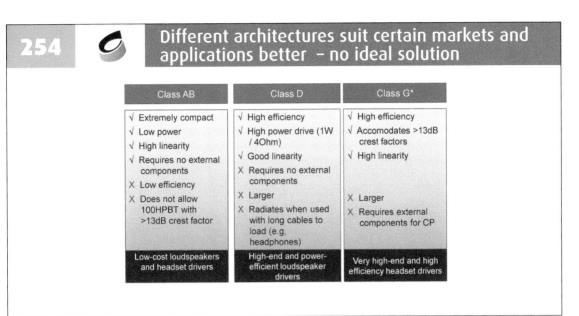

254 Different architectures suit certain markets and applications better – no ideal solution

Class AB	Class D	Class G*
√ Extremely compact √ Low power √ High linearity √ Requires no external components X Low efficiency X Does not allow 100HPBT with >13dB crest factor	√ High efficiency √ High power drive (1W / 4Ohm) √ Good linearity X Requires no external components X Larger X Radiates when used with long cables to load (e.g, headphones)	√ High efficiency √ Accomodates >13dB crest factors √ High linearity X Larger X Requires external components for CP
Low-cost loudspeakers and headset drivers	High-end and power-efficient loudspeaker drivers	Very high-end and high efficiency headset drivers

There is the comparison between Class AB, Class D and Class G. Class AB has the advantage in the small area, low power, good linearity and no external components. But it is low efficiency due to the crest factor effect. Class D has the higher efficiency, higher power to drive and good linearity, but it requires external components. Class G has the high efficiency, high linearity, but it requires external components for charge pump and the area is large. It is normally used in a very high-end and high efficiency headset.

255 Asynchronous Sample Rate Converter (ASRC) (I)

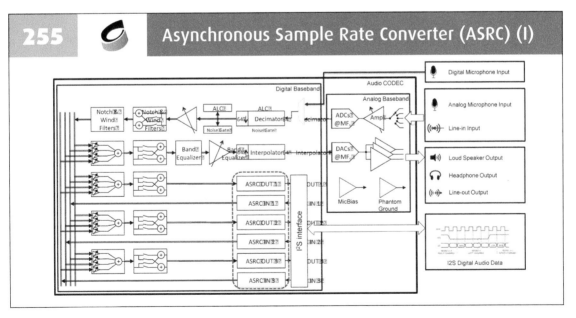

The next is the asynchronous sampling rate converters.

256 — Asynchronous Sample Rate Converter (ASRC) (II)

- The ASRCs used in the Audio Codec interface the internal sampling frequency, to the external audio sampling frequencies $F_{SX1..3}$.

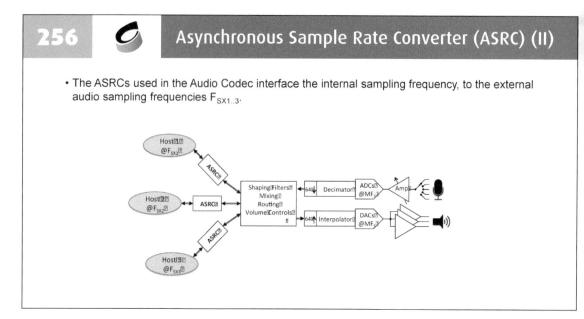

There are a different kinds of sampling rate signals coming from different sources, such as bluetooth, MP3, line in or a microphone in. If the signals with different sample rate are mixed together, the aasynchronous sample rate converter will be required.

257 — Asynchronous Sample Rate Converter (ASRC) (III)

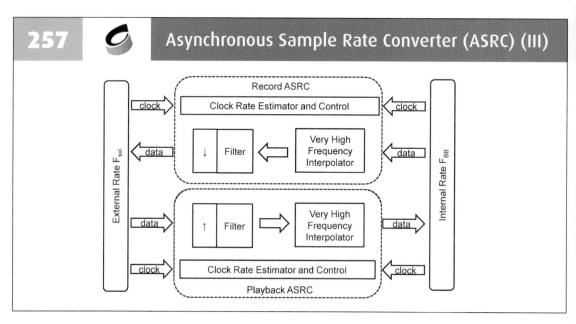

The ASRC requires a very high frequency to interpret, because the sample rate change is mostly a fractional ratio. For example, the 44.1kHz converts to 48kHz. Therefore, the 44.1kHz signal should be raised to a very frequency and then downsample to the 48kHz.

258 Asynchronous Sample Rate Converter (ASRC) (IV)

- **Clock Rate Estimator** receives as inputs the input and output rate clocks to the ASRC (CK_{IN} and CK_{OUT}). It basically computes the frequency ratio between the input and output samples of the VHF Interpolator, and adjusted according to the Halfband Filter in the signal path
- **Very High Frequency (VHF) Interpolator** receives samples at an input rate, interpolates them by a very large ratio, lowpass filters to remove the aliasing bands and resamples at an output rate

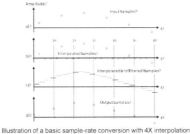

Illustration of a basic sample-rate conversion with 4X interpolation.

Clock rate estimator receives the CKin and CKout as inputs. It basically computes the frequency ratio between the input and output samples of the VHF interpolator, and adjusted according to the halfband filter in the signal path. Very High Frequency (VHF) interpolator receives samples at an input rate, interpolates them by a very large ratio, lowpass filters to remove the aliasing bands and resamples at an output rate.

259 I²S interface

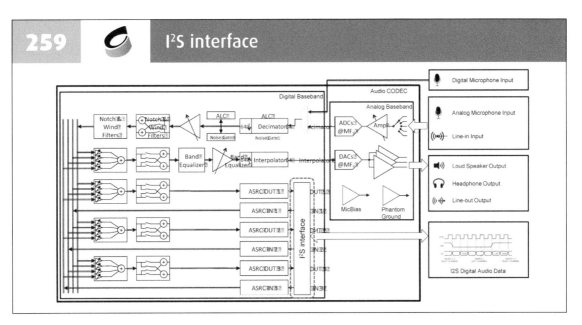

The I2S interface is the standard audio interface.

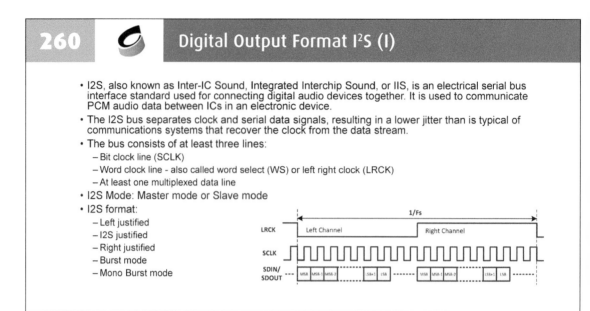

| 260 | | Digital Output Format I²S (I) |

- I2S, also known as Inter-IC Sound, Integrated Interchip Sound, or IIS, is an electrical serial bus interface standard used for connecting digital audio devices together. It is used to communicate PCM audio data between ICs in an electronic device.
- The I2S bus separates clock and serial data signals, resulting in a lower jitter than is typical of communications systems that recover the clock from the data stream.
- The bus consists of at least three lines:
  - Bit clock line (SCLK)
  - Word clock line - also called word select (WS) or left right clock (LRCK)
  - At least one multiplexed data line
- I2S Mode: Master mode or Slave mode
- I2S format:
  - Left justified
  - I2S justified
  - Right justified
  - Burst mode
  - Mono Burst mode

2S, also known as Inter-IC sound or Integrated Interchip Sound, is an electrical serial bus interface standard used for connecting digital audio devices together. The bus consists of at least three lines, bit clock line (SCLK), left right clock line (LRCK) and at least one multiplexed data line.

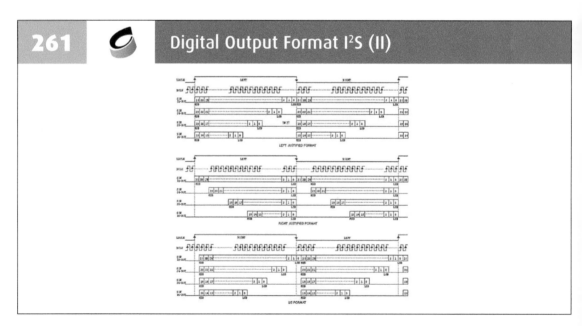

| 261 | | Digital Output Format I²S (II) |

The protocol is very simple. Basically, the first one is from the MSB and the last one is the LSB.

AUDIO CODEC DESIGN EXAMPLES

Two audio codec design examples will be showed.

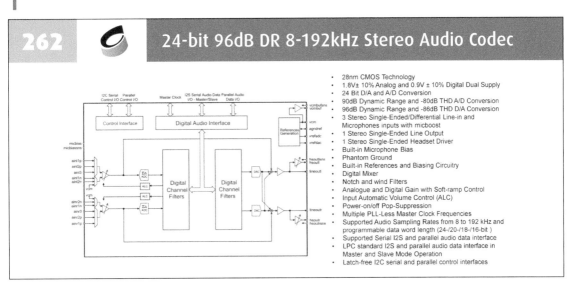

262 — 24-bit 96dB DR 8-192kHz Stereo Audio Codec

- 28nm CMOS Technology
- 1.8V± 10% Analog and 0.9V ± 10% Digital Dual Supply
- 24 Bit D/A and A/D Conversion
- 90dB Dynamic Range and -80dB THD A/D Conversion
- 96dB Dynamic Range and -86dB THD D/A Conversion
- 3 Stereo Single-Ended/Differential Line-in and Microphones inputs with micboost
- 1 Stereo Single-Ended Line Output
- 1 Stereo Single-Ended Headset Driver
- Built-in Microphone Bias
- Phantom Ground
- Built-in References and Biasing Circuitry
- Digital Mixer
- Notch and wind Filters
- Analogue and Digital Gain with Soft-ramp Control
- Input Automatic Volume Control (ALC)
- Power-on/off Pop-Suppression
- Multiple PLL-Less Master Clock Frequencies
- Supported Audio Sampling Rates from 8 to 192 kHz and programmable data word length (24-/20-/18-/16-bit)
- Supported Serial I2S and parallel audio data interface
- LPC standard I2S and parallel audio data interface in Master and Slave Mode Operation
- Latch-free I2C serial and parallel control interfaces

The first one is a 24-bit 96dB DR 8-192kHz stereo audio codec in 32nm CMOS technology. It includes the PGA with input mask, differential line in, microphone bias and boost, line out and headset driver out. It also has the buffer phantom ground, notch and wind filter, analog and digital gains soft ramping control, input automatic volume control, power-on/off pop suppression, multiple PLL-less master clock frequency. All of them are the standard audio codec specifications.

263 — Record Channel: Gain and Noise

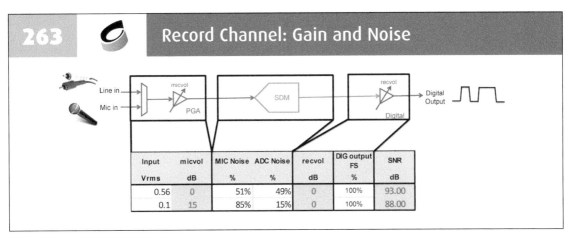

Input	micvol	MIC Noise	ADC Noise	recvol	DIG output FS	SNR
Vrms	dB	%	%	dB	%	dB
0.56	0	51%	49%	0	100%	93.00
0.1	15	85%	15%	0	100%	88.00

In the record channel, the 0.56V rms input with 0dB micvol can get a full scale output with 93dB SNR. The noise between ADC and PGA is almost the same. And a 0.1Vrms input with 15dB micvol can also get the same output with 88dB SNR, but the mic noise will dominate as 85%. Therefore, if the gian of PGA is large, the noise before ADC becomes more dominate and the overall performance is depleted.

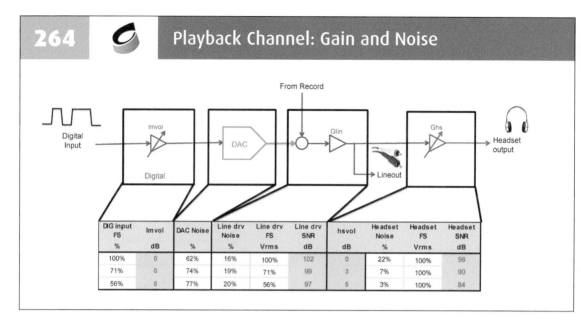

264 — Playback Channel: Gain and Noise

DIG input FS	Imvol	DAC Noise	Line drv Noise	Line drv FS	Line drv SNR	hsvol	Headset Noise	Headset FS	Headset SNR
%	dB	%	%	Vrms	dB	dB	%	Vrms	dB
100%	0	62%	16%	100%	102	0	22%	100%	98
71%	0	74%	19%	71%	99	3	7%	100%	90
56%	0	77%	20%	56%	97	5	3%	100%	84

In the playback channel, with the digital input signal becoming small, the line driver noise become large and the SNR in line driver decreases from 102dB to 97dB. Although with the gain of headset, a full scale output will be got, but the SNR is still lower than the case of 0dB hsvol.

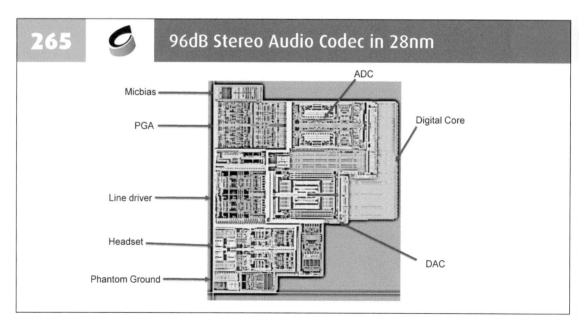

265 — 96dB Stereo Audio Codec in 28nm

This is the chip photograph and the shape is actually not quite regular. Because the audio codec is embedded inside a SoC, and there will be lots of different IPs around the codec. Besides, the design with unregular shape is like a append-able way and can be quickly adjusted according to user needs.

266 Audio Codec Test Setup Diagram

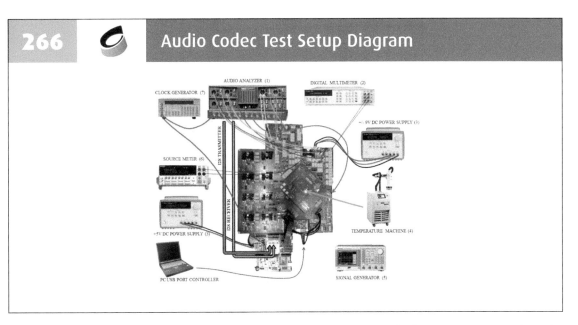

This is the test setup of the audio codec. Audio analyzer is the very standard testing equipment for audio application.

267 Record Path (I)

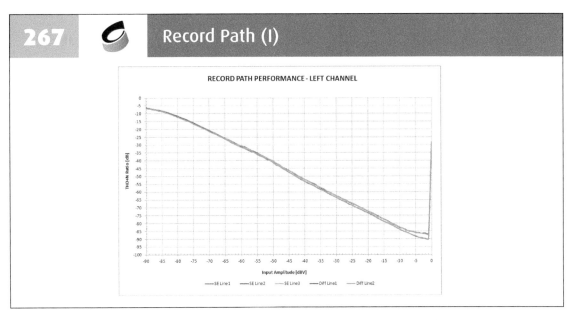

It is the performance of the record path, with single-end line in 1-3 and differential line in 1-2. The THD+N is around -37dB at -60dB full scale so that a 97dB dynamic range can be achieved.

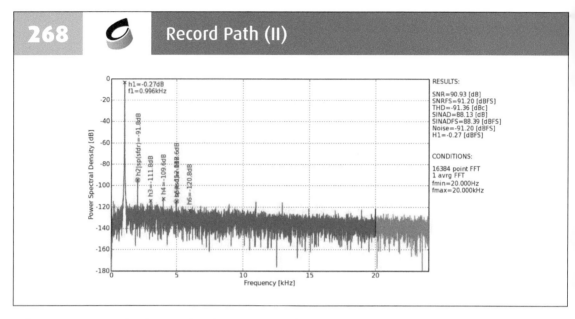

The THD is measured with a good performance at 91dB.

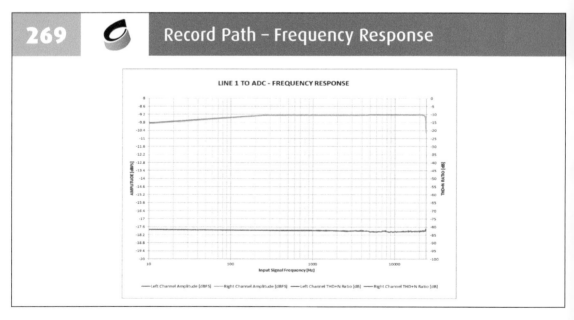

When the gain of PGA is changing, the resistance is changing, and the frequency response is also changing.

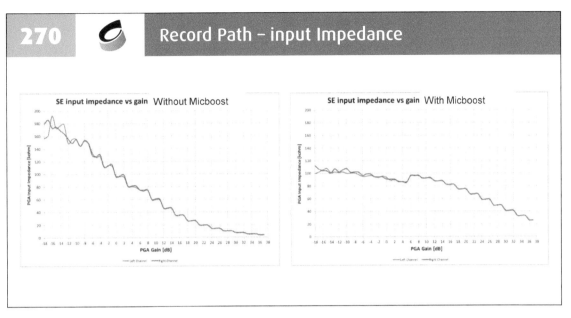

The impedance is actually reducing to only 2kΩ without the microphone boost. Therefore the gain will also drop several dB. But with the microphone boost, the input impedance measured is still more than 20kΩ.

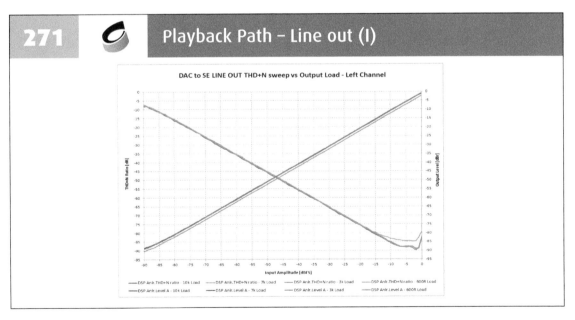

There are lots of measurements showing DAC to single line out with different kinds of loading, from 10kΩ to 600Ω.

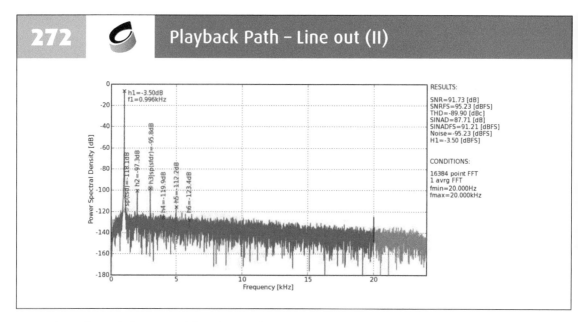

In playback channel, the SNR in full scale is 95dB and THD is -89dB.

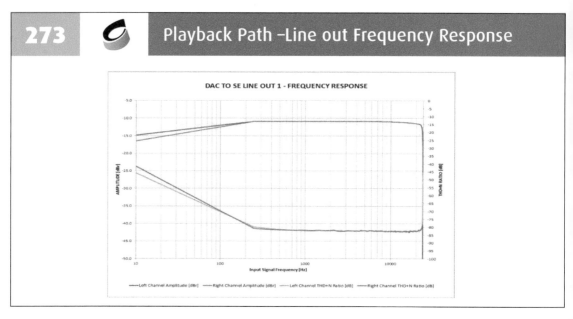

It is the output frequency response of the line out and the performance are reduced in the color frequency due to the ac coupled capacitor.

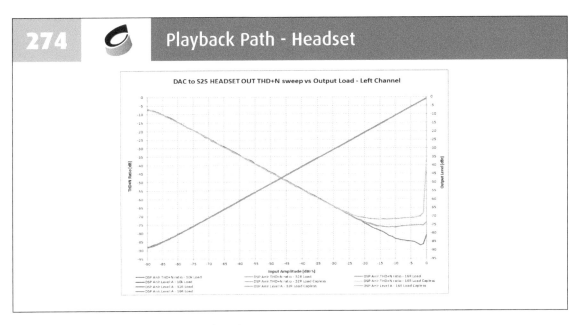

t shows the THD of DAC to single headset driver.

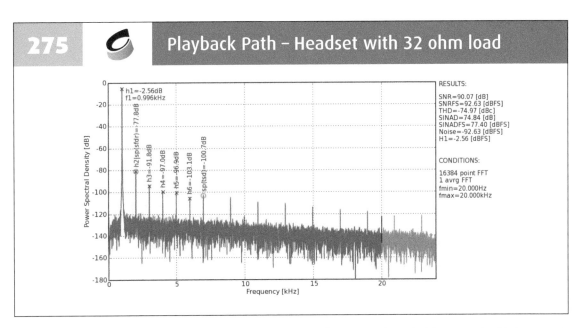

To drive the headset with 32Ω, the SNR is the same but the THD is degraded because the high power is delivered to the output. A -75dB THD is already very good.

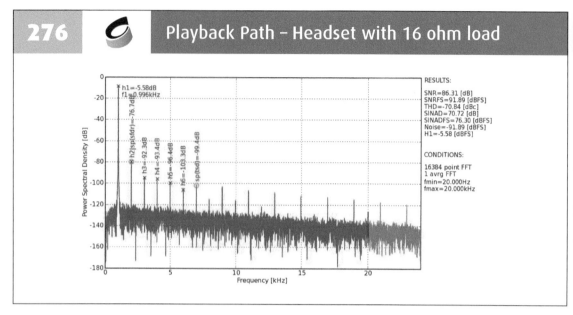

With 16Ω headset, the THD is -71dB.

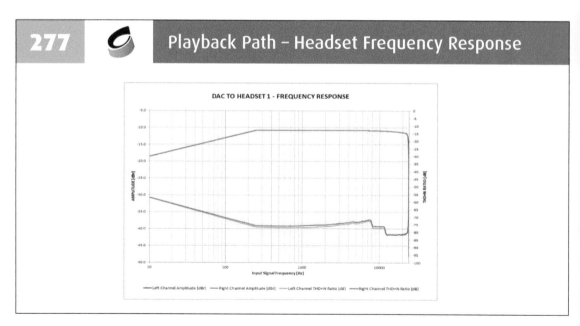

The high pass filter is also caused by the ac coupled capacitor.

278 Soft Ramp

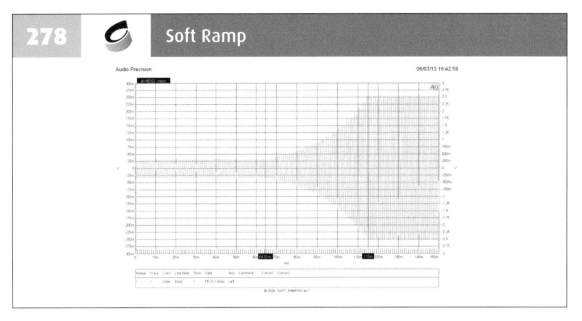

t is the soft ramping control.

279 Crosstalk

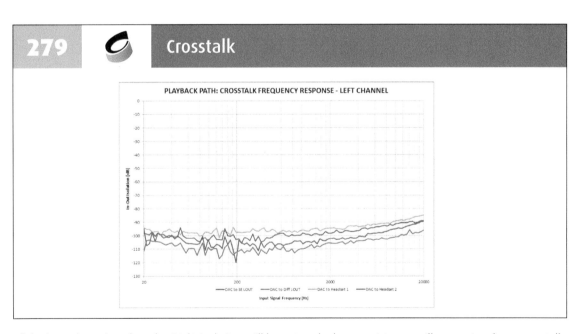

f the impedance is 10kΩ, the 90dB isolation will be got and a large resistance will generate a large crosstalk. Therefore, the audio layout and count parasitic resistance should be careful.

280 Microphone Bias

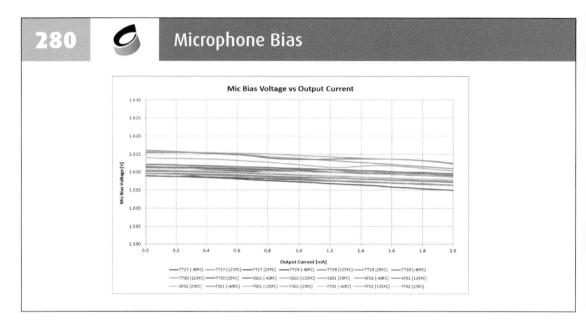

The micbias voltage can not vary a lot with different output current and temperature.

281 PSRR

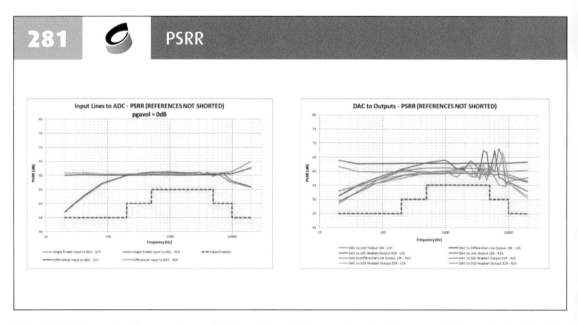

The PSRR is important because the audio is sensitive to the supply noise. There are different kinds of loading and path to measure.

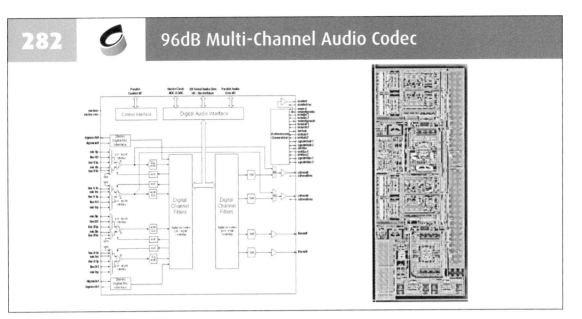

It is another multiple channel audio codec, which has multiple channel input ADC to stereo input and DAC for codec.

INTEGRATION OF AFE IN SOC

he next part is integration of AFE in SoC.

283   Integration of AFE in SoC

- The integration of AFE and Data Converters in aggressive SoC environment are essential and also challenged to maximize the AFE performance and avoid the pitfalls that can, sometimes, jeopardize overall system performance

- For example in a case of integrating a Data Converter, where :
  - 12-bit Data converter with Vref ~1V → Vfull-scale ~1.0 Vpp diff
  - Size of the LSB (min voltage to discriminate): LSB = 1.0 V /4096 ~ 250uV (!)
  - Internal digital aggressors use core voltage ~1V
    - 1.0 V / 250uV → >72dB isolation required such that LSB is not affected by aggressor
  - Digital I/O aggressors use I/O voltage ~1.8V – 2.5V
    - Larger isolation required

- Which signals must be protected?
  - Analog Input/Output and Reference: impact is direct
  - Clock: impact through jitter when sampling/delivering
  - Power and ground supply: impact through finite PSR

For example, the LSB is just 250uV for a 12-bit data converter with the reference 1V full scale. It is very sensitive to the noise and 72dB isolation is required so that the LSB is not affected by aggressors. However, it is hard to achieve such isolation with the digital I/O with 1.8V and 2.5V supply.

284 Data Converters – top-level view

- Analog input (in) / output (out) signals:
  - differential and/or single-ended
  - with one, two, or more channels

- Digital output/input (b) words

- Internal / external reference (ref):
  - internal reference generator or provided by an external reference input

- Conversion clock (clk):
  - driven from an internal phase-locked loop (PLL), or supplied externally from the chip

- Power and ground supply

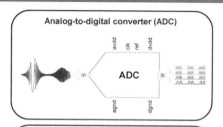

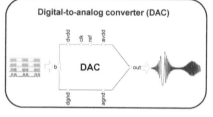

At the top level view, the input, output, internal and external reference, conversion clock and power supply should be considered.

DESIGN CONSIDERATIONS

285 A. Proper Placement (I)

- Excessive noise generated in other noisy blocks in the SoC can couple into the data converter block and impact its performance

- For the best possible isolation from other blocks, the first step in the physical integration process is to correctly place the data converter macro in the SoC

The first design consideration is the location. Excessive noise generated in other noisy blocks in the SoC can couple into the data converter block and impact its performance.

286 A. Proper Placement (II)

Create distance between active logic (aggressor) and the analog block (victim)

- Place the data converter macro away from sources of digital switching
- Position the digital interface of the data converter macro toward the noisier areas of the chip, and the analog interface toward the quieter areas of the chip.
- Place the clock source (e.g., PLL) as close as possible to the data converter macro.
- If the immediate data converter neighborhood includes strong digital switching routing or blocks, then maintain a keep-out area (that is, no metal, devices, or active area) separating the data converter macro from those blocks or routing.

It is required to create the distance between active logic and the analog block. Place the data converter macro away from sources of digital switching. Position the digital interface of the data converter macro toward the noisier areas of the chip and the analog interface toward the quieter areas of the chip. Place the clock source, e.g. PLL, as close as possible to the data converter macro. If the immediate data converter neighborhood includes strong digital switching routing or blocks, then maintain a keep-out area, which is no metal, devices or active area, separating the data converter macro from those blocks or routing.

287 A. Placement: Example (I)

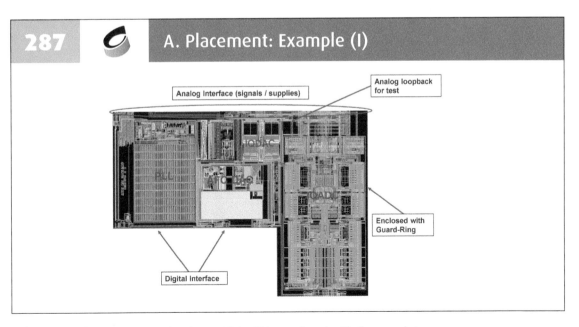

This is one placement example. The IQ ADC will be enclosed with the guard-ring.

288 · A. Proper Placement (III)

Place the data converter macro close to the analog I/O pads

Digital Interface

Analog Interface

IP deliverables

SYNOPSYS IP*- NS

Core poly orientation

Customer SoC (Integration Phase)

- No routing or filling allowed on top of the IP, except for RDL (only ADC signals allowed)

- North South Orientation

No routing or filling allowed on top of the IP, except for RDL (only ADC signals allowed). The metal filling on top of the capacitor are not fully matched, so that it can degraded the performance of the ADC.

289 · A. Placement: Example (II)

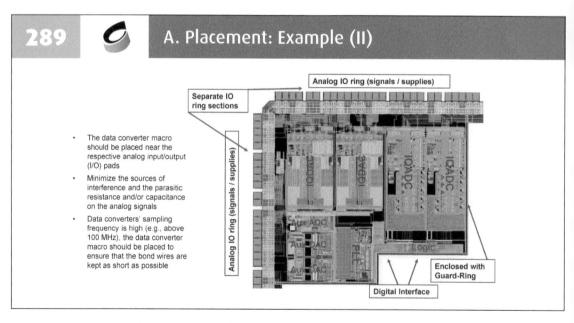

Analog IO ring (signals / supplies)

Separate IO ring sections

Analog IO ring (signals / supplies)

- The data converter macro should be placed near the respective analog input/output (I/O) pads

- Minimize the sources of interference and the parasitic resistance and/or capacitance on the analog signals

- Data converters' sampling frequency is high (e.g., above 100 MHz), the data converter macro should be placed to ensure that the bond wires are kept as short as possible

Enclosed with Guard-Ring

Digital Interface

The analog rings should not connect to the digital ring. The data converter macro should be close to the analog I/O pads to minimize the sources of interference and the parasitic resistance and capacitance on the analog signals. If the sampling frequency is higher than 100MHz, it should be ensured that the bond wires are kept as short as possible.

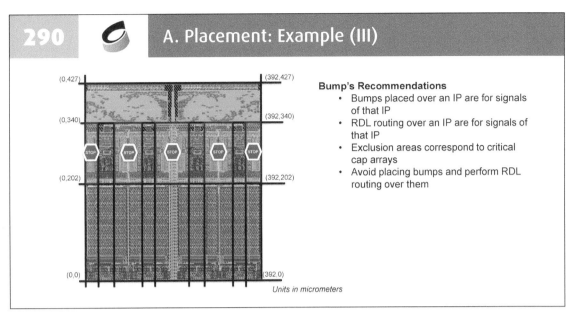

290 — A. Placement: Example (III)

Bump's Recommendations
- Bumps placed over an IP are for signals of that IP
- RDL routing over an IP are for signals of that IP
- Exclusion areas correspond to critical cap arrays
- Avoid placing bumps and perform RDL routing over them

Units in micrometers

The bump's recommendations are as followed. Bumps placed over an IP are for signals of that IP. RDL routing over an IP are for signals of that IP. Exclusion areas correspond to critical capacitor arrays. Avoid placing bumps and perform RDL routing over them.

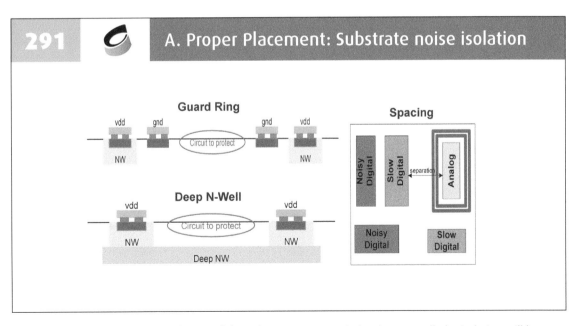

291 — A. Proper Placement: Substrate noise isolation

The guard ring should be placed to avoid the substrate noise. With the deep n-well, the isolation will be even better. Therefore, put the guard ring around the analog and so on.

292 B. Analog Signal Routing

- Any noise or unwanted signals coupling to the ADC input will be seen by the ADC as part of the "true" signal and, therefore, will also be present in its digital output

- ADCs with differential inputs, or DACs with differential outputs, have higher immunity to common mode noise because the aggressor is equally coupled to both positive and negative differential signals

- Noise or unwanted signals, if coupled to the reference, will also become part of the data converter output word

Any noise or unwanted signals coupling to the ADC input will be seen as part of the "true" signal, and will also be present in its digital output. Therefore, ADCs with differential inputs or DACs with differential outputs, should have higher immunity to common mode noise because the aggressor is equally coupled to both positive and negative differential signals. Noise or unwanted signals, if coupled to the reference, will also become part of the data converter ouput word and affect the ADC performance as well.

293 B. Analog Signal Routing: Reference Signals

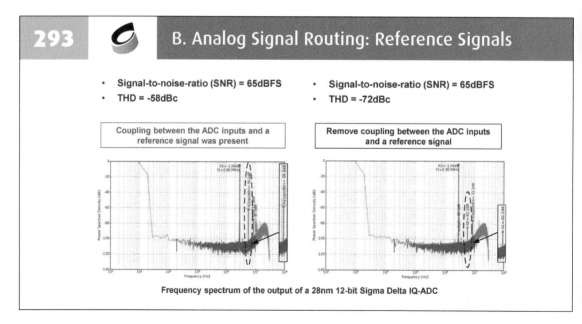

- Signal-to-noise-ratio (SNR) = 65dBFS
- THD = -58dBc

- Signal-to-noise-ratio (SNR) = 65dBFS
- THD = -72dBc

Coupling between the ADC inputs and a reference signal was present

Remove coupling between the ADC inputs and a reference signal

Frequency spectrum of the output of a 28nm 12-bit Sigma Delta IQ-ADC

This is an example that the reference signal coupling from ADC input, so that the reference signal has the input dependent noise. This will generate the distortion and the THD will decrease to -59dB. If the coupling between the inputs and the reference is removed, the THD will be -72dB.

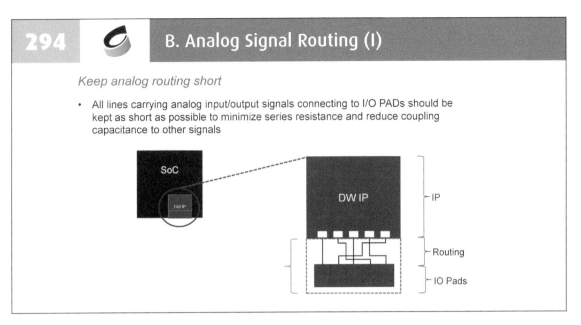

It is important to keep analog routing short and avoid the cross couple and the parasitic.

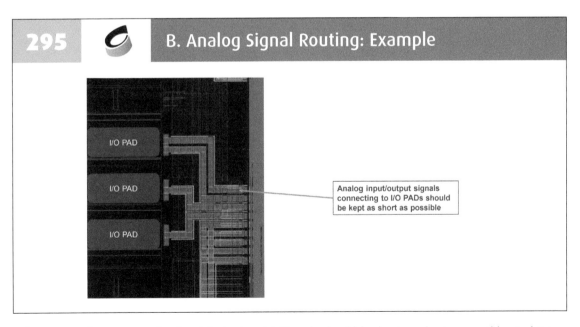

The connect between analog input/output and I/O pads should be kept as short as possible. And it is a better way to make it symmetric.

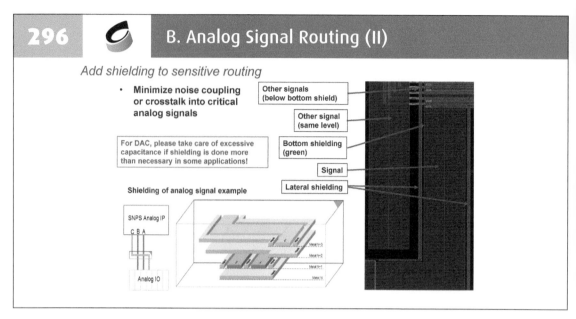

Shielding is actually necessary if the crosstalk can not be avoided. For example, the virtual ground of the opamp is high impedance and very sensitive. Therefore the routing should be short enough and the shielding should be added.

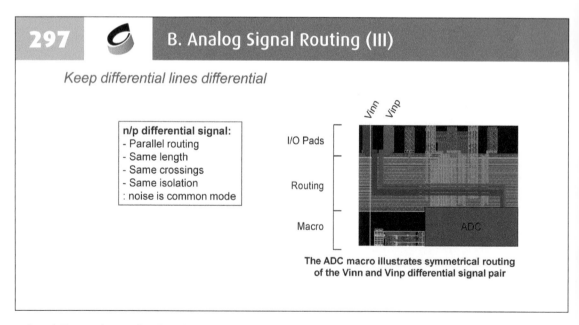

The differential signals should have the same length, crossing, isolation and parallel routing. The differential architecture should have independent common mode rejection ratio so that any noise from the common mode will not affect the differential signals a lot.

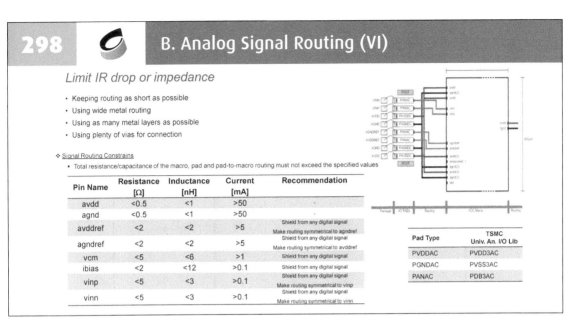

298   B. Analog Signal Routing (VI)

Limit IR drop or impedance

- Keeping routing as short as possible
- Using wide metal routing
- Using as many metal layers as possible
- Using plenty of vias for connection

❖ Signal Routing Constrains
- Total resistance/capacitance of the macro, pad and pad-to-macro routing must not exceed the specified values

Pin Name	Resistance [Ω]	Inductance [nH]	Current [mA]	Recommendation
avdd	<0.5	<1	>50	-
agnd	<0.5	<1	>50	-
avddref	<2	<2	>5	Shield from any digital signal Make routing symmetrical to agndref
agndref	<2	<2	>5	Shield from any digital signal Make routing symmetrical to avddref
vcm	<5	<6	>1	Shield from any digital signal
ibias	<2	<12	>0.1	Shield from any digital signal
vinp	<5	<3	>0.1	Shield from any digital signal Make routing symmetrical to vinp
vinn	<5	<3	>0.1	Shield from any digital signal Make routing symmetrical to vinn

Pad Type	TSMC Univ. An. I/O Lib
PVDDAC	PVDD3AC
PGNDAC	PVSS3AC
PANAC	PDB3AC

The IR drop or impedance should also be considered. In the simulation, the parasitic effect should be taken into account. Therefore the parasitic with 0.5Ω resistance and 1nH inductance is required to count to the supply line. And avddref and agndref have the more parasitics than the avdd. Vcm can be routed widely and there is less current drop because of the matching differential signals. As a result, the IR drop of vcm is not so critical. Input resistance is not very critical as well. However, the output impedance of the audio driver is only 8~32Ω so that only 5Ω routing resistance can affect the performance a lot.

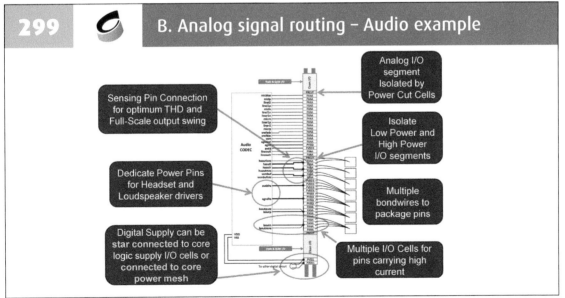

299   B. Analog signal routing – Audio example

Sensing Pin Connection for optimum THD and Full-Scale output swing

Analog I/O segment Isolated by Power Cut Cells

Isolate Low Power and High Power I/O segments

Dedicate Power Pins for Headset and Loudspeaker drivers

Multiple bondwires to package pins

Digital Supply can be star connected to core logic supply I/O cells or connected to core power mesh

Multiple I/O Cells for pins carrying high current

How to handle with the sensing pins? For example, the amplifiers have the output and the feedback loop. These two nets are connected at the I/O pads so that the routing resistance will be put in the feedback loop and will not affect the output. The digital supply can be star connected to core logic supply I/O cells or connected to core power mesh. As a result, the noise from the digital switchings can be avoided. Besides, for loudspeakers with 8 or 4Ω, multiple bondwires to package pins are also required.

300 C. Clock Jitter Consideration (I)

- **Clock jitter impact in SNR**
- Clock jitter adds noise during signal sampling

$$SNR_{total} = -20\log\sqrt{10^{\left(\frac{-SNR_{ADC}}{10}\right)} + 10^{\left(\frac{-SNR_{jitter}}{10}\right)}}$$

- SNR_{jitter} decreases with signal frequency

$$SNR_{jitter} = -20\log\left(2\pi f_{in}\sigma_{t,jk}\right)$$

- **Example**
- PLL clock with 8ps$_{rms}$ jitter
- Clock path containing 9 clock buffers, each adding 6ps$_{rms}$ jitter
- Then, clock signal at the ADC input has ~20ps$_{rms}$ jitter
- A 12 bit ADC with a 5MHz input signal will lose 1.5bit of ENOB

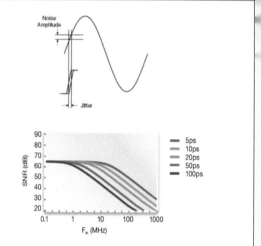

The clock jitter is very sensitive at the high frequency. For example, the PLL clock has 8ps rms jitter and the clock path contains 9 clock buffers and each will add 6ps. The clock signal has 20ps rms jitter and a 12 bit ADC with 5MHz input signal will lose 1.5bit of ENOB. Therefore, the clock of ADC should be designed carefully.

301 C. Clock Jitter Consideration (II)

Place clock source close to the data converter macro

Place the PLL close to data converter to reduce likelihood of extraneous signals coupling to the clock line and contributing to clock jitter

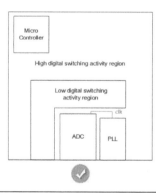

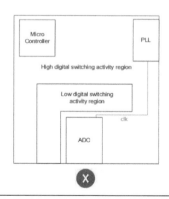

Placing the PLL close to data converter can reduce the likelihood signal coupling to the clock line and contribute to the clock jitter. If the PLL is shared with others, the design should be careful and adding the buffer image as well as possible.

302   C. Clock Jitter Consideration (III)

Check the transition times

- **Clock signal slope degradation**
  - Increased time uncertainty (dt2) = more jitter
  - Duty cycle degradation

- **Add adequate buffering in clock tree**
- **Avoid supply domain transitions**
- **Use clean supply**

- **Rule of thumb: edge < 100ps**

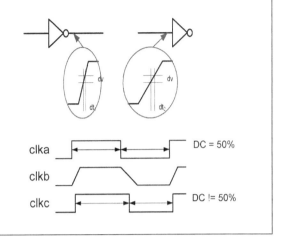

Clock signal slope degradation can increase the time uncertainty and couse the duty cycle degradation. It is necessary to add adequate buffering in clock tree, avoid supply domain transitions and use clean supply. The rule of thumb is to keep the rising edge or falling edge less than 100ps.

303   C. Clock Jitter Consideration (IV)

Minimize supply transitions

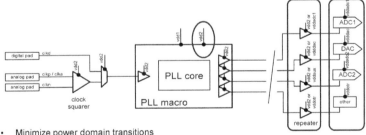

- Minimize power domain transitions
- Buffered using only one supply
- Shielding from other signals
- Selective clock disable
- Place the clock source close to the data converter macro

To minimize supply transitions, the buffer should use only one supply, which is not PLL or shared with others. Place the clock source close to the data converter macro and the clock can be selective disable when it doesn't work.

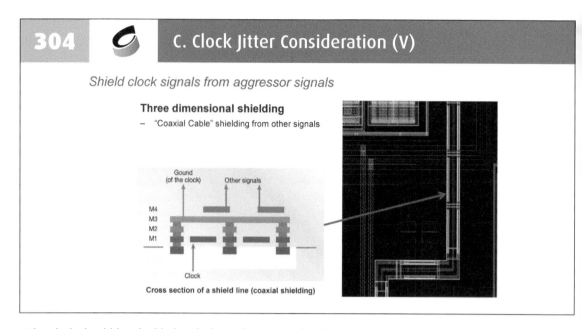

304 · C. Clock Jitter Consideration (V)

Shield clock signals from aggressor signals

Three dimensional shielding
− "Coaxial Cable" shielding from other signals

Cross section of a shield line (coaxial shielding)

The clock should be shielded with three dimension shielding, which it puts the ground and the via metal around the clock. It can protect the clock well in a high resolution or high speed design.

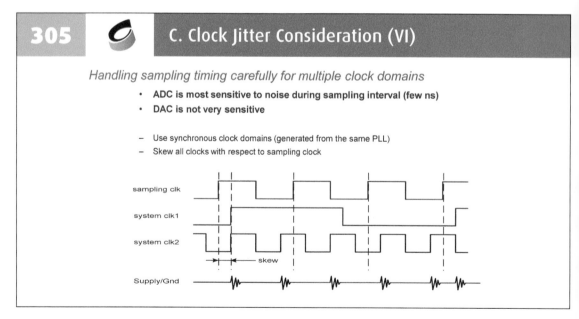

305 · C. Clock Jitter Consideration (VI)

Handling sampling timing carefully for multiple clock domains

• **ADC is most sensitive to noise during sampling interval (few ns)**
• **DAC is not very sensitive**

− Use synchronous clock domains (generated from the same PLL)
− Skew all clocks with respect to sampling clock

ADC is most sensitive to noise during sampling interval. If multiple clock domains are used, it is necessary to avoid the big switching happening before the sampling inverval. Because the big switching can introduce the noise to the supply or ground, which can also be sampled by the ADC. Therefore, with the synchronous frequency, it should be avoided that switching happening before the sampling of ADC. However, there is no any choice in the asynchronous ADC due to the uncertain sampling.

306 D. Power Routing (I)

- Any analog circuit will have a finite power supply rejection ratio (PSRR); Excessive noise injected in the power and ground supplies may affect its performance.

- This is especially true if it is processing broadband signals,

- Analog supplies should be clean and proper decoupling used

- Additional effects, such as excessive routing resistance that could lead to a DC voltage drop outside the data converter operating range, and that could also cause a slow AC response

Any analog circuit will hive a finite power supply rejection ratio and excessive noise injected in the power and ground supplies may affect the performance. Therefore, the analog supply should be clean and proper decoupling used.

307 D. Power Routing (II)

Isolate noisy supply source

- –DC-DC in-band spurious tone – LDO isolation
- –Example of a possible power management scheme for 2 AFE supplies and 2 digital core supplies

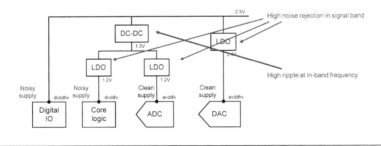

It is important to isolate noisy supply source. For example, DC-DC power actually generates lots of spurious tones so that the LDO should be used for isolation. It should be well controlled to separate the noise domain and the clean domain.

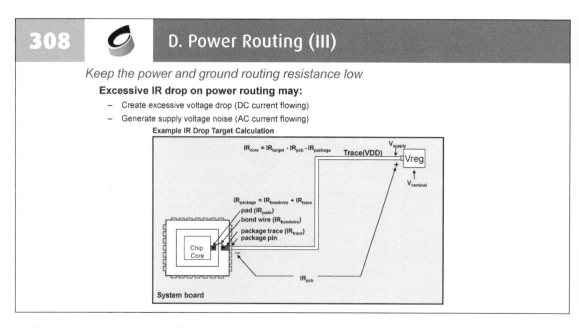

308 — D. Power Routing (III)

Keep the power and ground routing resistance low

Excessive IR drop on power routing may:
- Create excessive voltage drop (DC current flowing)
- Generate supply voltage noise (AC current flowing)

Example IR Drop Target Calculation

$IR_{core} = IR_{target} - IR_{pcb} - IR_{package}$

Trace(VDD)

V_{supply}

Vreg

$V_{nominal}$

$IR_{package} = IR_{bondwire} + IR_{trace}$
- pad (IR_{pads})
- bond wire ($IR_{bondwire}$)
- package trace (IR_{trace})
- package pin

Chip Core

IR_{pcb}

System board

R drop on the power supply will create excessive voltage drop and noise. The IR drop from IP core, PCB and package will count to the design, which can degrade the performance as well.

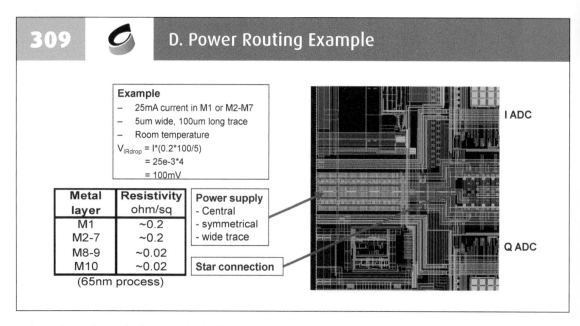

309 — D. Power Routing Example

Example
- 25mA current in M1 or M2-M7
- 5um wide, 100um long trace
- Room temperature

$V_{IRdrop} = I*(0.2*100/5)$
$= 25e-3*4$
$= 100mV$

Metal layer	Resistivity ohm/sq
M1	~0.2
M2-7	~0.2
M8-9	~0.02
M10	~0.02
(65nm process)	

Power supply
- Central
- symmetrical
- wide trace

Star connection

I ADC

Q ADC

This is the IR drop calculations, which shows IR drop in this long strip at the room temperature. It also shows that routing in the top metal will have less resistance. Besides, if the resistance is not avoided, the star connection is still the better choice.

310   D. Power Routing (IV)

Use dedicated power routing

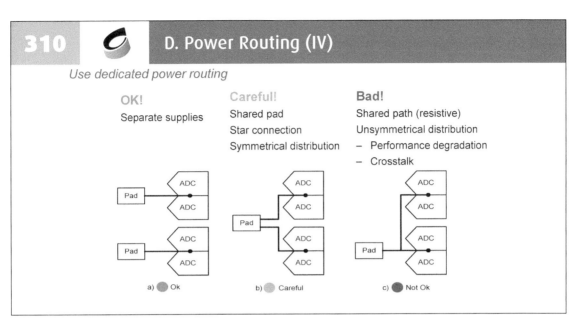

The separate supply is the perfect choice. But if no dedicate pad for ADC, it should be designed carefully with star connection and symmetricala distribution. Therefore, the last one is a very bad design.

311   D. Power Routing (V)

Flip Chip vs Wire Bond

- Wire bond package has power pins only on the edge, making it more difficult to meet IR drop requirements
  - Smaller IR drop budget because of higher package IR drop
  - Larger IR drop for the core since current has to travel from the edge of the die to the center
  - More routing resources required in the PG grid to keep IR drop within budget
- Flip Chip package has distributed power bumps, resulting in smaller, local IR drop. However, the flip chip methodology is more complex
  - RDL routing
  - Hierarchical blocks are more difficult

Wire bond package has power pins only on the edge, making it more difficult to meet IR drop requirements. It has smaller IR drop budget because of higher package IR drop. There is larger IR drop for the core since current has to travel from the edge of the die to the center. The flip chip package has distributed power bumps, resulting in smaller, local IR drop. Although the flip chip methodology is more difficult, it still is the better choice for power routing.

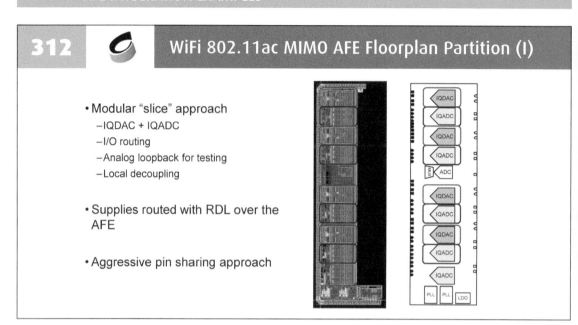

312 **WiFi 802.11ac MIMO AFE Floorplan Partition (I)**

- Modular "slice" approach
  - IQDAC + IQADC
  - I/O routing
  - Analog loopback for testing
  - Local decoupling

- Supplies routed with RDL over the AFE

- Aggressive pin sharing approach

This is an example of WiFi 802.11ac MIMO AFE floorplan. Due to the requirement of MIMO, the routing becomes very complex and it is required to share the pins in order to reduce the pins.

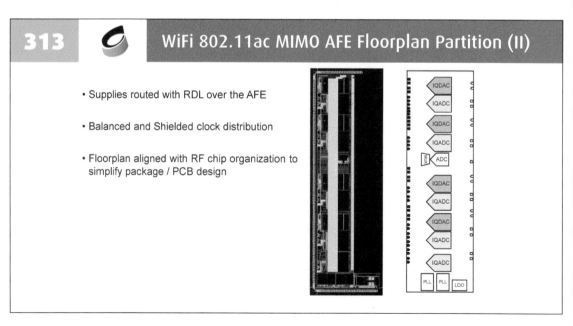

313 **WiFi 802.11ac MIMO AFE Floorplan Partition (II)**

- Supplies routed with RDL over the AFE

- Balanced and Shielded clock distribution

- Floorplan aligned with RF chip organization to simplify package / PCB design

The LDO supply routing is symmetrical in the AFE.

314   AFE Modular Integration Example (I)

Module = IQADC + IQDAC

- Modular "slice" approach
  - –IQDAC + IQADC
  - –I/O routing
  - –Analog loopback for testing
  - –Local decoupling

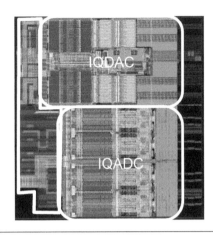

The module puts the IQ ADC together and there are some shared common blocks between them. Analog loopback is a loop that the ADC digital bit streams come to the DAC and than back to ADC. It is the test method to measure the analog signal directly. The local decoupling are very important.

315   AFE Modular Integration Example (II)

Module = IQADC + IQDAC

- Modular "slice" approach
  - –IQDAC + IQADC
  - –I/O routing
  - –Analog loopback for testing
  - –Local decoupling

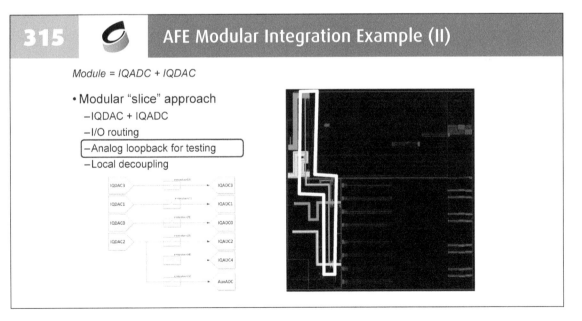

This is analog loopback for testing. There are lots of loopback between ADC and DAC.

316 AFE Modular Integration Example (III)

Module = IQADC + IQDAC

• Modular "slice" approach
  - IQDAC + IQADC
  - I/O routing
  - Analog loopback for testing
  - Local decoupling

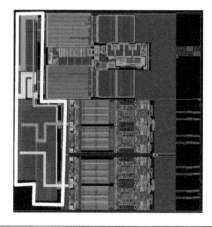

Remerber to put the decouping as much as possible. The decoupling effect will be reduced if the the resistance of the decoupling routing is not so small.

SUMMARY

317 Summary

• Analog Front-Ends (AFEs) and data converters are key components of Communication and Multimedia SoC

• To properly interpret the AFE components' electrical specifications and how they determine the overall performance are meaningful to the SoC system design

• For broadband applications, digital calibrated SAR ADCs with also interleaving techniques is a prominent scaling architecture with respect to the most advanced process for wide-range data converters with high-efficiency in terms of power and area as well as design cycle

• Growing complexity in audio from multiple formats, channels, sampling rate, features and sound processing for customer electronics drives the dedicated and configurable audio subsystem with audio codec and audio processor for multimedia SoCs

• Successful SoC relies not only the high quality, robust and modularized AFEs but also considerate and careful integration design

Analog Front Ends and data converters are key components of communication and multimedia SoCs. To properly interpret the AFE components' electrical specifications and how they determine the overall performance are meaningful to the SoC system design. For broadband applications, digital calibrated SAR ADCs with also interleaving techniques is a prominent scaling architecture with respect to the most advanced process for wide-range data converters with high-efficiency in terms of power and area as well as design cycle. Growing complexity in audio from multiple formats, channels, sampling rate, features and sound processing for customer electronics drives the dedicated and configurable audio subsystem with audio codec and audio processor for multimedia SoCs. Successful SoC relies not only the high quality, robust and modularized AFEs, but also considerate and careful integration design.

Acknowledgement:

Manual Mota, Pedro Figueiredo, Louis Lao, William Lai, Roberto Guerreiro and other colleagues for their contributions

CHAPTER 03

Low Power Digital Design for Mobile Computing

Alice Wang

Assistant General Manager of
High-performance Processor Technology,
MediaTek

Driven by consumer demand, mobile devices such as smartphones and tablets are offering more desktop-like capabilities. High-performance processors, which handle compute-intensive tasks, are key to enhancing the user experience in applications such as 3D gaming, high-definition video and internet browsing. The user experience is judged by the ability of the processors to deliver maximum performance under a thermal limit and able to sustain the performance with a limited battery capacity. This talk will discuss circuit techniques to maximize power while keeping low power used in mobile SoC's. The talk will focus on the system level considerations that require close collaboration between hardware and software and also span across process nodes from 28nm down to 10nm.

1. LOW POWER FOR MOBILE COMPUTING

1 **Did you know?**

- Today's cell phone has millions of time more computing power than all of the NASA computers that put two astronauts on the moon

THEN   NOW

So just imagine in 1960 how much computing has improved from then to now. We have done a lot in terms of power and size. And what we will do is exploring some of those techniques that went from then to now.

2 **The Dawn of Low-Power**

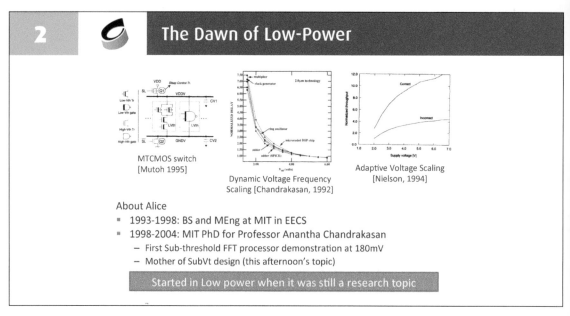

MTCMOS switch
[Mutoh 1995]

Dynamic Voltage Frequency Scaling [Chandrakasan, 1992]

Adaptive Voltage Scaling
[Nielson, 1994]

About Alice
- 1993-1998: BS and MEng at MIT in EECS
- 1998-2004: MIT PhD for Professor Anantha Chandrakasan
  - First Sub-threshold FFT processor demonstration at 180mV
  - Mother of SubVt design (this afternoon's topic)

Started in Low power when it was still a research topic

The MTCMOS switch. It is a powerful switch in 1995 by Mutoh. Dynamic Voltage Frequency Scaling by professor Chandrakasan in 1992, first Sub-Vt design. And Adaptive Voltage Scaling by Nielson in 1994.

I developed this sub-threshold FFT processor demonstration and it works at 180Mv.

first joined Texas Instruments. At that time, Texas Instruments was doing a lot of feature phones, like candy bar phones and it was very popular. Candy bar phones for Nokia. Texas Instruments in the smart phone and picture phone vision doing low power. And Then after a few years, I left Texas Instruments for MediaTek, and right now I am an assistant general manager there and leading High-performance Technology Unit Manage R&D groups in Taiwan and Singapore doing High-performance CPU design for application processor, std cell library and memory compilers.

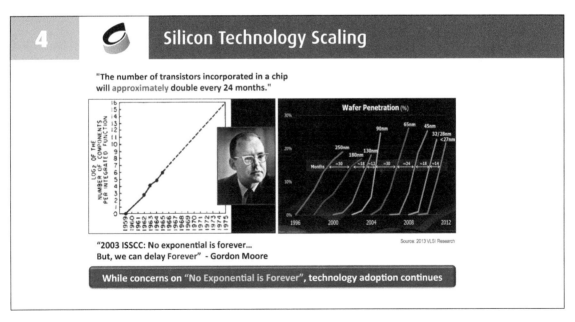

No exponential is forever or we can keep delaying forever. But I firmly believe engineers can solve any problems. So that's why we're still tracking on the Moore's Law curve you see and this part here shows the truth of that cost.

The 250 nanometer in each generation and actually got 30 months here but even has call a tow line so the years to two years and then maybe you hit a difficult year.

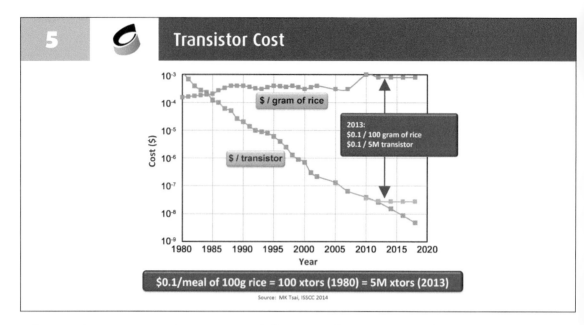

The cost of rice per gram of rice pretty much flat and not changed from 1980 to 2017. But the cost of transistors literally gone down so that even today time fly we can use only 10 cents for hundreds of rice and 10 cents you can exchange for 5 billion transistors.

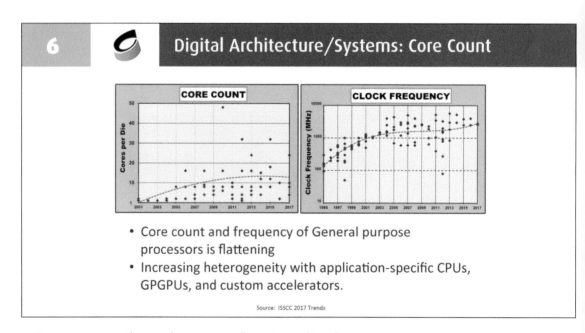

You can see now the trend curve starts flattening now. it was growing linearly in the last decades. So not adding too many new general purpose processors over the last decade and also we were seeing that the clock frequency also was not growing linearly anymore.

We have more graphics processors and custom accelerators instead of just putting general purpose processors.

7 Communication trends

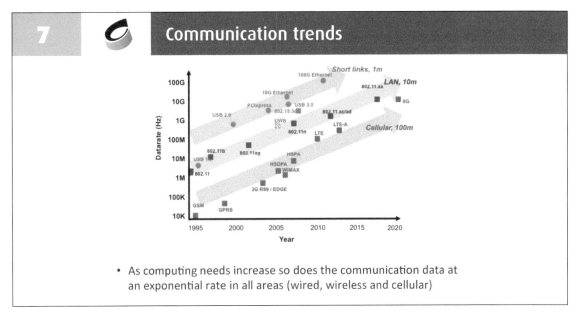

- As computing needs increase so does the communication data at an exponential rate in all areas (wired, wireless and cellular)

Computing needs increase and also the communication data rate. And this shows a few different cellular here and the wireless and short links. All of them are increasing over time at an exponential rate, so wired, wireless and cellular all showing it is the same for communication increasing and computing.

8 Digital Architecture: Power

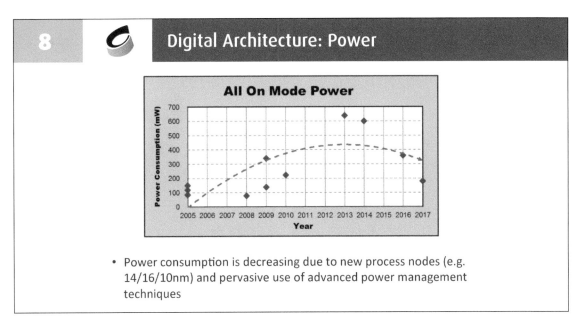

- Power consumption is decreasing due to new process nodes (e.g. 14/16/10nm) and pervasive use of advanced power management techniques

You can see technology helps to lower the power without sacrificing performance. And also we will see a lot of new advanced management technique introduced in order to reduce the power.

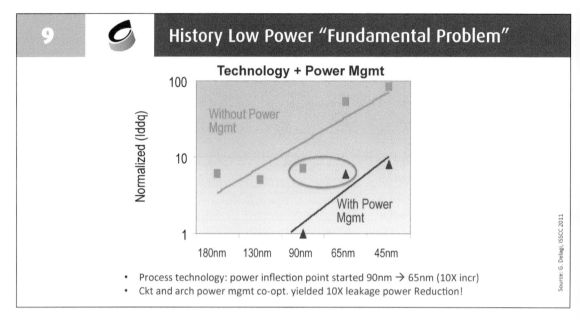

The green line is the Power without Power Mgmt, and the blue one is the Power with Power Mgmt. My job is doing the Mgmt in 65nm in TI to reduce the leakage current as the 90nm. Yielded 10X leakage power reduction.

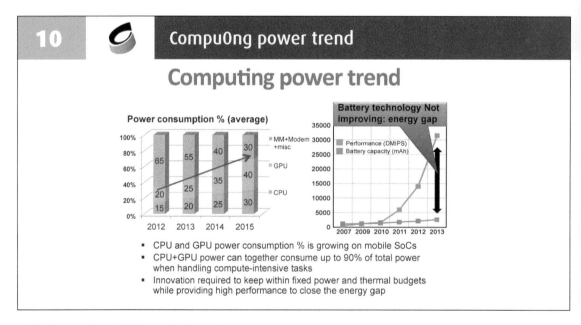

This shows the breakdown of where the compute go on an average.

11 — Smartphone/tablet User-reality

Total battery usage/day

Standby mode	**Performance:** Gaming, Web browsing
Games	
Email	
Web Browser	**Power:** All use cases
Social Networking	
Text	
Voice	

(Battery chart with y-axis: 0%, 10%, 20%, 30%, 40%, 50%, 60%, 70%, 80%, 90%, 100%)

For best user experience, need to optimize both power and performance

12 — Today's Topics for discussion

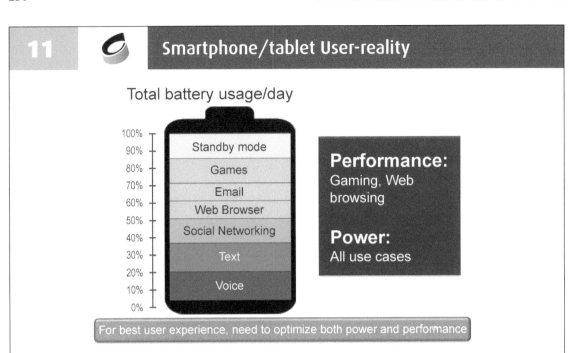

Techniques are demonstrated in Silicon on the following CPU's and process nodes

28nm Planar → 28nm HiK → 20nm HiK → 16/10nm FinFet
ISSCC'14   ISSCC'15   ISSCC'16   ISSCC'17

- What's going on around me? (Adaptive techniques for power/perf opt)
- Sipping power! (Power delivery network)
- I feel the need, the need for speed (Clocking)
- Lay a good foundation (Std cell/SRAM ckt techniques)
- Shipping 100's Millions of IC's (Advanced test/debug techniques)
- Up the food chain (ckt → system → S/W → Humanties?)

2. WHAT'S GOING ON AROUND ME?

13

Performance, Thermal & Power (PTP) Technology (I)

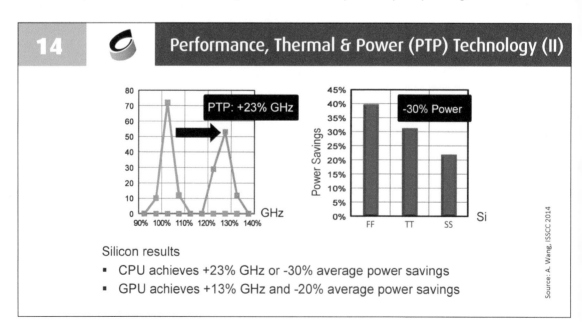

PTP concept
- Issue: To ensure consistent Power/Performance for an SoC the design is closed at a PVT (process, voltage, temperature) which is worst case across all possible conditions
- Result: Over-design and significant power overhead
- Solution: Distribute Performance, Thermal and Process (PTP) sensors on critical IP's (CPU, GPU) to adaptively manage the voltage to the minimum possible Vdd

Source: A. Wang, ISSCC 2014

The first topic is about the adaptive techniques for power/perf optimization.

We distribute the PTP sensors---performance, thermal and process sensors on critical IP's, like CPU and GPU. And from this data that we collect in the CPUs, then we can actively manage the voltage they provide to the IP and to the minimum possible VDD that they can adaptively manage.

14

Performance, Thermal & Power (PTP) Technology (II)

Silicon results
- CPU achieves +23% GHz or -30% average power savings
- GPU achieves +13% GHz and -20% average power savings

Source: A. Wang, ISSCC 2014

The power and performance are like two, can be interchange pretty much. Imagine that CPU GPU to take up you know to 90% of the total power percentage and that means your battery can last 30% longer if you can do this kind of adaptive voltage managements.

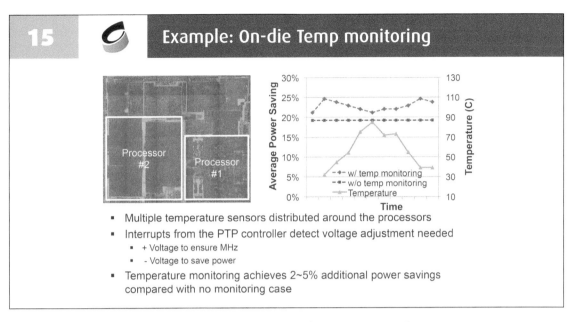

15 **Example: On-die Temp monitoring**

- Multiple temperature sensors distributed around the processors
- Interrupts from the PTP controller detect voltage adjustment needed
  - + Voltage to ensure MHz
  - - Voltage to save power
- Temperature monitoring achieves 2~5% additional power savings compared with no monitoring case

The second kind of adaptive behavior is for temperature. We distribute multiple temperature sensors all around the processor and measure the temperature, at the time we judge the voltage in order to compensate for temperature. So you can get more power saving to 25 percent more power saving compared to non-monitoring temperature.

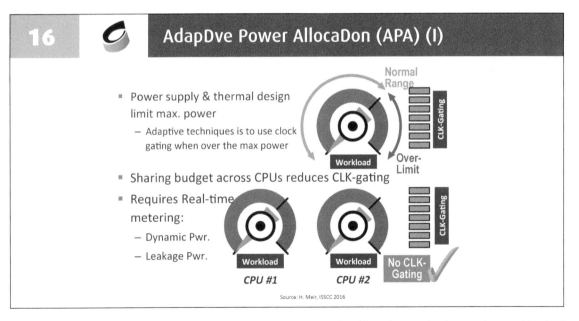

16 **AdapDve Power AllocaDon (APA) (I)**

- Power supply & thermal design limit max. power
  - Adaptive techniques is to use clock gating when over the max power
- Sharing budget across CPUs reduces CLK-gating
- Requires Real-time metering:
  - Dynamic Pwr.
  - Leakage Pwr.

Source: H. Mair, ISSCC 2016

Thermal design is really important for mobile SOC and because of the thermal limit that we have to make sure that our CPUs, that our SOCs is not exceeding this thermal limit. So with the CLK-gating, the CPUs can come back into the normal range not even overheat the design.

We should understand what is the workload of each one of them and how do we maximally to do the clock gating when it is the right time. So we want to share this thermal budget across different CPUs and then we can reduce the amount of clock gating that we have to do.

17 AdapDve Power AllocaDon (APA) (II)

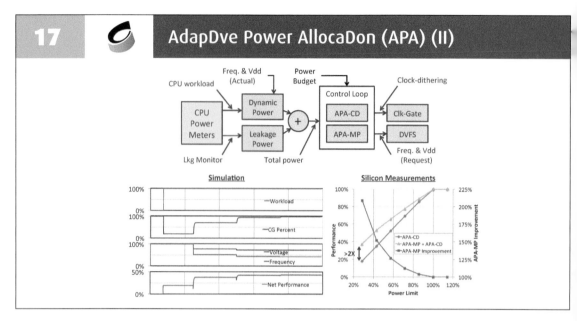

Clock Gating, DVFS and Clock Dithering are all techniques for the APA.

The silicon measurements of this technology and it shows if we were to do just a clock gating, as you have a very very low power limit, your performance is pretty bad only 20% performance when the power limit is very low. But if we can do the clock dithering and MP maximal performance together then you can increase this by 2X and have even more performance at a very very low power limit.

18 Dynamic Power Meter (I)

- **Event-Accumulator estimation:**
  - **Estimation based on Block-Level Events, e.g. SIMD instruction decode**
  - **Good correlation at high-power** ✓
  - **Poor correlation at low-power** ✗

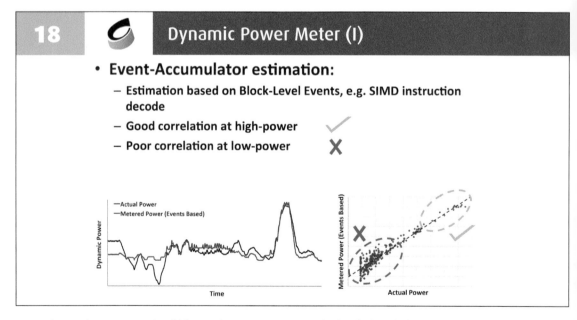

To achieve the APA, we should know the Dynamic Power & Leakage Power. The first one is a dynamic power meter.

The most popular way is to do the event accumulator estimation. We could do we can break out the kind of work that CPU does which is a different kind of event like a SIMD instruction, decode or a little store for example. You can see the low-power event is all over the place. The correlation is not very good.

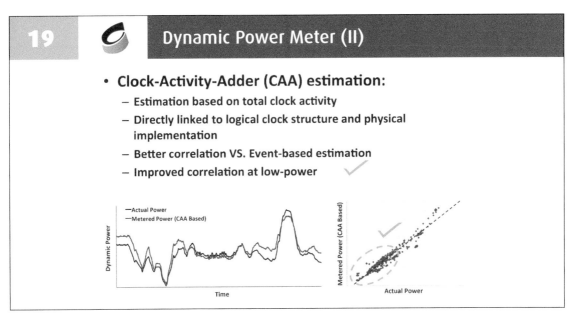

19   Dynamic Power Meter (II)

- **Clock-Activity-Adder (CAA) estimation:**
  - Estimation based on total clock activity
  - Directly linked to logical clock structure and physical implementation
  - Better correlation VS. Event-based estimation
  - Improved correlation at low-power

The second way is looking to it called a clock activity adder estimation. You know how many devices you have switching is pretty much a good indication of power. So you can see that if you look at just a clock then you have a pretty good correlation all over the place here.

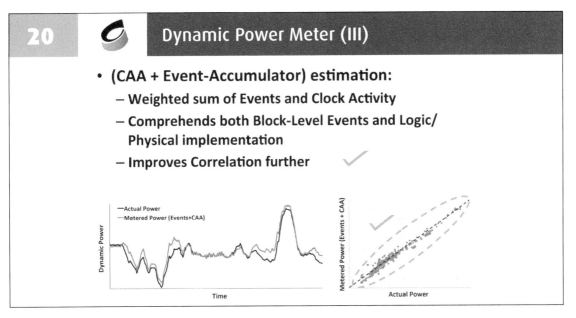

20   Dynamic Power Meter (III)

- **(CAA + Event-Accumulator) estimation:**
  - Weighted sum of Events and Clock Activity
  - Comprehends both Block-Level Events and Logic/ Physical implementation
  - Improves Correlation further

We do combination of both the clock activity adder and event accumulator estimation. And we take away to some of both the events and the clock activity, so that we have both good correlation at the high power area and also good correlation at the low power area and get even better correlation in either one separately.

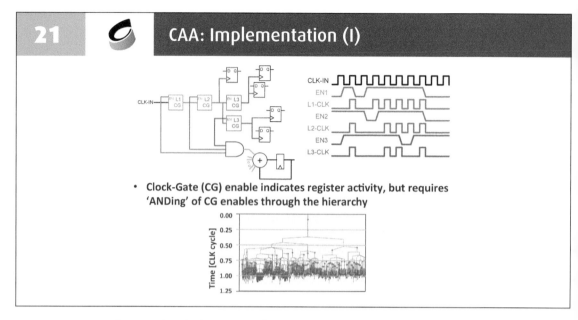

- **Clock-Gate (CG) enable indicates register activity, but requires 'ANDing' of CG enables through the hierarchy**

However, to implement the clock activity adder is quite difficult. A better way is to monitor the clock enable signal. You can imagine a clock enable are telling you how often those clocks are being toggled and the clock being toggled is most likely everything else being toggled.

But this part is really difficult to close the timing at the kind of speed that network trying to do. It is a good idea but it is very hard to implement.

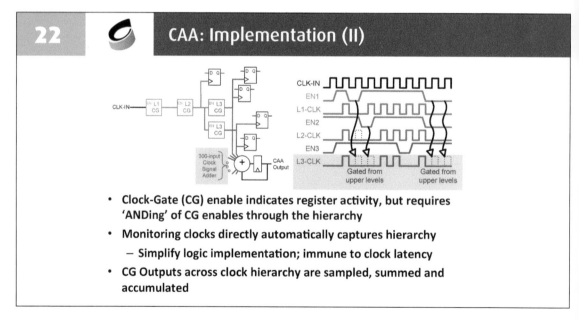

- **Clock-Gate (CG) enable indicates register activity, but requires 'ANDing' of CG enables through the hierarchy**
- **Monitoring clocks directly automatically captures hierarchy**
  - **Simplify logic implementation; immune to clock latency**
- **CG Outputs across clock hierarchy are sampled, summed and accumulated**

If we look at the last enable clock instead of enables there we see all of these enabling features from the previous stages so it's a much simpler implementation of this technology because these clock signals are much more distributed in time when we do clock balancing and so we just sample the CG outputs. The outputs of CG are summed and turn out to do this kind of estimation is much easier for the physical designer than the previous one which is monitoring the enable signals.

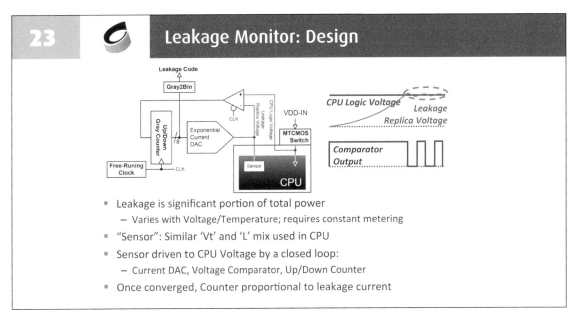

23 Leakage Monitor: Design

- Leakage is significant portion of total power
  - Varies with Voltage/Temperature; requires constant metering
- "Sensor": Similar 'Vt' and 'L' mix used in CPU
- Sensor driven to CPU Voltage by a closed loop:
  - Current DAC, Voltage Comparator, Up/Down Counter
- Once converged, Counter proportional to leakage current

We want to monitor leakage to give our close loop the most accurate amount of power that it dissipated at a given time. The reason why we want to do is that the leakage is actually very very significant portion of total power.

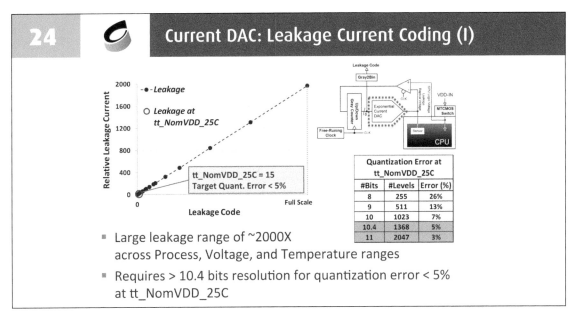

24 Current DAC: Leakage Current Coding (I)

Quantization Error at tt_NomVDD_25C		
#Bits	#Levels	Error (%)
8	255	26%
9	511	13%
10	1023	7%
10.4	1368	5%
11	2047	3%

tt_NomVDD_25C = 15
Target Quant. Error < 5%

- Large leakage range of ~2000X across Process, Voltage, and Temperature ranges
- Requires > 10.4 bits resolution for quantization error < 5% at tt_NomVDD_25C

But it is very hard to monitor actually. And what we want to do is to drive this sensors voltage to the same voltage as the CPU is running and we use a closed loop in order to do that. The closed loop has a current DAC, a voltage comparator and an up/ down counter. If you have an accurate current DAC, this counter value here is now proportional to the leakage current. You can visit it is just a tiny replica of the CPU.

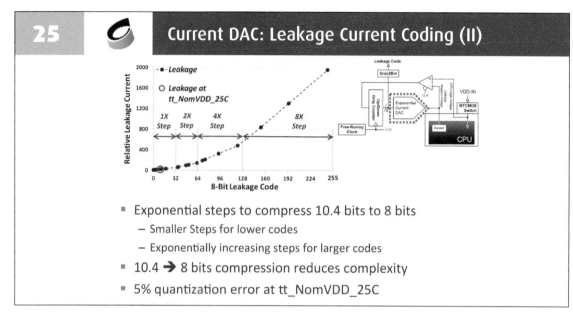

25 Current DAC: Leakage Current Coding (II)

- Exponential steps to compress 10.4 bits to 8 bits
  - Smaller Steps for lower codes
  - Exponentially increasing steps for larger codes
- 10.4 ➔ 8 bits compression reduces complexity
- 5% quantization error at tt_NomVDD_25C

So die at the backside and the leakage could be about 2,000X. You can get for process slow-slow versus fast-fast that voltages may be down from point five volt to one volt and temperature ranges between 0 degree and 125 degrees. So you have all these ranges and the leakage monitor needs to be able to take this huge range. And then the exponential current DVC also needs to support this huge range.

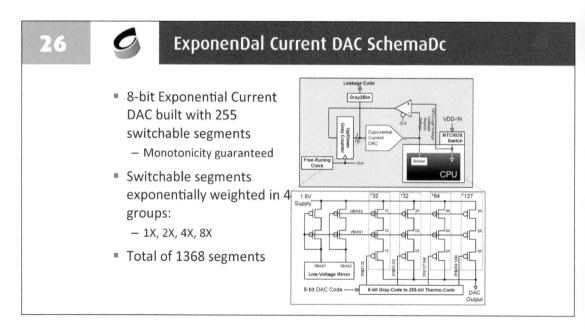

26 ExponenDal Current DAC SchemaDc

- 8-bit Exponential Current DAC built with 255 switchable segments
  - Monotonicity guaranteed
- Switchable segments exponentially weighted in 4 groups:
  - 1X, 2X, 4X, 8X
- Total of 1368 segments

This is the schematic of the exponential current DAC. Here's our 1X, and 2 X, then to 8X to make sure that it's covering in exponential way and then it's a total of 1,368 segments in the exponential DAC.

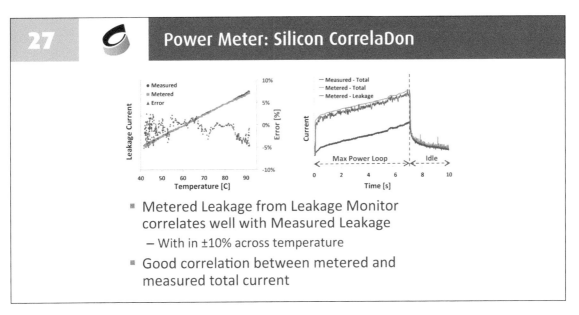

27 — Power Meter: Silicon CorrelaDon

- Metered Leakage from Leakage Monitor correlates well with Measured Leakage
  - With in ±10% across temperature
- Good correlation between metered and measured total current

This is the correlation of this power meter versus silicon. The leakage monitor say about the leakage is quite well correlated across temperature. And the error is about within +/- 10% across temperature. As you keep running the CPU, it gets hotter and hotter. The temperature increases and the leakage also increases, so you see the total power increases over time.

3. SIPPING POWER!

28 — Power Delivery Network

- Getting the power to where it needs to go is the difficult part of the power delivery network
- Involves PMIC efficiency, PCB consideration and package/on-die co-design
- For Mobile products, strict thermal and cost constraints means new innovation is needed on power delivery

The most difficult part is designing your power delivery network. It involves a lot of components not just the IC, the PMIC, the power management IC. We need to have innovation where we trade off the cost of the PDN and also the thermal on the delivery. There are a few of the considerations that we do in our current mobile SOC and we are going to trade off and to do a much smarter PDN.

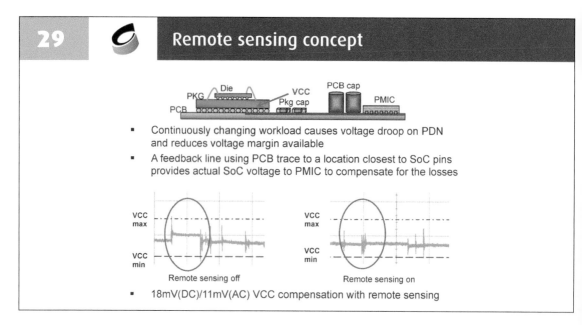

29 — Remote sensing concept

- Continuously changing workload causes voltage droop on PDN and reduces voltage margin available
- A feedback line using PCB trace to a location closest to SoC pins provides actual SoC voltage to PMIC to compensate for the losses

Remote sensing off

Remote sensing on

- 18mV(DC)/11mV(AC) VCC compensation with remote sensing

Remote sensing is a way to reduce the voltage drop. We take a trace in the PCB on the package, although that's the closest to SOC pins and we take this trace all the way back to the PMIC and then we can monitor this droop and try to compensate for the droop as fast as possible. So what we see is that going from this side here to this side here you can see the improvement is 18mV(DC)/11mV(AC) VCC compensation with the technique of remote sensing.

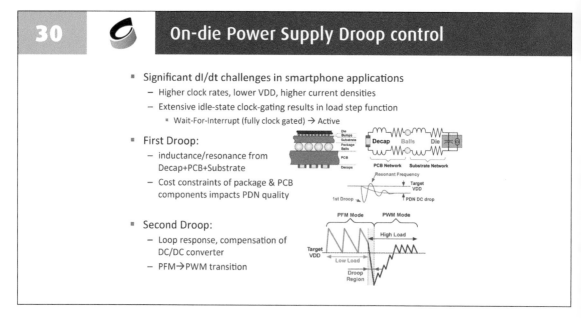

30 — On-die Power Supply Droop control

- Significant dI/dt challenges in smartphone applications
  - Higher clock rates, lower VDD, higher current densities
  - Extensive idle-state clock-gating results in load step function
    - Wait-For-Interrupt (fully clock gated) → Active
- First Droop:
  - inductance/resonance from Decap+PCB+Substrate
  - Cost constraints of package & PCB components impacts PDN quality
- Second Droop:
  - Loop response, compensation of DC/DC converter
  - PFM→PWM transition

Droop control of the power supply on the die is important for smartphone applications. When you come out of idle state, the power goes from zero to max within one or two cycles. There are two reasons induced the droop.

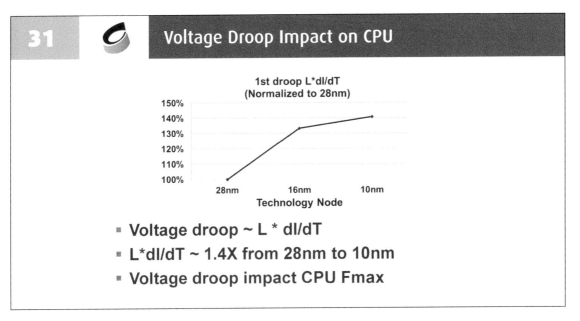

31 — **Voltage Droop Impact on CPU**

- **Voltage droop ~ L * dI/dT**
- **L*dI/dT ~ 1.4X from 28nm to 10nm**
- **Voltage droop impact CPU Fmax**

As it shows that the L*dI/dT is actually getting worse from technology node to technology node. This big drop will impact the CPU Fmax. To maintain the Fmax , we have to increase the DCDC voltage and cause larger current.

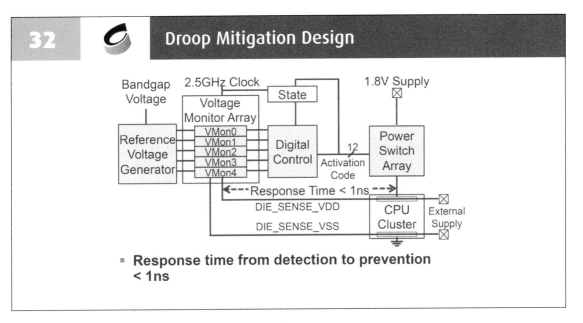

32 — **Droop Mitigation Design**

- **Response time from detection to prevention < 1ns**

So we need a very quick response to mitigate the droop.

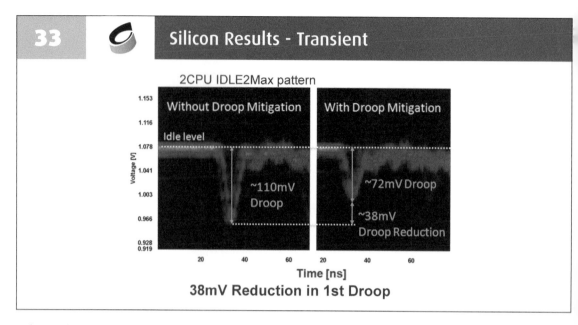

That is the silicon test to show the mitigation of VDD Droop. With Droop Mitigation 1st Droop will reduce 38mV from 110mV to 72mV.

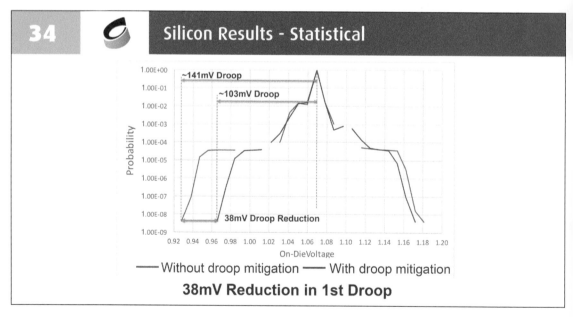

We have to see statistically for many cycles what is the reduction improvement. Basically running for millions of cycles and the worst droop is down here. So this technique shows that for this part which typically will determine Fmax of CPU, you can improve like 30 millivolts and that may be translated back to frequency improvement.

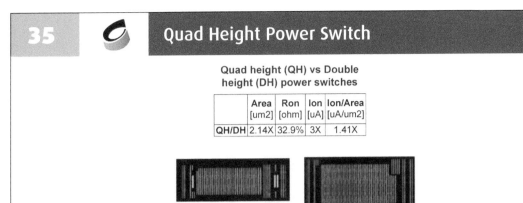

35　Quad Height Power Switch

Quad height (QH) vs Double height (DH) power switches

	Area [um2]	Ron [ohm]	Ion [uA]	Ion/Area [uA/um2]
QH/DH	2.14X	32.9%	3X	1.41X

Double-height (DH) power switch

Quad-height (QH) power switch

- QH PS improves 41% uA/um2 and saves CPU switch area by 2X while meeting the same IR drop spec

Source: A. Wang, ISSCC 2014

Also one important part of our PDN design is power switch. When the block is not if we can disable the power switches and make the leakage very close to zero then you can save a lot of power.

By increasing the quad height means you get much more ion purses with the area. The QH will save the Switch Area by 2X.

36　Hybrid Power Switch

- **The hybrid switch performs 3 functions:**
    1. Main power switch for normal operation
    2. Diode connected switch for retention
    3. NWELL bias switch for Forward Body Bias (FBB)

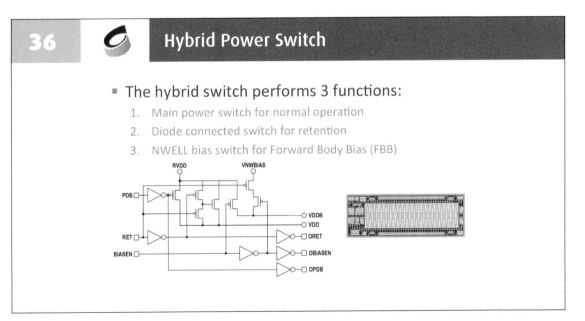

We have 20 nanometer with a hybrid power switch that can perform three different functions. The first function is the main power switch and we use a HVT PMOS for the main power switch which is a very high VT to make sure that the total leakage of power switches is very low. Secondly we

built into the power switch a diode connected switch for retention. Finally you use the switch in order to enable forward body bias. Forward body bias in the concept of changing the voltage to the back gate of the device you speed up.

37   Power Control Sequencing

- Sequencing of the control signals is required in order to prevent short-circuit current between supplies due to distributed switch control

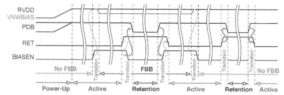

1. BIASEN should be asserted while the externally applied FBB voltage is equal to VDD.
2. RET should be asserted prior to de-assertion of BIASEN when entering retention mode.
3. RET asserted before PDB de-asserted

When you have such a complicated power switch at doing on these different functions. Actually the most important thing is making sure that the sequencing of the control signals to the switches correct to prevent any short circuit currents.

4. "I FEEL THE NEED, THE NEED FOR SPEED"

The sentence is from Top Gun. Speed in CPU is all about performance, what is the gigahertz you can need. That is kind of the big selling point. The next topic is about clocking.

38   Clock Design for High Performance

- A structured clock is developed to minimize clock variation and insertion delay.
    - But still take advantage of beneficial skew.

- Must realize these benefits without impacting Area, Productivity (schedule), or <u>Power</u>

- Develop hybrid mesh/CTS approach, 'Fishbone'

- Implementation Steps:
    1. Restructure clocking
    2. Implement physical clock structure
    3. Clock Gate (CG) cloning/optimization
    4. Traditional CTS of lower-level clock tree

We want to make a structured clock is to minimize clock variation and insertion delay. But to gain these benefits will impacting Area, Productivity or Power. So we develop the hybrid mesh/CTS approach, and named it "Fishbone".

39   Clock Structure Change

- CPU design utilizes high-level functional CG's
  - Small reduction in insertion delay (15%) due to few end-points.
- Restructuring:
  1. Collapse 1st/2nd level CG's; more end-points on fishbone
  2. Add pre-gating with OR of all first level CG's to eliminate dynamic power when all groups are gated-off

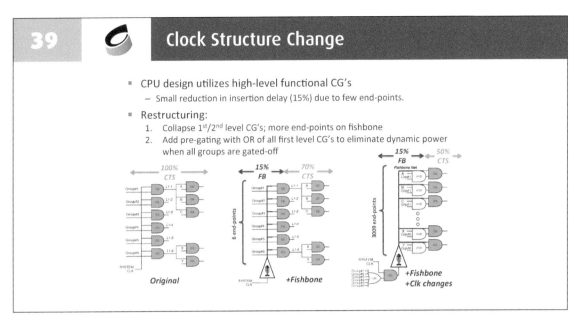

There are two steps to implementing the Fishbone structure. The first step is to do a structure change to better fit our hybrid mesh in CG's. There is small reduction in insertion delay about fifteen percent due to few end-point.

40   Physical Clock Structure

- **Only 2 Low-R metals available (M7, M8)**
  - Both primarily used for power distribution
  - M8 is also bump landing
    - Not practical to build a mesh
- **Adopt fishbone style distribution**
  - ✓ Vertical zero-skew buffer along Spine
  - ✓ 'Spine' on M8, in gap between bump targets
  - ✓ 'Ribs' on M7
  - ✓ Tap point buffers
  - ✓ Uni-directional 'mesh' reduces power

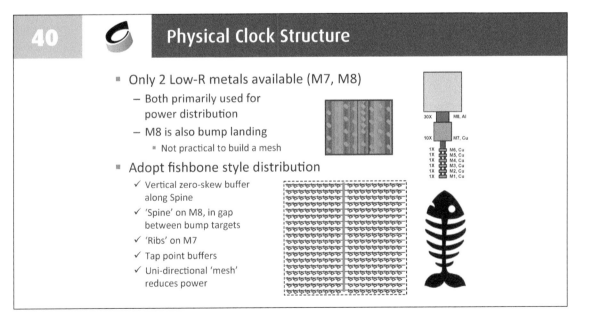

Look at the physical cock structure, how to create this fishbone? In order to make the clock very very fast, we need to use the lowest resistance metal. And we have to share the TOP low resistance Metals with the Power Grid. We put this spine on metal 8, in the gap between bump targets. And then we put the ribs on metal 7 across the whole CPU. It looks like a fishbone.

41 Clock Timing Results

- Timing graph view of clock tree

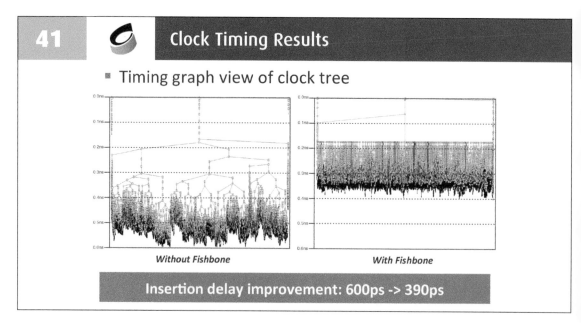

Without Fishbone   *With Fishbone*

Insertion delay improvement: 600ps -> 390ps

t shows some of the clock timing results of the fishbone or without the fishbone. Insertion delay can improve from 600ps to 390ps.

42 Duty Cycle Sensitivity

- Delay-skewed SRAM clocking is used in CPU
  - Skew achieved through negative edge clocking

- Results in clock duty cycle dependency.
  - Narrower pulses stresses SRAM input timing
  - Wider pulses stresses SRAM output timing

We have to care about the clock duty cycle when we use delay-skewed SRAM clocking. And if the pulse is narrower, then it will stress the timing of the SRAM on input. If you have a duty cycle gets wider and it will stress the output timing. Overall means you have to add margin to compensate with this big variation.

43 Duty Cycle Corrector (DCC) Design

- Two identical delay elements are placed in a cross-coupled clock-generator topology
  1. Delay +ve edge, pulse narrows
  2. Delay -ve edge, pulse widens
- Use two delay elements to minimize delay for 50% DC
- Each delay element comprised 3 stages, each with 5-bit thermo-coded switched capacitor loads

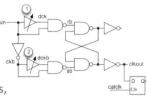

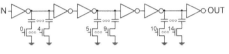

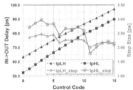

- Delay adjustment resolution: <3ps
  - <1% of clock period (400ps clock period)

A duty cycle corrector circuit on the CPU will correct the duty cycle. How we corrected per-die? We use two identical delay elements in a cross-coupled clock generator topology. The first one is more for delay positive edge and the pulse was narrow. And then the second one is a delay negative edge the pulse was widen. You can get 3ps resolution and the reason for this is to target about 1% of the clock period.

44 DCC Calibration

- Auto-calibration is performed at power-up to provide per-die correction

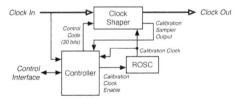

- Operation:
  1. A pre-programmed 'target' duty cycle is set
  2. 'Error to target' is determined by un-correlated under-sampling of the shaped clock.
     - Un-correlated sample clock from internal oscillator
  3. An FSM performs a linear search of code space
     - Starts at 0 (no adjustment), range of +15 to -15
     - increment/decrement based on sign of error to target

We will do the auto-calibration in order to correct it automatically. Add a ROSC to create a free running clock and in order to do the calibration closed loop during the power of stage. We sampled using the internal oscillator and then the controller will perform a linear search of the code space. It starts at zero which means no adjustments from the current clock and ranges from +15 to -15. And increment is based on sign of error until it gets to the target.

5. LAY A GOOD FOUNDATION

The next topic is about having good foundation which is the standard cell and SRAM you choose and the architecture you choose for designing.

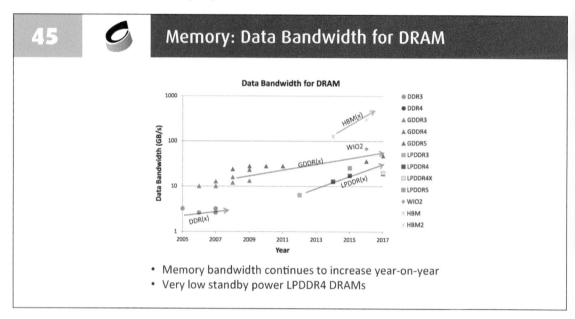

45 **Memory: Data Bandwidth for DRAM**

- Memory bandwidth continues to increase year-on-year
- Very low standby power LPDDR4 DRAMs

This is the data of Data Bandwidth of DRAM from ISSCC. From year on year, memory bandwidth continues increasing. Also the different standards are pushing towards lower and lower power.

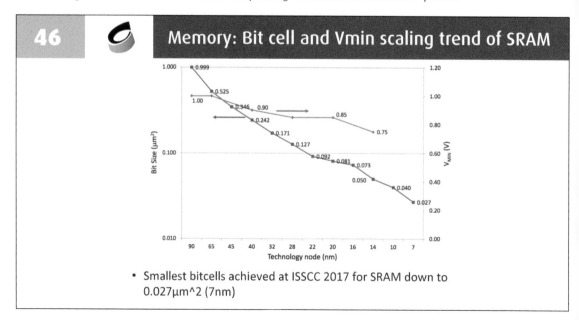

46 **Memory: Bit cell and Vmin scaling trend of SRAM**

- Smallest bitcells achieved at ISSCC 2017 for SRAM down to 0.027μm^2 (7nm)

At the same time, memory bit cell keeps scaling down. Bit- cell scaling is a good indicator of Moore's law scaling. And basically 2017 ISSCC showed the smallest bit-cells achieved is with the 7nm (0.027 um^2).

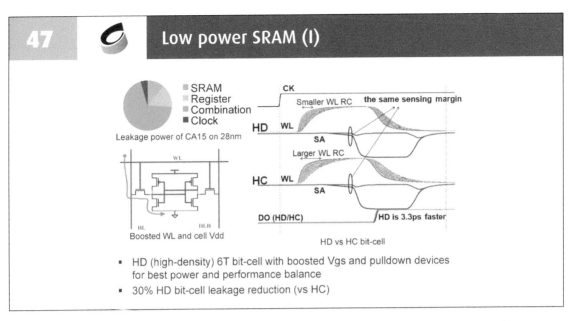

47   Low power SRAM (I)

- HD (high-density) 6T bit-cell with boosted Vgs and pulldown devices for best power and performance balance
- 30% HD bit-cell leakage reduction (vs HC)

Now will show some considerations on choosing SRAM, designing SRAM and designing standard cells. CPUs typically have to have the fastest SRAM on the whole chip.

This is a high-density 6T bit-cell with boosted Vgs and pulldown devices. Because due to the parasitic, it has a smaller word-line RC than the high current which has a larger WL RC. The HD one is 3.3ps faster and 30% leakage reduction VS HC.

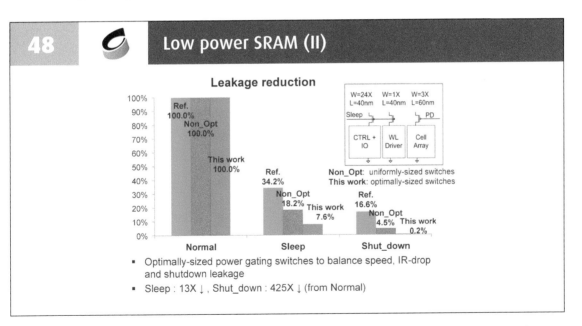

48   Low power SRAM (II)

- Optimally-sized power gating switches to balance speed, IR-drop and shutdown leakage
- Sleep : 13X ↓ , Shut_down : 425X ↓ (from Normal)

Other thing you have to consider is the low power capability of the SRAM. So how do we design the power gating switches to balance between speed, IR drop and shutdown leakage? We can use a uniformly-sized power switches for the different parts like control, word-line driver, cell array. This the test results of two types design.

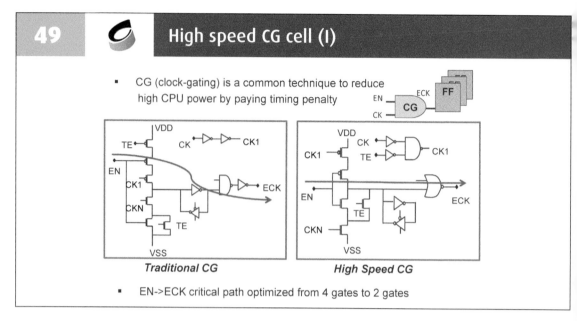

49 High speed CG cell (I)

- CG (clock-gating) is a common technique to reduce high CPU power by paying timing penalty

Traditional CG   *High Speed CG*

- EN->ECK critical path optimized from 4 gates to 2 gates

Designing new and improving standard cells to further boost the performance. Clock gating is commonly used to provide low-power, and actually in many cases the clock gating can also become the timing critical signal when you are closing the timing. We're able to change the critical path to two from four gates.

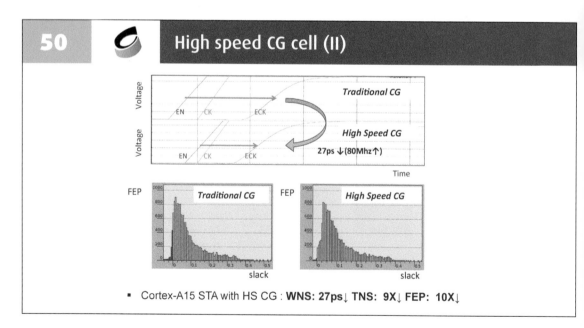

50 High speed CG cell (II)

Traditional CG

High Speed CG

27ps ↓(80Mhz↑)

Traditional CG   *High Speed CG*

- Cortex-A15 STA with HS CG : **WNS: 27ps↓ TNS: 9X↓ FEP: 10X↓**

This shows the simulation of the traditional clock gating and high-speed clock gating basically done by 27ps improve 80Mhz with just single standard cell. So we run the STA with or without the clock gating, we saw WNS-worst negative slack improve by 27ps, the TNS improve 9X and FEP improve 10X.

51   L1$ SRAMs PPA Trade-off

- 3 key design options were explored for the L1 cache
    1. High-Density (HD) vs. High-Current (HC) bit-cell
    2. 64 bits/bitline (b/bl) vs. 32b/bl
    3. SVT vs. LVT periphery transistors

- Limit to +25% Area (HC-32b/bl too large)

- LVT eliminated due to leakage power and mask cost

- HD-64b/bl did not provide enough performance

- Compare (similar area): HD-32b/bl & HC-64b/bl

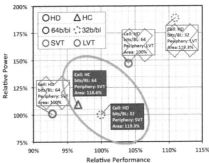

For the L1 SRAM, do we want to have 64bits per bit-line or 32 bits per bit-line. The 32 typically are faster but burn more power. What kind of VT do we need to use? SVT vs LVT. LVT would be faster but then although the leakage would be higher. The first guideline is the limit to +25% area. The second one we eliminated is LVT. LVT needs extra mask cost. So One is the HD-32b/bl with SVT periphery, the second one is Hc-64b/bl with SVT.

52   Comparing HD-32b/bl & HC-64b/bl

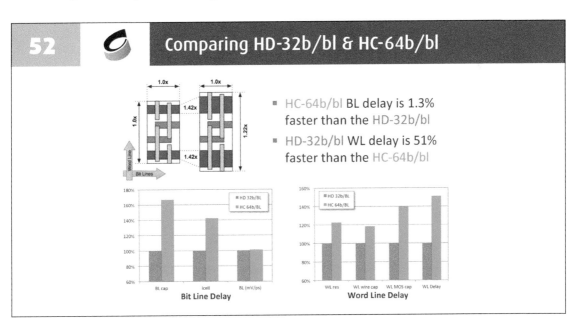

- HC-64b/bl BL delay is 1.3% faster than the HD-32b/bl

- HD-32b/bl WL delay is 51% faster than the HC-64b/bl

This shows comparison of the bit-cells of HD and HC. The bit line is horizontal and is pretty much similar between the two bit cells. But the word line is vertical so you see the world line is here for the HC is 1.22X. For bit-line delay, the high density will win on the BL cap vs high current. And then the icell is worse for the high density vs high current. These two will combine, pretty much the same of BL(mV/ps). For word line delay, it is the opposite. The high current has more resistance, more caps. So compounded means the word line delay is much better for the high density than the high current.

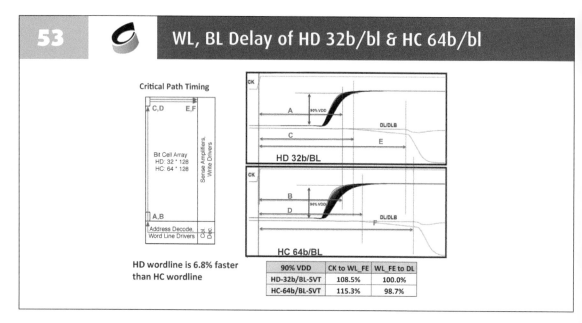

53 WL, BL Delay of HD 32b/bl & HC 64b/bl

Critical Path Timing

HD 32b/BL

HC 64b/BL

HD wordline is 6.8% faster than HC wordline

90% VDD	CK to WL_FE	WL_FE to DL
HD-32b/BL-SVT	108.5%	100.0%
HC-64b/BL-SVT	115.3%	98.7%

This is an actual timing of a real SRAM of these two different architectures. Analyzing this whole critical path timing, you can see the top one is high density, the bottom one is high current. So the high density in the first part-the clock to the word line is 108.5% with the high current is 115%.

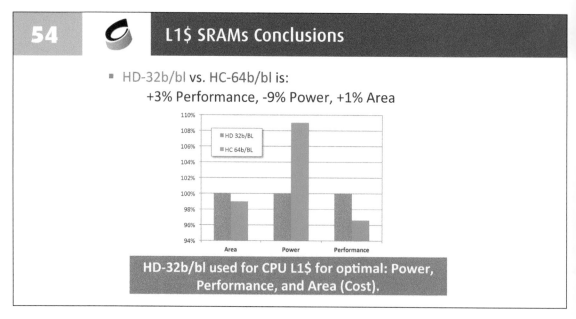

54 L1$ SRAMs Conclusions

- HD-32b/bl vs. HC-64b/bl is:
  +3% Performance, -9% Power, +1% Area

HD-32b/bl used for CPU L1$ for optimal: Power, Performance, and Area (Cost).

So the conclusion of course is very clear after doing all of the analysis is that the two architectures are very similar in area. Using a HD-bit-cell is very good for bit line but CPU L1 cache and then you can get the optimal power, performance, and area trade-off.

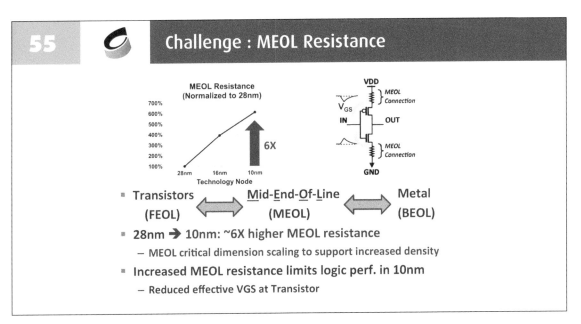

I think this is quite a final one in foundation IP analysis and some of the very big challenge we face is as you go to MEOL's node mid-end-of-line resistance of the technology. So this chart here shows mid-end-of-line resistance is normalized 28nm. 28nm is okay, then they want to go to 16nm. That resistance gets 4X larger and in 10nm, it gets 6X larger. Just because the dimensions are scaling, we have to get the impact on the metal resistance. So we did see in 10nm the huge logic limitation of performance due to this MEOL resistance and basically reduced our effective VGS at transistor.

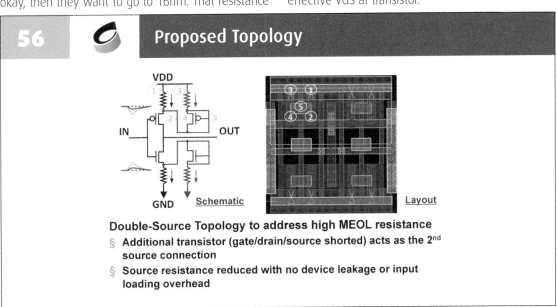

We are trying to design a better standard cell library to reduce the effects of MEOL. What we did was basically adding additional transistor that shorted. It shorts between the gate, the drain and the source. This source resistance reduced and does not have any device leakage or input loading overhead. So it can really provide much better connection to the device.

57 Simulation Results

	FO4 Delay [ps]
Standard 2x INV	9.1
Double-Source 2x INV	8.6
Frequency Benefit	5.5%

FO4 Delay at 0.75V, TT, 25°C

- This technique is applied to inverters, buffers, and commonly used combinational cells
- The double-source cells are over 5% faster, and 18% larger on average than conventional counterparts
- Area impact from the double-source cells managed by restricting usage to critical timing paths

Compared to a regular inverter, while you get a better performance from 9.1ps to 8.6ps. It's a very nice benefit in performance about 5.5%. So in order to manage this area impact from the double-source cells we have to restrict the usage to only the critical timing paths.

6. SHIPPING 100'S MILLIONS OF IC'S

This topic is about advanced test and debug techniques. It is really about how we can make sure we're shipping hundreds of millions of IC's.

58 Motivation for Design for Debug

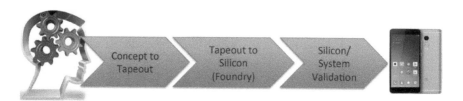

- Time to Market is Key to any business!
- Any Silicon Bug can cause millions of dollars and time wasted...

In the industry, you have to get millions of IC's working and working reliably well and having good yields. So a lot of work that we do in my group is what we called design for debug. It may not be the most exciting in sexy research topic to be working on but actually it is really important for industry and the idea of design for debug is that we might have a concept that we will take the tape out. This is in general.

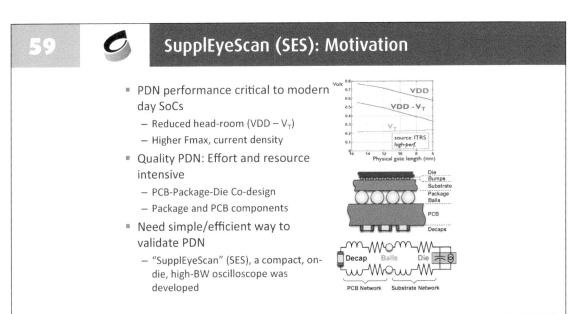

One of them we call our SupplEyeScan and the idea is I talked a lot about our PDN having a very robust power delivery network, important and critical for our SOCs to work. So how can we from the inside know the quality of power delivery network. We build basically on-die high-BW oscilloscope to validate. And that's why we call SupplEyeScan.

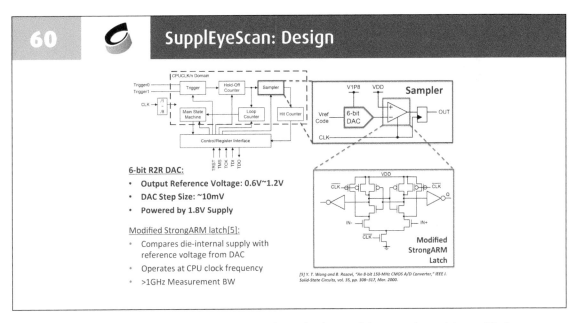

This is a block diagram of our SupplEyeScan. It's basically 6-bit R2R DAC. We were sampling the delivery network at regular time and it is based on output reference voltage : 0.6V~1.2V with the DAC step size of 10 millivolts, powered by 1.8 volts supply. The heart of this sampler is this modified strongARM latch here from this paper. This operates at CPU clock frequency, so we're taking samples as fast as the clock of the CPU and it has a greater than 1 gigahertz measurement bandwidth.

61 SES: Histogram Mode

- Histogram Mode captures higher probability portion of the distribution through <u>sub-sampling</u> of VDD

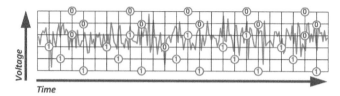

- **Up to 255 samples of (VDD – VREF) captured:**
  - Spaced to capture low-frequency VDD variation
- **DAC reference (VREF) swept from min to max**
- **Probability of (VDD>VREF) measured at each VREF**

You can see that this voltage supply is very noisy as a function of time. And what we do is we sample up to 255 samples of (VDD –VREF) captured and it's spaced to capture the low-frequency VDD variation. We're doing like a statistical sampling of the voltage supply and we captured the probability that VDD is higher than VREF measured at each VREF.

62 Histogram Mode: Silicon results

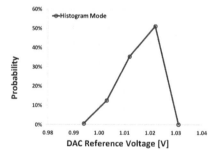

- Histogram mode captures high probability statistics of supply voltage
  - Infrequent voltage excursions not captured

And what we get is basically the histogram of the voltage at different voltage VREFs. So different voltage VREFs, this is a probability of the voltage being higher than that reference voltage. And this is the high probability statistics of supply voltage.

SupplEyeScan Peak-Detect Mode

63

- Peak-Detect Modes capture "tail" of the distribution through <u>continuous-sampling</u> of VDD

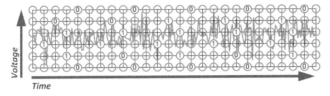

Voltage

Time

- **Within each window, record any occurrence of:**
  - **— Peak-Low: VDD < VREF**
  - **— Peak-High: VDD > VREF**
- **Capture worst-case min/max voltages**

We have a second mode to do that. That's called peak-detect mode. Instead of capturing the high probability level voltages, it's capturing the low probability of the voltage and the tail of the voltage distribution. Keep sweeping the VREF like that and now it's like all of the time it will go below the ref until get to the high point. And then we do peak high which is to figure out when this is the VDD go above the VREF.

Peak-Detect Mode: Silicon results

64

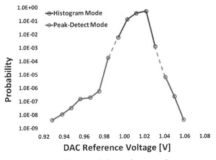

- Peak-Detect (high and low) modes capture "tail" statistics of supply voltage
- Histogram and Peak-Detect modes combined provide overall PDN quality

Put the two together and then you have a whole histogram of the voltage and what it looks like from the inside. And so the peak-detect part captures the high and low tail. In histogram part, it captures the middle part. It is a clever way to do on-die voltage characterization of our power delivery network.

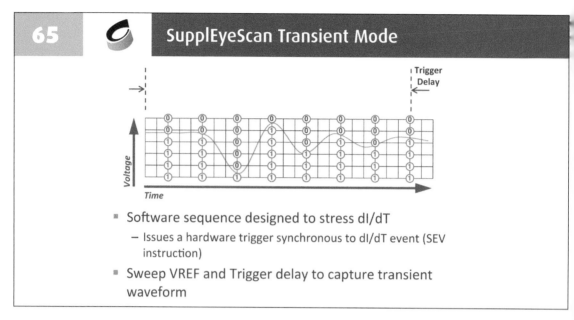

65 — **SupplEyeScan Transient Mode**

- Software sequence designed to stress dI/dT
  - Issues a hardware trigger synchronous to dI/dT event (SEV instruction)
- Sweep VREF and Trigger delay to capture transient waveform

Another mode of our on-die oscilloscope is transient mode. And if you just want to see the function of time what's going on inside. For example, if there's a very big dI/dT, we want to see this droop that we were trying to monitor. So we use the on-die oscilloscope to understand the droop behavior of the PDN. And what this system does is a trigger synchronous to dI/dT event and then basically after a certain amount the trigger delay, it measures the voltage. What you capture is the probability of the function of time of the voltage.

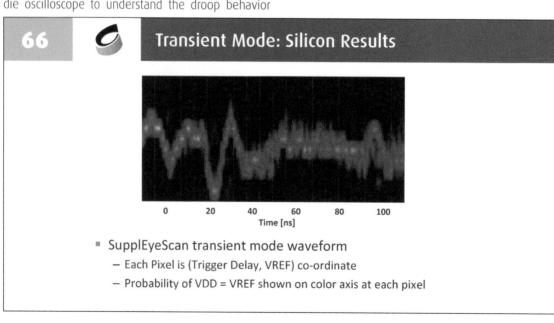

66 — **Transient Mode: Silicon Results**

- SupplEyeScan transient mode waveform
  - Each Pixel is (Trigger Delay, VREF) co-ordinate
  - Probability of VDD = VREF shown on color axis at each pixel

You can see that this is a transient mode waveform of the droop and each of these pixels is a trigger delay and VREF co-ordinate.

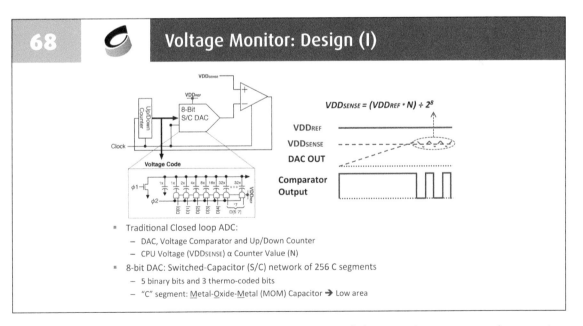

67 **Voltage Monitor: Motivation**

Prior Work:

- **Continuous power monitoring to maximize performance with in fixed power budget**
- **Software Complexity: Needs voltage update after every DVFS change**

ISSCC 2016: A 20nm 2.5GHz Ultra-Low-Power Tri-Cluster CPU Subsystem with Adaptive Power Allocation for Optimal Mobile SoC Performance

This Work:

- **Voltage Monitor: On-die CPU Voltage monitoring**
  - **Removes Software Complexity**
  - **Power efficient ADC with low area overhead**
  - **Accurate envelope tracker: High frequency transients ignored**

It's not as so fine grain monitor of the voltage but it's something that the software can continually ping to know the voltage quality or the voltage level of the CPU. And so what I had described earlier was this apa clock loop and one of the annoying things about when we implemented it is that for the software it had to keep feeding the frequency and voltage into this block.

68 **Voltage Monitor: Design (I)**

$$VDD_{SENSE} = (VDD_{REF} * N) \div 2^8$$

- Traditional Closed loop ADC:
  - DAC, Voltage Comparator and Up/Down Counter
  - CPU Voltage (VDDSENSE) α Counter Value (N)
- 8-bit DAC: Switched-Capacitor (S/C) network of 256 C segments
  - 5 binary bits and 3 thermo-coded bits
  - "C" segment: Metal-Oxide-Metal (MOM) Capacitor → Low area

This voltage monitor design is a big DAC and it's actually really similar to the leakage monitor design that I described earlier. You have this VDD sense and then you have a VREF when you just basically count up/down counter until the two voltages are the same.

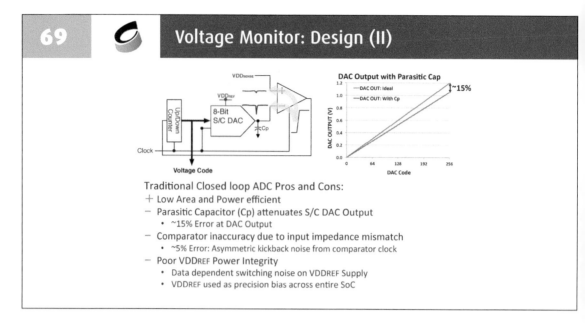

69 — Voltage Monitor: Design (II)

Traditional Closed loop ADC Pros and Cons:

+ Low Area and Power efficient
− Parasitic Capacitor (Cp) attenuates S/C DAC Output
  • ~15% Error at DAC Output
− Comparator inaccuracy due to input impedance mismatch
  • ~5% Error: Asymmetric kickback noise from comparator clock
− Poor VDDREF Power Integrity
  • Data dependent switching noise on VDDREF Supply
  • VDDREF used as precision bias across entire SoC

Traditional closed loop ADC Pros and Cons:
It is area and power efficient, but its accuracy is affected by
1, The parasitic capacitor induce about 15% Error at DAC output.
2, The Comparator inaccuracy induce about 5% Error.
3, Need a precision VDDRef bias across entire SOC.

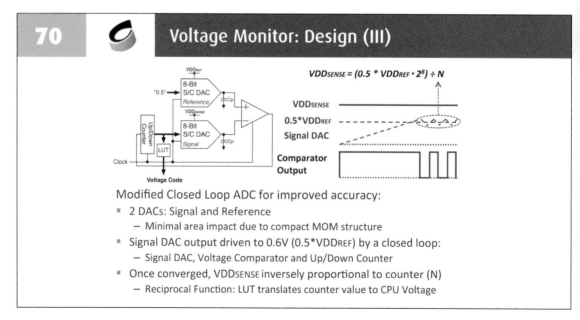

70 — Voltage Monitor: Design (III)

$$VDD_{SENSE} = (0.5 * VDD_{REF} * 2^8) \div N$$

Modified Closed Loop ADC for improved accuracy:

- 2 DACs: Signal and Reference
  − Minimal area impact due to compact MOM structure
- Signal DAC output driven to 0.6V (0.5*VDDREF) by a closed loop:
  − Signal DAC, Voltage Comparator and Up/Down Counter
- Once converged, VDDSENSE inversely proportional to counter (N)
  − Reciprocal Function: LUT translates counter value to CPU Voltage

To improve the accuracy, we can use two DAC in the closed loop ADC.
The Main update is that the signal and reference both use the same type DAC to reduce the parasitic capacitor problem.
It is easy to calculate the value of VDDSense.

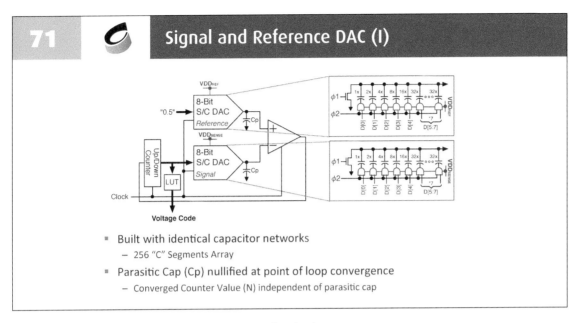

71   Signal and Reference DAC (I)

- Built with identical capacitor networks
  - 256 "C" Segments Array
- Parasitic Cap (Cp) nullified at point of loop convergence
  - Converged Counter Value (N) independent of parasitic cap

Now you can see the detail of the 8bit DAC used in the loop.

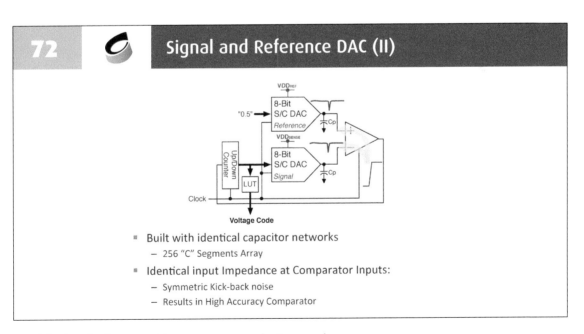

72   Signal and Reference DAC (II)

- Built with identical capacitor networks
  - 256 "C" Segments Array
- Identical input Impedance at Comparator Inputs:
  - Symmetric Kick-back noise
  - Results in High Accuracy Comparator

Kick-back Noise is symmetric, so accuracy can be improved.

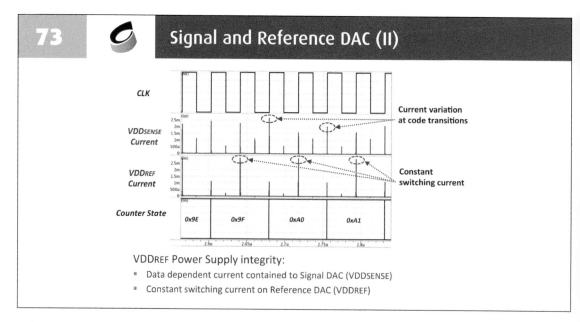

73 **Signal and Reference DAC (II)**

VDDREF Power Supply integrity:

- Data dependent current contained to Signal DAC (VDDSENSE)
- Constant switching current on Reference DAC (VDDREF)

The Sense Current is different between signal and reference. One is data dependent, and one is constant switching current.

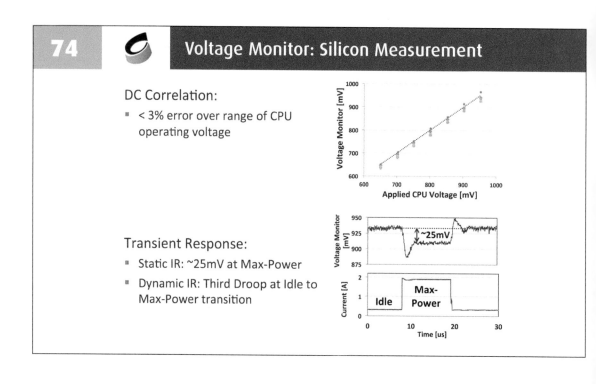

74 **Voltage Monitor: Silicon Measurement**

DC Correlation:

- < 3% error over range of CPU operating voltage

Transient Response:

- Static IR: ~25mV at Max-Power
- Dynamic IR: Third Droop at Idle to Max-Power transition

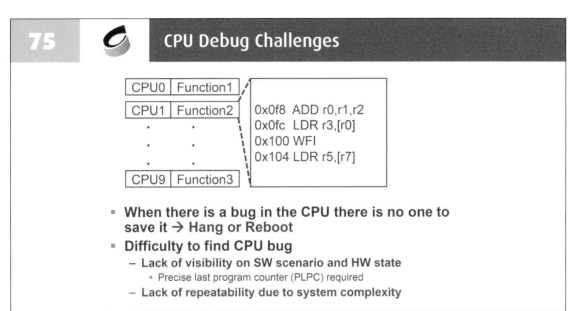

75   **CPU Debug Challenges**

CPU0	Function1		0x0f8 ADD r0,r1,r2
CPU1	Function2		0x0fc LDR r3,[r0]
.	.		0x100 WFI
.	.		0x104 LDR r5,[r7]
CPU9	Function3		

- **When there is a bug in the CPU there is no one to save it → Hang or Reboot**
- **Difficulty to find CPU bug**
  - **Lack of visibility on SW scenario and HW state**
    - Precise last program counter (PLPC) required
  - **Lack of repeatability due to system complexity**

It is easy to debug software because we can save the CPU state. But if CPU has a bug, there is no one to save it. The CPU will hang or Reboot. So it lacks of visibility on SW scenario and HW state. Even we can save something, we can't repeat the bug because of the system complexity. So to debug the CPU, we need to get the precise last PC.

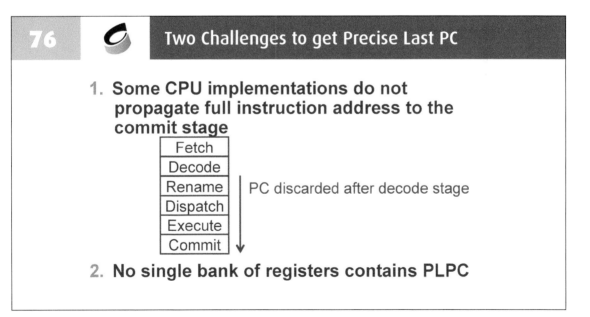

76   **Two Challenges to get Precise Last PC**

1. **Some CPU implementations do not propagate full instruction address to the commit stage**

| Fetch |
| Decode |
| Rename |
| Dispatch |
| Execute |
| Commit |

PC discarded after decode stage

2. **No single bank of registers contains PLPC**

There are two challenges to get the Precise Last PC. One is some CPU implementations do not propagate the full instruction address to the commit stage. For example, after the Decode stage, the address will be discarded. So when you need it, it is gone!

Two is no single bank of registers contains PLPC.

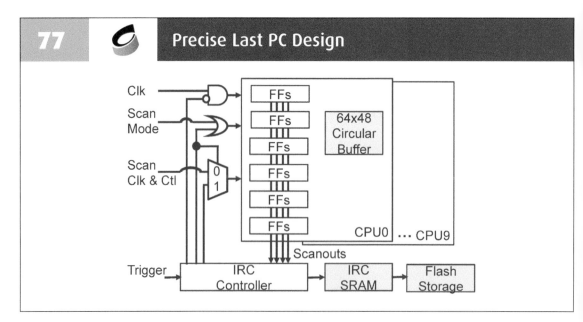

To get the PLPC, we need to add some hardware and software. When finding the CPU is hang, then send a trigger to activate the IRC. First is gating the CPU clock to stop the CPU from further damage. And then scan out all the FFs to a special SRAM. Finally dump the SRAM data to the Flash Storage.

We can debug the CPU bug by the data in Flash Storage in anytime.

7. UP THE FOOD CHAIN

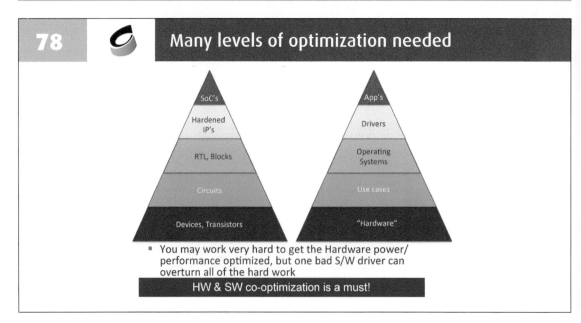

One bad software driver can make all that hard work goes on nothing. And it's just too easy for a software to introduce bugs or just take away our power performance benefits. The most important thing is how do we work together.

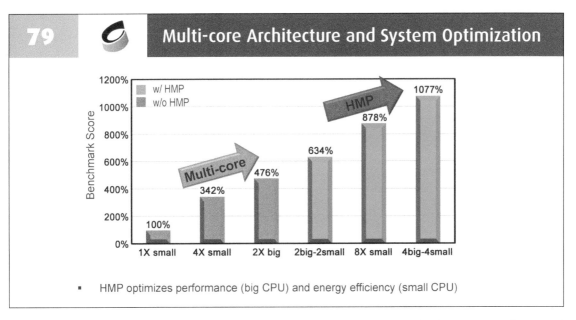

79 Multi-core Architecture and System Optimization

- HMP optimizes performance (big CPU) and energy efficiency (small CPU)

There are some examples of what we did is looking at how to take advantage of a multi core architecture and do some system optimization. If you need performance, use the big CPUs. But if you need power efficiency, then you use small CPU. That's the concept of heterogeneous multiprocessing.

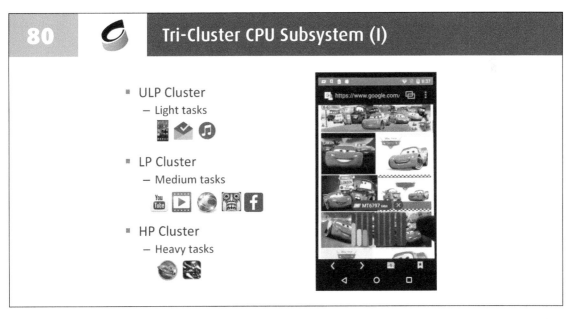

80 Tri-Cluster CPU Subsystem (I)

- ULP Cluster
  - Light tasks

- LP Cluster
  - Medium tasks

- HP Cluster
  - Heavy tasks

In our most current high end processor, we actually have three different types of CPUs. So it's really important that we find to give all the knobs to our software guys. Whatever application needs they have, they can choose the right CPU.

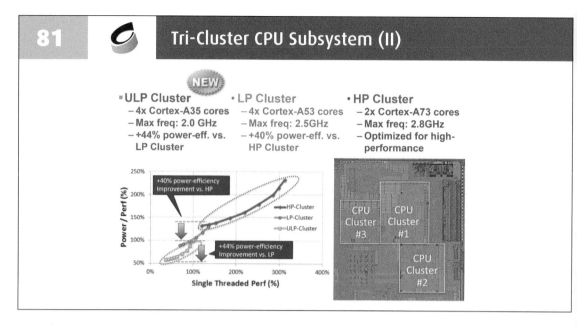

81 · **Tri-Cluster CPU Subsystem (II)**

And this is again a tri-cluster CPU system was a high performance the low power and the ultra-low power showing the power efficiency of the CPU as a function of performance.

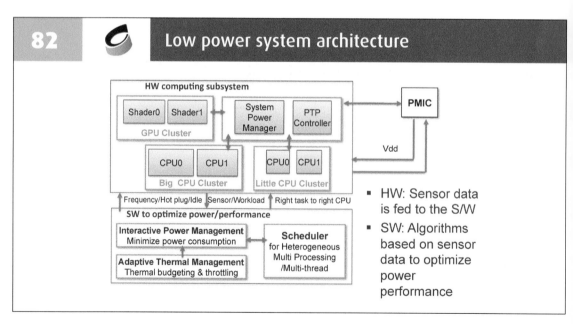

82 · **Low power system architecture**

Also important is having a low power system architecture and so what I described here for the most part are hardware which has CPUs, GPUs and sensors and everything. But what is the software architecture that can really take advantage of this system?

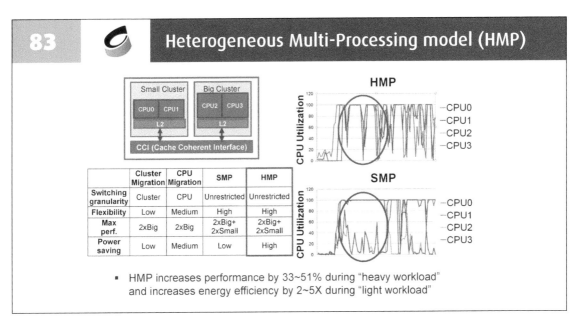

83 Heterogeneous Multi-Processing model (HMP)

	Cluster Migration	CPU Migration	SMP	HMP
Switching granularity	Cluster	CPU	Unrestricted	Unrestricted
Flexibility	Low	Medium	High	High
Max perf.	2xBig	2xBig	2xBig+ 2xSmall	2xBig+ 2xSmall
Power saving	Low	Medium	Low	High

- HMP increases performance by 33~51% during "heavy workload" and increases energy efficiency by 2~5X during "light workload"

One of the big challenges when going to HMP model-heterogeneous multi-processing is how the software can really take advantage of this new scheme. Beforehand, we pretty much always had SMP. All the CPUs are the same. And now we have different kinds of CPUs, so the software has to change in order to really take advantage. The real benefit comes when you can recognize what the capability of big is, what the capability of small is, and how do you schedule your tests correctly. So what the HMP software does is increasing the performance by 33 to 51 percent during heavy workload.

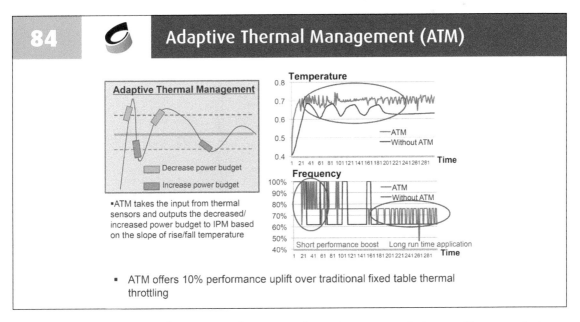

84 Adaptive Thermal Management (ATM)

- ATM takes the input from thermal sensors and outputs the decreased/increased power budget to IPM based on the slope of rise/fall temperature

- ATM offers 10% performance uplift over traditional fixed table thermal throttling

Thermal management is really important, because the temperature is changing all the time and you don't want to keep exceeding a certain thermal budget. What ATM does is taking the input from thermal sensors and then making sure that you can look at the slope of how fast the temperature is increasing in order to know how best to manage the clocking.

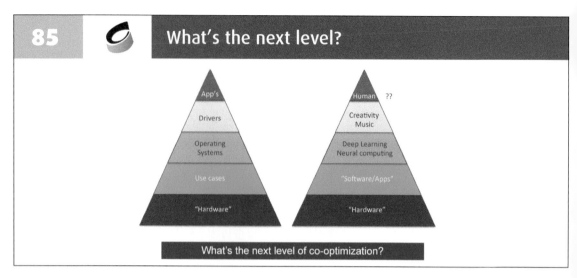

85 — **What's the next level?**

What's the next level of co-optimization?

Encourage you to work with software guys. What's the next level of co-optimization that we should be working on? You can see that maybe it will be the deep learning, neural computing. How does the human mind think? Or maybe we can think about how to be more creative like music or art and humanities. Is that the ultimate optimization of our hardware and software of world? So I think we'll just continue to co-optimize up the next level in order to really get the best performance.

10 YRS OF SUBVT – WHAT'S NEXT?

What do you think are the next big growth engines for semiconductors?

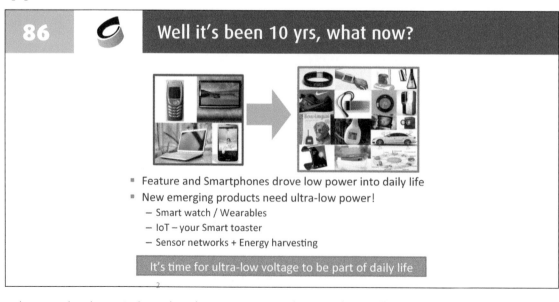

86 — **Well it's been 10 yrs, what now?**

- Feature and Smartphones drove low power into daily life
- New emerging products need ultra-low power!
  - Smart watch / Wearables
  - IoT – your Smart toaster
  - Sensor networks + Energy harvesting

It's time for ultra-low voltage to be part of daily life

There are the electronical products between 10yrs before and now. So what's in store for the future? And what we saw is that feature and smart phone really drove low power and low power is in all of our devices. It's actually time for ultra-low voltage to be part of the daily life, kind of similar to the way that low power became in our daily life after ten years.

87 Mobile Impact to Emerging Countries

China: Helping farmers for real-time pricing   Pakistan: Smartphones to fight Dengue

AGRICULTURE   HEALTH CARE
EDUCATION   SERVICE

KHAN ACADEMY
A free, world-class education for anyone, anywhere

3   Khan Academy's vision   Kenya: M-PESA branch-less banking

[MK Tsai, ISSCC 2014]

There are some pictures that were taken from ISSCC, just showing how mobile can really help the economy in many different countries. Such as in China, helping farmers for real time pricing is on the field. Or in Pakistan, using smartphones to fight Dengue fever. Education, you can see the Khan Academy's photo, you know this boy doesn't even have clothes but he has a laptop. So he can have his education and the servicing. So in Africa, being able to use your phone to do banking instead of having to go to a branch. I mean those things just really energize the next level of economy for the world.

88 Energy Challenge of Emerging Markets

- 1.3 billion people (20% of world population) are without access to electricity, 84% in rural areas
- 7% of world's total electricity production today could cover all basic human needs (World Energy Council)

Energy in emerging markets is critical and lacking

And energy is a key challenge in all of these different emerging markets. If you look at this like the G20 from a few years ago our energy sustainability working meeting, but the statistics are that one point three billion people which is twenty percent of our world population are without access to electricity and eighty-four percent of them are in rural areas. And seven percent of the world's total electricity production today could cover all basic human needs and that's the statement from the world energy council. So indeed energy is really important metric in order to make new economies in emerging markets.

89 Renewable Energy in Emerging Markets

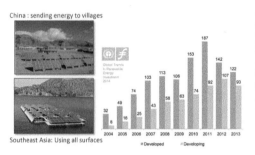

China : sending energy to villages

Southeast Asia: Using all surfaces

India: Goal to generate 15% of domestic power by renewable sources

- In 2011, for the first time renewable energy investment in emerging markets overtook US and Europe
- Emerging economies are fastest growing energy consumers and high demand causes a need to invest in new capacity

Here just shows the function of the year that the investment in renewable energy. That the darker green one is the developed countries and it is actually peak here but then going down.

90 Even in Africa : Creative ways to generate energy!

1 minute pedaling → 400 minutes of light

Children can do their homework at night

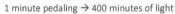

Energy in rural areas is so precious, not to be squandered

http://www.wbir.com/video/1627313404001/43961430001/Pedal-Power-Lights-up-Rural-Rwanda

And just a few cases in Africa, they are just finding creative ways to generate energy. There's a fellow here who takes a minute of peddling and then she can generate four hundred minutes of light. And I think actually over here is like a cell phone charger.

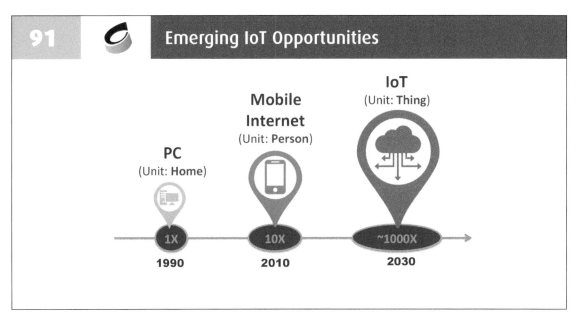

So as we know IoT is one of the big growth opportunities in the future. In the past, it was the PC and the unit was home, so 1X in the 1990. In mobile internet where we are now, the unit is the person and so that's about 10X for mobile internet. And actually I have many for one person, so it's even growing per person. But we know IoT will really be about each thing, each electronic in your house. We will have mobile capability so that the prediction is by 2030 will have 100X of the IoT things.

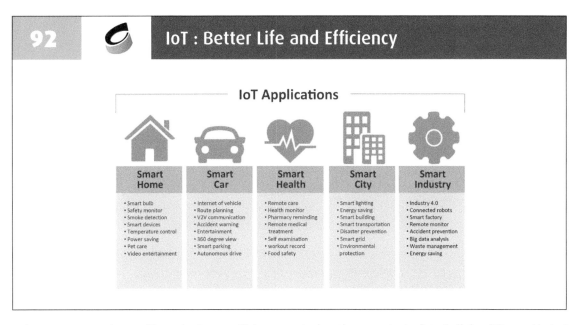

The IoT means a better life and a better efficiency and so these IoT applications will make our home, our cars, our medical, our city and industry smarter. And so these are just a list of all the different kind of things or we can expect to see. IoT will be enabling in the future.

93 · Are We There Yet?

"How many guys d'you know with
a solar-powered wristwatch?"

But the question is that are we there yet from a technology point of view in order to see this happening. And these are just some two cute cartoons. Like this guy with his wearable smart watch but he has a huge solar power energy hovering on his back in order to power his smart watch. And then this guy with his wearable technology you know all strapped. That's kind of the feeling of where we are today.

94 · Circuit and System for IoT Applications

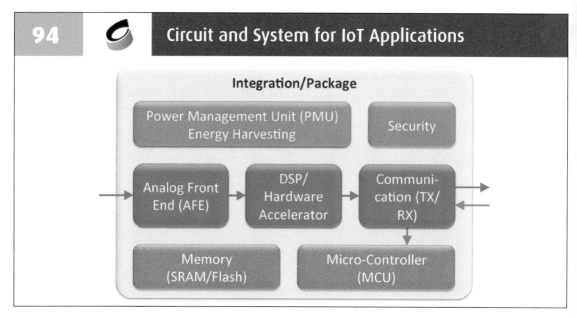

You have the analog front end that acquires the data and then you have a DSP hardware accelerator that processes the data and finally some kind of communication to transmit and receive the data. So that is the hardware view of IoT. Of course there are other components very important such as like a micro controller(MCU) that helps to do the control of all this data and communication, a memory component, we need to store the memory somewhere and also we have the power management unit possibly energy harvesting and security is also really important.

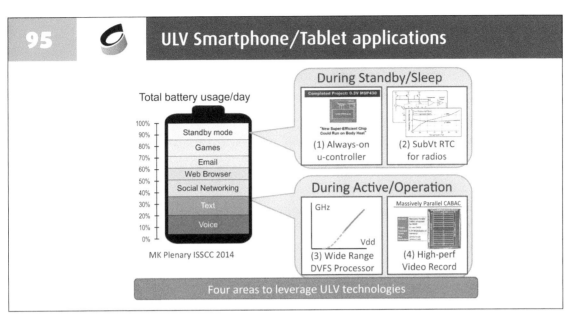

The time standby/sleep is really important. Most of the time, your device will not be operating. So how can we make standby/sleep mode even more low power. And then secondly, once the device is doing something, it's in active doing operation.

These are the four areas that I believe can leverage our ultra-low voltage techniques.

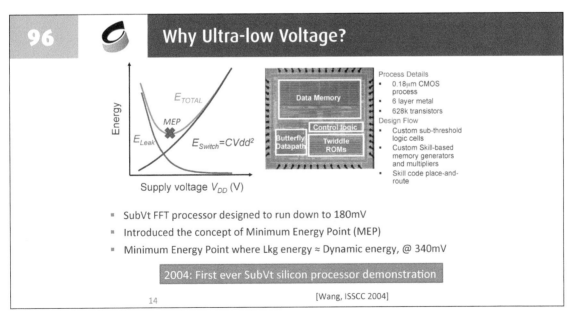

If you lower the supply voltage, you get lower dynamic power because you have CVDD2. However, as you lower supply voltage, the clock gets very very slow and then the leakage energy is increasing like this.

This is the minimum energy of this particular function and it was demonstrated in the subVT processor although the design can run as low as 180 millivolts actually the minimum energy point was at 340 millivolts. And that's approximately where leakage energy and dynamic energy will cross over.

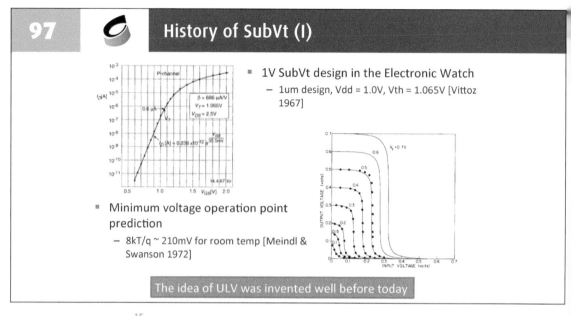

97 ⬤ History of SubVt (I)

- 1V SubVt design in the Electronic Watch
  - 1um design, Vdd = 1.0V, Vth = 1.065V [Vittoz 1967]

- Minimum voltage operation point prediction
  - 8kT/q ~ 210mV for room temp [Meindl & Swanson 1972]

The idea of ULV was invented well before today

The very first that I could find in publication was in 1967 by professor Eric Vittoz. He actually did a micron design and his VDD was one volt. This plot here of Vgs and Id showing that 1V is able to operate his electronic watch at that time. In 1972 by professor Meindl and Swanson? they predicted that the lowest operating voltage possible is 8KT/q or about 210mV at room temp.

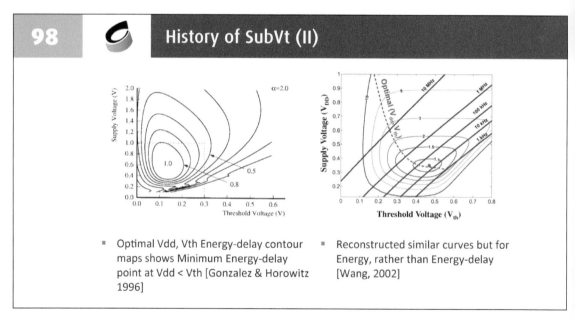

98 ⬤ History of SubVt (II)

- Optimal Vdd, Vth Energy-delay contour maps shows Minimum Energy-delay point at Vdd < Vth [Gonzalez & Horowitz 1996]

- Reconstructed similar curves but for Energy, rather than Energy-delay [Wang, 2002]

In 1996, professor Horowitz had a paper where he showed the optimal VDD and Vth, energy delay contours. The plot shows that the minimum energy delayed point is where VDD is less than the threshold.

So that was the paper I published in 2002 showing that there is an optimal VDD and threshold curve at that point.

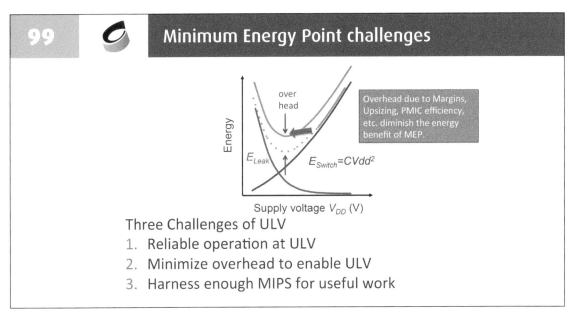

So I want to talk about some of the challenges of actually making this minimum energy point really happen in production. So there are three key challenges for ultra-low voltage.

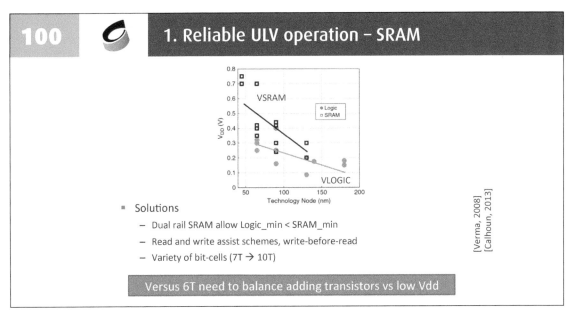

One of the first killers in reliable low voltages is the SRAM. And as I mentioned before, the SRAM is typically the smallest device, the smallest cell in the whole SOC or CPU. Actually the SRAM supply is always higher than the logic supply.

Solutions are using Dual rail SRAM to meet the dual voltage, or adding read and write assist schemes, or using 7T-10T bit-cells. There is a balance from adding transistors to low vdd.

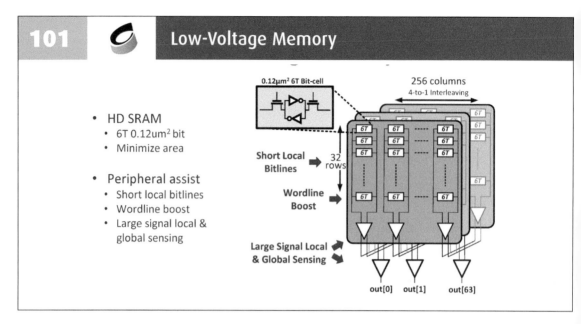

This shows a 28nm 6T SRAM can look at using short local bitlines, wordline boost or large signal local and global sensing. All those techniques can help to lower the voltage of the SRAM.

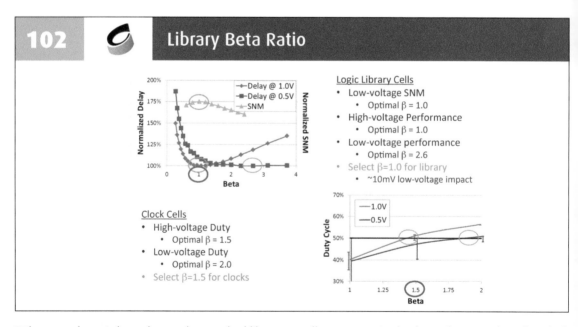

The second one is how change the standard library to adaptive the low vdd. The PMOS versus NMOS ratio is different at different vdd. To the Logic Library Cells Beta=1.0 is the best choice and to the Clock Cells Beta=1.5 is the best choice.

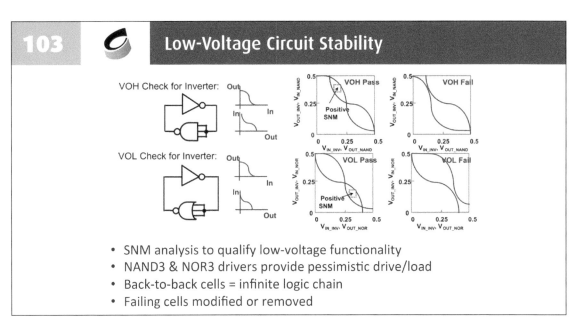

103 **Low-Voltage Circuit Stability**

- SNM analysis to qualify low-voltage functionality
- NAND3 & NOR3 drivers provide pessimistic drive/load
- Back-to-back cells = infinite logic chain
- Failing cells modified or removed

We can use SNM analysis to qualify low-voltage functionality like SRAM. As we called butterfly curve. Nand3&Nor3 drivers provide pessimistic dirve/load. Positive SNM means PASS both of VOH and VOL check.

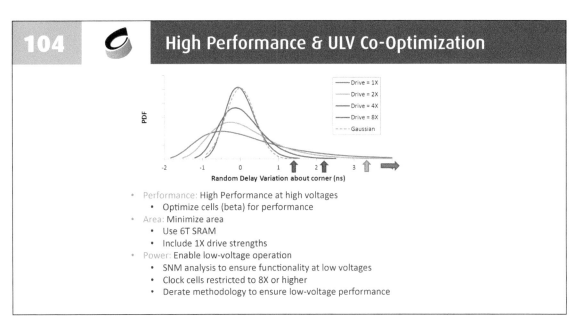

104 **High Performance & ULV Co-Optimization**

- Performance: High Performance at high voltages
  - Optimize cells (beta) for performance
- Area: Minimize area
  - Use 6T SRAM
  - Include 1X drive strengths
- Power: Enable low-voltage operation
  - SNM analysis to ensure functionality at low voltages
  - Clock cells restricted to 8X or higher
  - Derate methodology to ensure low-voltage performance

To co-optimize high performance and low voltage. To minimize area, we use 6T SRAM & 1X drive strengths. To keep the steady performance we should consider the standard cells's random delay variation about corners. 8X is good , so clock cells restricted to 8X or higher.

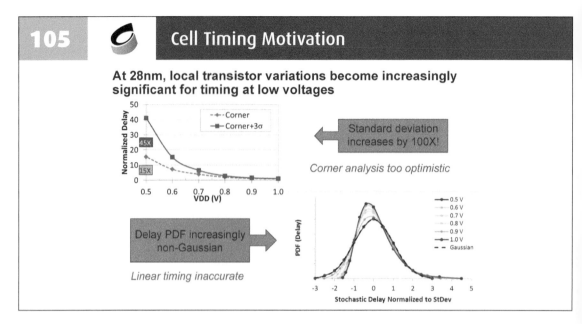

At 28nm the cell timing has two special features. One is local transistor variations is significant at low vdd, 45X vs 15X. Another is the Delay PDF is non-Gaussian. All the model need updates.

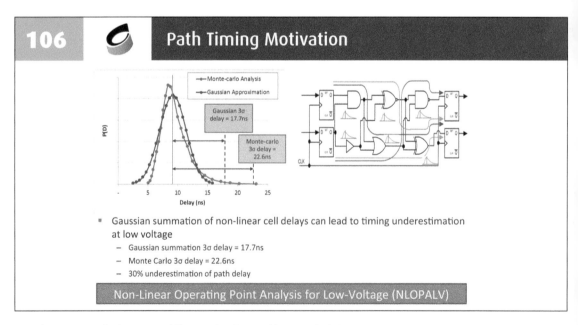

Path Timing Calculation is different for low vdd. Gaussian summation 3Sigma delya is 17.7ns and Monte-Carlo 3Sigma delay is 22.6ns. 30% worse of path delay.

Non-linear Operating Point Analysis is important for low-voltage design.

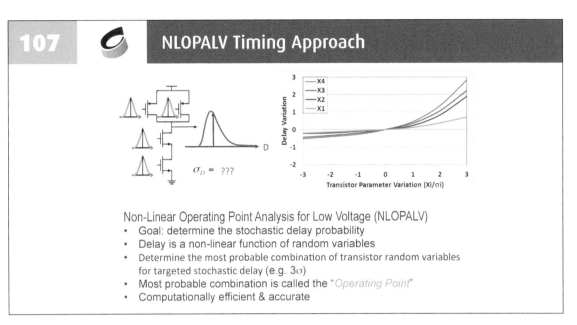

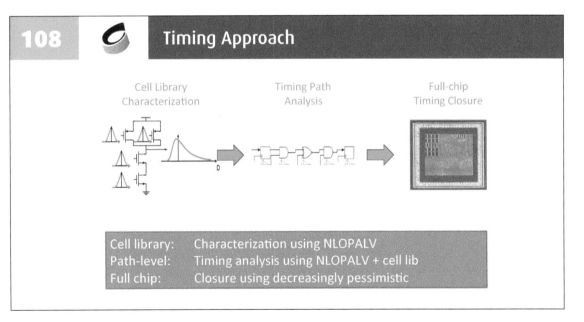

NLOPALV is a new approach for timing calculation compare to the normal Gaussian model. The Goal is determine the new stochastic delay probability. Random variables are also Gaussian but the cell delay is non-linear because of the low voltage.

The Timing Approach is three steps, First Step is Cell library characterization. Second Step is Timing Path Analysis. The last Step is Full-chip Timing Closure.

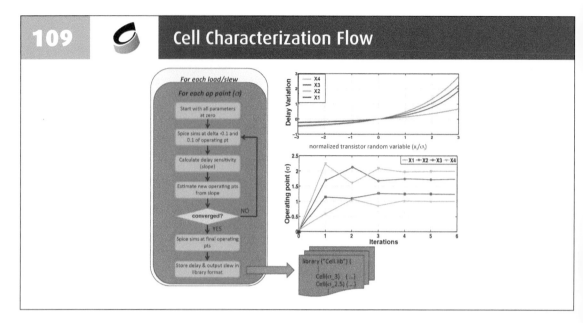

That is the Flow of Cell Characterization. To get the LIB, we must calculate for each load/slew and for each op point. In each calculate, we start with all parameters at zero, then Spice sim and updates the new operating points. When simulation converged, we can get the delay & slew in library format.

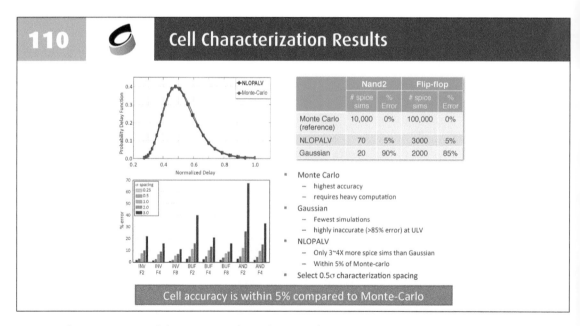

Using the NLOPALV Model we can reduce the simulations to get the accurate results. Only 3~4X more spice sims than Gaussian within 5% of Monte-Carlo.

The sigma step is 0.5 Sigma for the balance of accurate and speed.

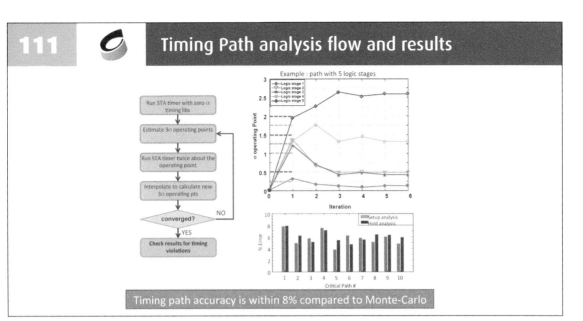

This is the Timing Path analysis flow and compared results between NLOPALV and the Monte-Carlo. Timing path accuracy is within 8% compared to Monte-Carlo.

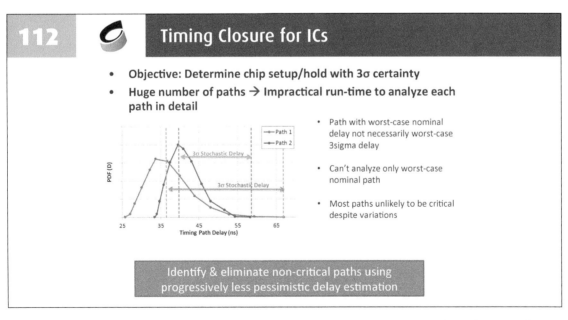

Now we have a cell characterization, we have a path characterization. The difficult thing is how to do this. Because the number of paths on your chip is millions and millions, so it's really impractical to analyze every single path and details. So we devised and said it is a way to identify and eliminate non critical paths using progressively less pessimistic delay estimation.

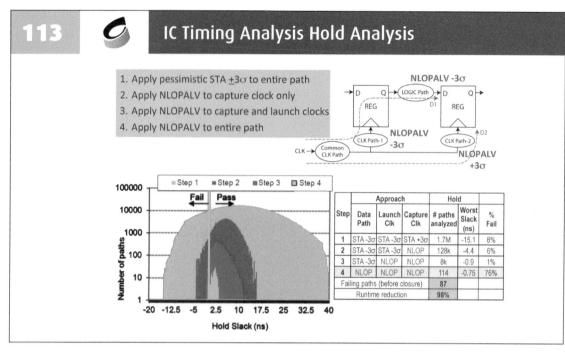

This is the Hold Analysis flow in IC Timing Analysis using NLOP.

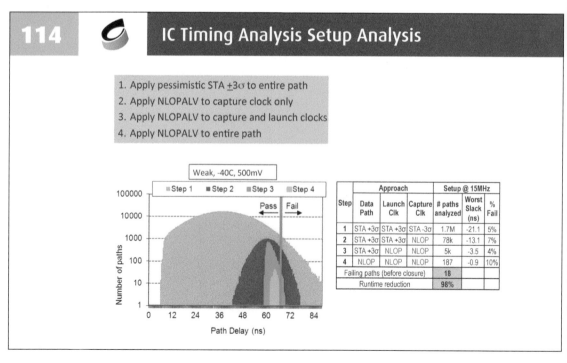

This is the Setup Analysis flow in IC Timing Analysis using NLOP.

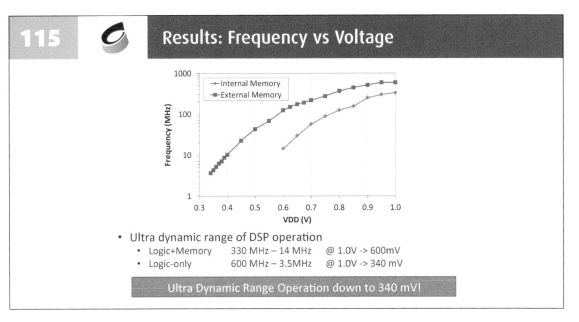

Consider the internal memory, we can reduce the vdd to 0.6 volt. If we use external Memory, the vdd can be 0.34 volt. And the clk can reduce from 600 Mhz to 3.5 MHz.

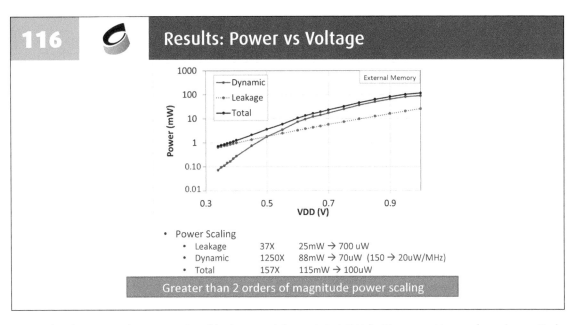

Consider the power, the cross point of leakage and dynamic is 0.5 Volt. We can get two orders of magnitude power scaling by enabling 0.5V operation.

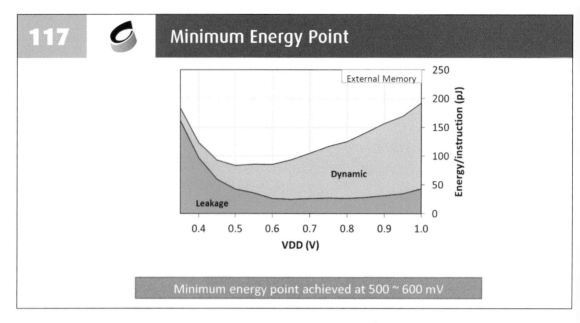

The minimum energy point is between 0.5and 0.6 around here of energy for instruction.

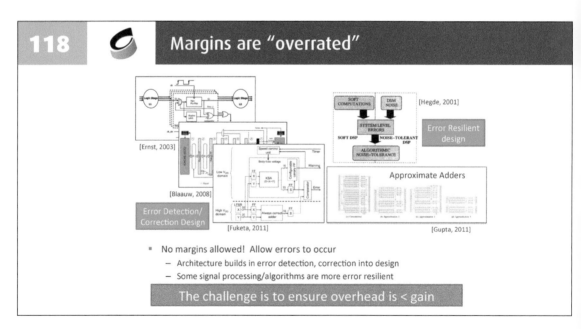

When you reduce the voltage, you reduce the margins. When no margins left, we have to face the errors. Two way to solve the error problem. One is add error detection and correction into design. The other one is changing signal processing/algorithms more error resilient. For example , we can't find a little distortion in your signal or in your image.

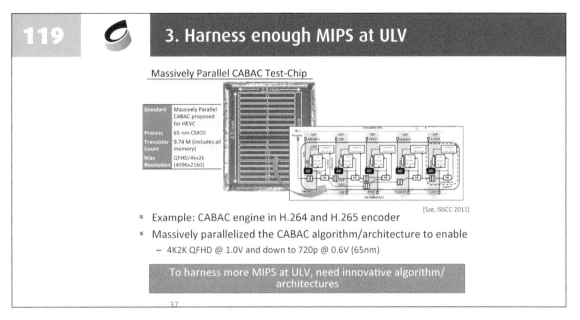

This is one concept done by professors Sze at MIT which is how do I massively paralyze this particular function and that allow me to go to very low voltage and still operate at a very good high quality. The example here is the CABAC engine which is found in video encoding H.264 and H.265/ what we show is that by doing this massively paralyzed CABAC she can still operate this very high definition 4k2k at 1V and she drops down 0.6V then she can still do 720p.

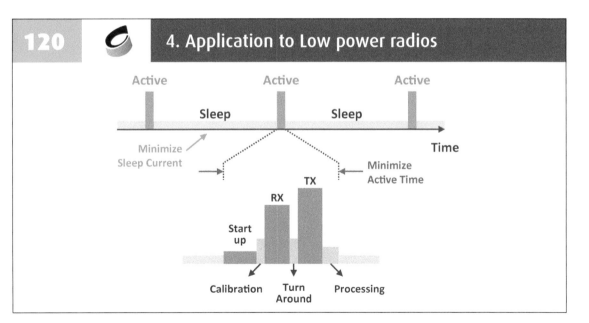

For low power radio, basically most of the time the radio is in sleep. And it will wake up for a certain amount of time and do some start up, transmit, receive and transmit and then it basically goes back to sleep. So minimize the sleep current is very important.

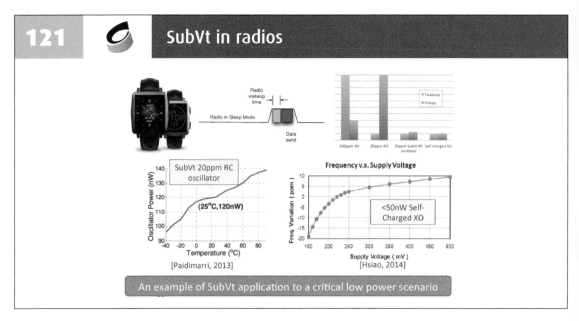

121 — SubVt in radios

An example of SubVt application to a critical low power scenario

Sleep power is mainly contributed by the Oscillator Power. There is two examples in SubVt Oscillator. One is RC oscillator in 2013, 120nW at 20ppm accuracy. One is Self-Charged XO in 2014, <50nW at 20ppm accuracy.

122 — Conclusions

- ULV: From cool idea → To solving real problems
  - Emerging applications and markets need ULV solutions
- Challenges : need "Everyday Genius"
  - Circuit design: Reliable operation at ULV
  - Design margins: Minimize overhead to enable ULV
  - H.265 CABAC: Harness enough MIPS for useful work
  - Low Power Radio : Need to solve the right problems!

10 yrs later, so now is the time!

ULV is a wonderful solution for solving real problems from cool idea.
We have to face the challenges now and solve them in ten years.

CHAPTER 04

Mobile Embedded DNN and AI SoCs

Hoi-Jun Yoo

KAIST

This chapter will first introduce the importance of deep learning based on the many core processor architecture, followed by an explanation on the mobile embedded neural network processor from theoretical basics to SoC examples, i.e. deep learning network, attention based item recognition, object recognition, the hot issues of virtual reality (VR) and augmented reality (AR).

Human being is the most intelligent in the world to my knowledge. Artificial intelligence (AI) is a mimic of human brain, a manmade intelligence. Nowadays, AI is still far less intelligent than human brain, only in very limited applications, such as object recognition, face recognition or cat recognition. There are many people are doing research on artificial intelligence. Someone even claimed that in 2045, artificial intelligence may outperform the human intelligence. They call it singularity. Nobody knows when it will happen, but anyway I believe it will happen sometime.

Contents

1. MANY-CORE PROCESSOR ARCHITECTURES

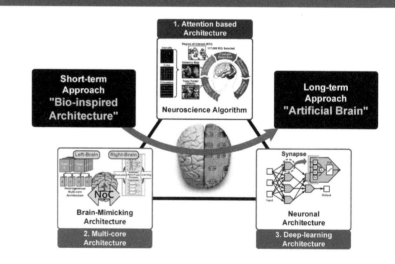

Our brain is quite small, only 1.35 liter and 1.35 kg. And it contains 1011 neuros, each neuron has about thousands of synapsis, so the number of synapses is 1014. And the speed is quite slow about million seconds, but because there are so many processes, so we can do high-speed operation. And memory is 1020 ~ 1021. However, compare to the computer disk, it's quite big. The reliability is a very interesting thing, every day about 10,000 neurons in your brain are dying out. New neuros come after. It is a very fortune, because of death of neuros, we found your intelligence and we found your memory. So, this is a necessary phenomenon. Plasticity mean learning, bring our brain many things, like actions, words, images. Plasticity means like plastic. It strengthens some parts of the neurons.

Brain consists of billions of neurons. Different approaches have been applied to study the inside of the brain, i.e. neural biology, neural physiology and psychology. Artificial intelligence is a mimic of human brain from different aspects. This chapter introduce the integration of a psychological effect, the attention architecture, and a neural network activity.

3

Trends of Process Technology

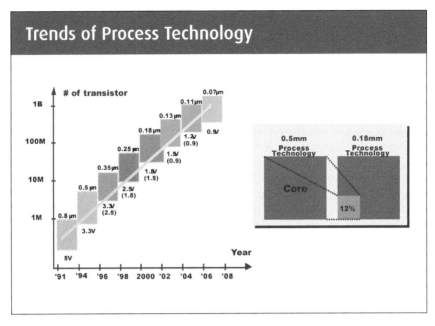

In 1996, the Sandia Lab in Monica, developed a super computer, consisting of 1.6T flops. Ten years later, in 2006, play station, integrated all of the functions with a much smaller size. It is the silicon chip, the System-on-Chip (SoC) design that makes it possible. And Moore's law pushes this trend in the last several decades. Nowadays, heterogeneous cores system integrated i.e. GPU, CPU with artificial intelligence is widely used in wearable devices. However, the power consumption is still too high for wearable applications, and the integrated intelligent function is limited.

4

Total Design

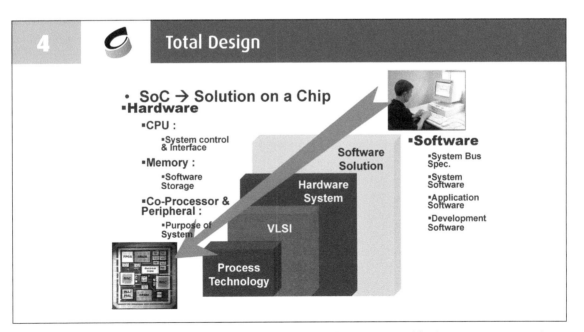

The solution is SoC, not System on-Chip, but Solution-on-Chip, which integrates both the software, and the hardware. When we talk about hardware, fabrication technology is taken into consideration. It is important especially when we go beyond 28nm. Wearable devices require to solve a problem on chip rather than using the software to reduce the power consumption as well as to improve the processing speed.

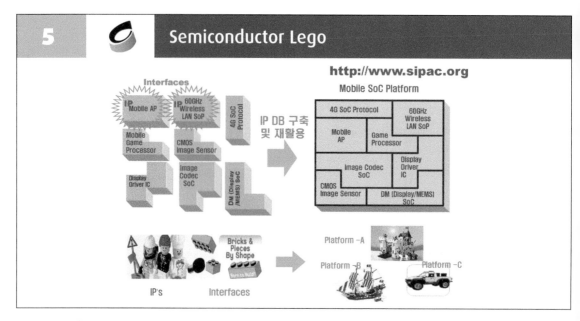

S olution-on-Chip designs requires vertical engineer rather than a horizontal engineer. A horizontal engineer assembles different parts from other experts to form a system, while Solution-on-Chip for AI application requires disciplinary expertise from software, to OS, to hardware to the optimization skills. IP-based design or platform-based design is used.

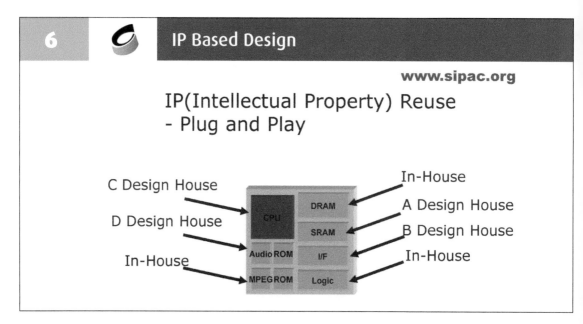

T his slide shows the architecture of our chip. The MPEG ROM and logic circuits are customized designs, while the SRAM and interface logic are standard IPs. The integration of the customized modules and standard IPs is a great challenge to interface design. It is important to optimize the interface to increase the design capability. The most efficient method is to adapt On-Chip Network (OCN), which connects all the IPs with independent addresses, which is similar to the IP network. The design of the IP interface becomes the key of the entire design.

7   BONE-2

- **Multimedia SoC with low-power on-chip networks**

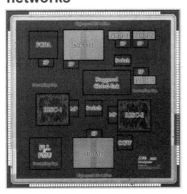

- 0.18μm 6metal CMOS technology
- Max. freq. = 1.6GHz
- Max. B/W = 11.2GB/s
- 5mm x 5mm
- Integrated IPs
  - 32bit RISC x 2
  - 64kb SRAM x 2
  - On-chip FPGA
  - PLL & Off-chip gateway
  - Peripheral memory x3 via small swing link
- Low-Power Techniques
  - less than 51mW @ full bandwidth

Kangmin Lee, et al., "A 51mW 1.6GHz On-Chip Network …" ISSCC-2004

Our first OCN design was proposed in 2004 [1]. It integrated two RISC chips, two SRAMs, one FPGA, one PLL and many other functional modules. All the modules are connected through the OCN. Experimental results show it operated as expected successfully. The proposed design can be configured by programming the RISC.

8   RAMP & MobileGL

- **Optimized Embedded S/W for Mobile 3D Graphics Application**
  - **Limited Computing Power of RISC Processor for 3D Graphics**
    - **Hand-optimized Assembly Code for Performance Improvement**
    - **Math Function Optimization for 3D Graphics Pipeline**

MobileGL

http://ssl.kaist.ac.kr

Since the area of memory dominates, Processor in memory (PiM) becomes popular. PiM integrating computational processor with memory on a single chip. DRAM technology instead of logic technology is used. The capability of the logic functions is reduced due to the complexity of the technology caused from the vertical stacking 3D cell structure of the DRAM. In order to further optimize the logic functions with hand-optimized assembly code. The platform, the OS, and software are integrated as well.

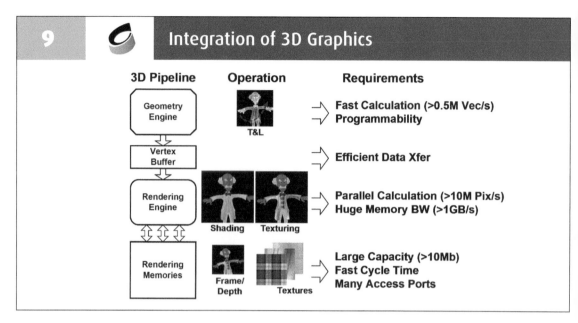

We proposed the first portable 3D graphics chip enabling 3D games on cell phone in 1999 [2]. The procedure of 3D graphics begins from using key points to reduce the calculation workload of the geometry engine during rotation or translation. The target is colored and shaded with special effect after its position changes by the rendering engine, which requires a very high power consumption.

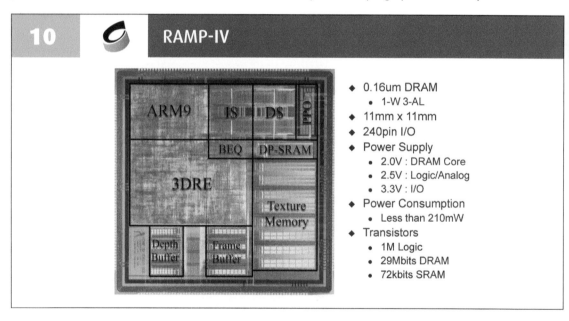

It was fabricated using a low cost 160nm technology on a single chip, integrating all the functions, texture memories, and 3DRE to calculate the translation and rotation. The proposed design featured the lowest power consumption (210mW) at that time with the highest performance of 3D graphics (66Mpixels/s and 264Mtexels/s for texture processing) on cellphone. It integrated a 32-bit RISC, a 3D rendering engine, a slimshader for shading effect, a 3D accelerator, and 29Mb DRAM, a power management unit. Software is designed as well. The optimization of the assembly code can improve the performance by more than 50%. Nivida and Sumsang used this technology for their first embedded GPU design.

11 Moore's Law

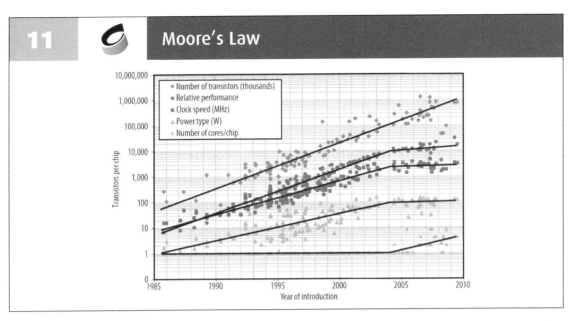

According to Moore's Law. Clock speed doubles every three years. However, in about 2005, even though the transistor size kept shrinking, the CPU speed stopped increasing. The CPU speed was stuck at about 2GHz due to power consumption, the capability limitation of the air cooling system. This is called power wall. In addition, there is another wall, the memory wall caused from the lack of large amount of fast speed memories. Moreover, there is performance wall due to the complexity of programming on multiple-core CPU.

12 Current CPU Research Trend

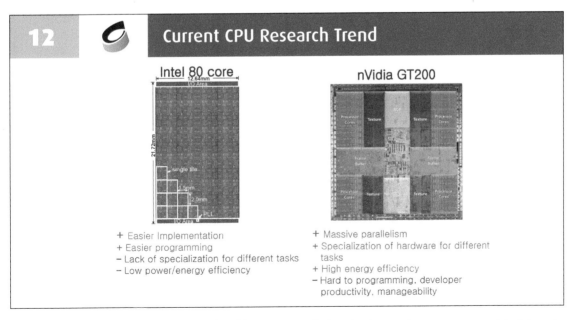

For portable and wearable devices, the dilemma between battery size and functionality causes a big gap. Intel integrate many homogeneous CPUs in a signal chip as a solution. In order to future optimize the integration, we proposed to integrate heterogeneous CPUs, i.e. graphics processor, DSP, neural network processor, featuring a better performance than homogeneous approach.

Recent Trends in IT

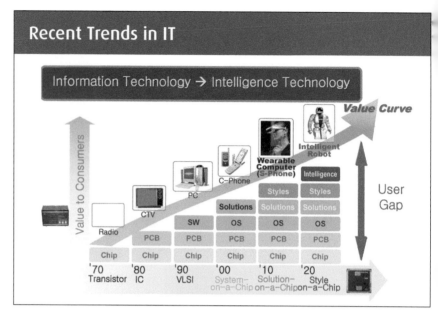

In the past half centu-ry, we experience the changes from informa-tion technology to intel-ligence technology. The designers' attention moved from transistors to software in PC era, to application in the cell phone era, to maybe intelligence in the next era. The gap between the designers and the users become larger and larger. However, the system designer, or SoC designer must understand the end user, or at least the application.

Multicore CPU vs Neuromorphic HW

There are two approaches for the multicore CPU design, Von Neumann architecture and non Von Neumann architecture, or Neuromorphic hardware, or deep neural network. There is not a clear boundary between logic and memory in non Von Neumann architecture. A lot of efforts have been invested into non

♦ **Multicore CPU**
- Performance Saturation

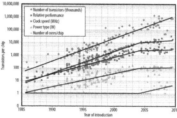

- **Limitations in** Von Neumann arch.
- **No optimized parallel processing arch.**
- → **Inefficient parallel processing**
- → **Saturation in performance scaling**

♦ **Neuromorphic Hardware**
- Implementation Inefficiency

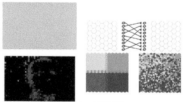

Stanford (2010)

- **Only naive pattern recognition**
- → **Limited applications**
- **Complex training / control flow**
- → **Hardware overhead**
- → **No SW optimization**

Von Neumann architecture to extend its applications. nVidia and AMD proposed GPU based design [3], but consumes too much power. Intel proposed the heterogeneous multicore processor. The power consumption was improved, but still high. IBM proposed their neuromorphic chip, TrueNorth or SyNAPSE [4]. The power consumption is very low, but the processing capability is poor. Stanford proposed a neuromorphic chip mimic the brain activity for brain research with very low processing speed [5]. The LeCun group in New York University designed a deep neural network chip in 45nm technology, consuming 570mW with limited functionality [6].

15   KAIST Approach: MC Processor + M/E DNN

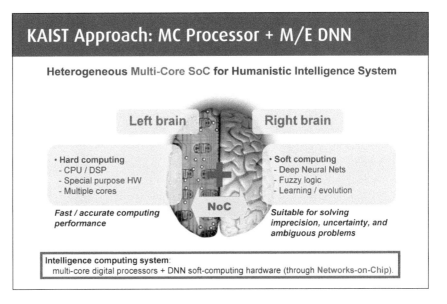

Heterogeneous Multi-Core SoC for Humanistic Intelligence System

Left brain   **Right brain**

- **Hard computing**
  - CPU / DSP
  - Special purpose HW
  - Multiple cores

- **Soft computing**
  - Deep Neural Nets
  - Fuzzy logic
  - Learning / evolution

NoC

Fast / accurate computing performance

Suitable for solving imprecision, uncertainty, and ambiguous problems

Intelligence computing system:
   multi-core digital processors + DNN soft-computing hardware (through Networks-on-Chip).

KAIST proposed to integrate the neuromorphic style chip with a logical processor, i.e. CPU, or DSP. The logical processor enables mathematical processing with high precision, just like our logic brain, the left brain, while the neuromorphic style chip performs AI functionalities, just like our emotional brain, the right brain. The integration is challenge. OCN is applied.

2. MOBILE/EMBEDDED DEEP LEARNING PROCESSOR

2.1 BASIC OF DEEP NEURAL NETWORK

16   Analogy between Biology and ANN

Biological Neural Network	Artificial Neural Network
Soma	Neuron
Dendrite	Input
Axon	Output
Synapse	Weight

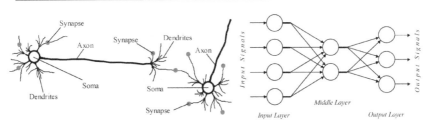

Nowadays, AI applications can be found everywhere, i.e. eye recognition and auto driving. AI is trying to resemble the neurons. Neuron cells receive signals through dendrites, while sending signal out through axon. Artificial neural network (ANN) is designed to mimic the neural system, in which each neuron receives multiple input signals, and generates one output signal after processing. The output signal can be send to multiple neurons.

17

Brief History: Neural Networks

The idea was invented 1943 by McCulloc and Pitts. They proposed a mathematical model on neural activity [7]. In the middle of the 1950s, Rosenblatt invented a system [8], consisting of two layers of many neurons to identify letters or make decisions for the unknown input according to the known output. 1950s was the first bloom of AI. However, in 1970s, the AI boom was ended by the conclusion from Minsky and the MIT group, saying that AI cannot solve the XOR problem at all [9]. In 1980, Rumelhart developed a new circuit implementing more layers with back propagation for training [10]. However, the complexity of neural network is limited to five by using back propagation training.

The 1990s and early 2000s was a winter time for neural network research. In 2006, the University of Toronto, Hinton develop a new algorithm to train even hundreds of neuron layers [11]. People applied this concept to perform image recognition achieving a higher accuracy than human being. The society are so excited about this results, which made AI is popular again.

18

Pattern Recognition & Deep Learning

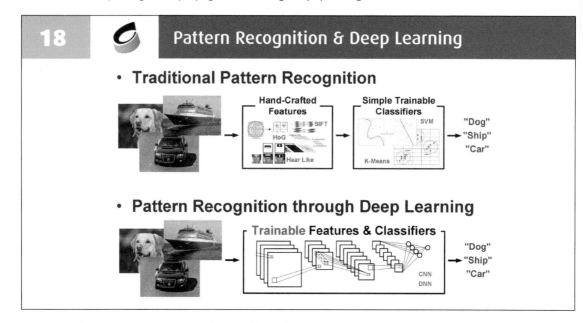

Traditional image recognition was designed based on pattern recognition using features, i.e. HoG, SIFT, Haar. Training was performed at circuit level, not the network level, i.e. SVM, K-Means. On the other hand, the training of DNN does not require background knowledge. The progress is quite simple, that even high school students can train and use it.

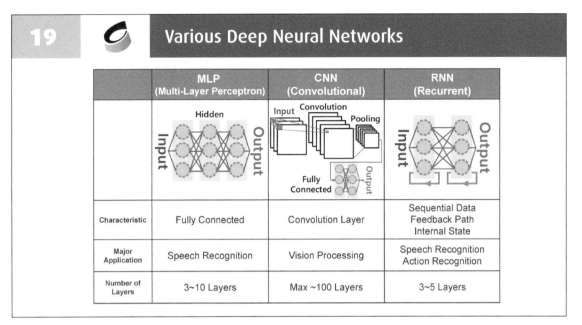

NN features higher accurate than any other object classification. Deep learning is part of machine learning, while machine learning is part of AI. All these are mathematical approach, and neural network. In the early 2000, natural intelligence or brain inspired is still smarter than AI. There are three key networks: multi-layer perceptron, CNN and recurrent neural network (RNN).

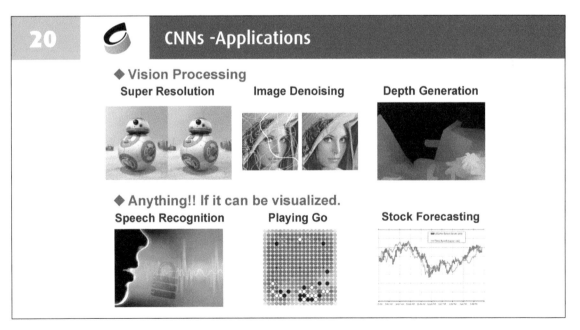

NN is widely used for image recognition, i.e. face recognition, object classification, and sign recognition, and gesture recognition, pedestrian detection, and car detection, while RNN is good for the sequential data analysis, such as moving feature analysis. CNN is also useful for de-blurring of image with high resolution. The depth can be generated while removing the noise. It is also utilized for speech recognition, go game, and many trading companies for prediction of the stock price.

2.2. M/E-DNN MOBILE/EMBEDDED DEEP NEURAL NETWORK

21

Compare to local feature based operation, DNN requires heavy hardware, including high performance CPU, large memory, high power consumption, as well as big database. Recently, many companies collect a big amount of data through the web. In addition, the usage of GPU, make it possible for the deep learning applications. The big data, and GPU are all very important to DNN implementation.

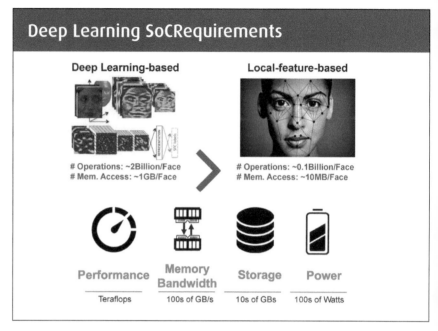

There are two different approaches in implementation, cloud-oriented and mobile/edge-oriented. Many companies, i.e. Amazon, Intel, Google, Facebook, IBM, are developing AI chips/hardware for their database or data center. The cloud-oriented approach is to collect, to store as well as to analyze all the data in the data center. On the other hand, smart phone, drone and Google Glasses require fast decision maker performed in the local device. And the power consumption is limited by the battery.

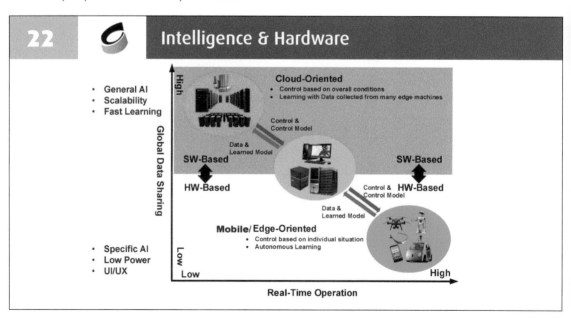

23

Mobile/Embedded DNN Requirements

Mobile embedded DNN is very difference from the cloud based DNN. Cloud-oriented DNN features more layers, sometimes hundreds or even a thousand. Each layer consists of more than thousands, sometimes more than ten thousand of neurons. The learning is very complicated. In addition, there is communication latency. The power consumption is high, but the precision is also high. On the other hand, for mobile embedded DNN, real time operation is very important. Power

Cloud-based DNN	Mobile/Embedded DNN
• Cloud Intelligence	• On-device Intelligence

• General AI
  - Very deep network
	• Application-specific AI
	  - Specific DNN network

• Communication latency	• Real-time operation
• High Number Precision	• Reduced Number Precision
• High power / cooling cost	• Low-power consumption
• Large memory capacity	• Efficient memory arch.

consumption must be low. The number of layers and calculation precision is limited. One very interesting fact is that accuracy is not decreased at all.

24

M/E-DNN Optimization Approaches

Hierarchy	Example Techniques
Algorithm	• Application-Specific DNN Architecture • Binary/Modulus Number • Reduced Number Precision • Tensor Decomposition (Separable Filter)* • Sparsity (ReLU, Pruning), Reusability • Approximate Computation – Skipping, Loop Perforation
Architecture	• PE & PE Array Optimization • PE Network Optimization • Exploiting Sparsity, Reusability • Efficient Memory Hierarchy
Circuit	• Near-threshold Operations • LUT based Operations • Customized SRAM Circuit

[1] Kyeongryeol Bong et al, ISSCC 2017

Most of the deep learning network research has been done by software engineers, with GPU or CPU hardware. However, recently people realized the importance of hardware. In order to realize a rapid, low power consumption operation, a customized DNN silicon chip is required. Many useful ideas are proposed, i.e. application specified DNN architect, tensor decomposition [12], sparsity [13], [14], reusability [15] and approximate computation [15], the use of PE array [12]. In addition, in order to reduce power consumption, near threshold operation [12], or LUT based operation [16], or new SRAM circuit [12] are proposed as well.

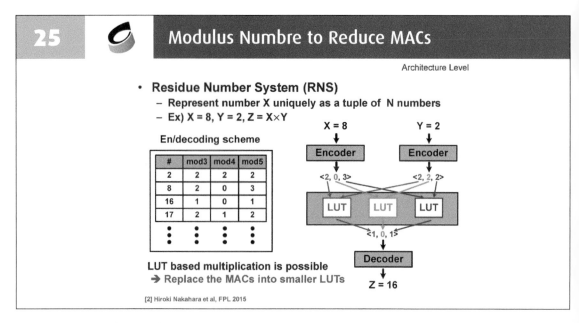

For example, residue number system (RNS). Every member can be uniquely specified by three modulus numbers. With mod3, mod4 and mod5, 2 is <2,2,2>, and 8 is <2,0,3>. By feeding <2,2,2> and <2,0,3> into a specified LUT, a result of <1,0,1> is achieved. <1,0,1> is 16 with mod3, 4 and 5. Such a modulus system is very simple, thus hardware requirement is low. The LUT can be optimized.

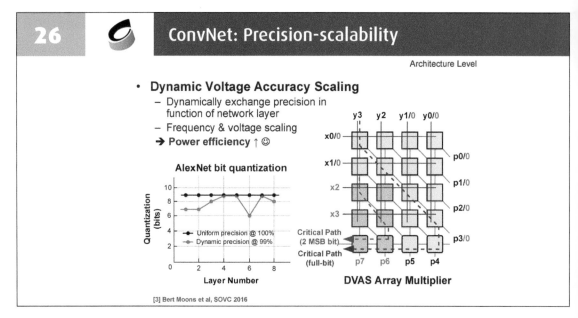

With dynamic voltage accuracy scaling, sometimes full 8-bit number is not required, 6-bit or even 2-bit is enough, the voltage can be reduced as well as the frequency.

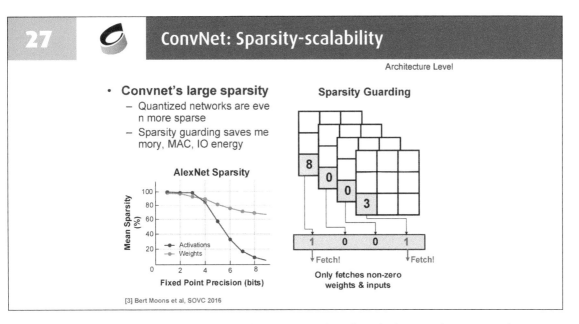

With sparsity-scalability for the convolutional operation, it skips the calculation with zeros to reduce power consumption.

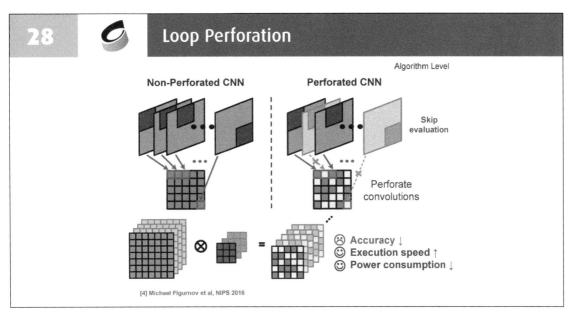

Loop perforation is also widely used for the convolutional operation. Convolutional operation is required in every pixel in CNN. However, the process in perforated CNN is randomly, which reduces the hardware complexity dramatically. However, the accuracy is sacrificed. There is a trade-off between the energy and accuracy.

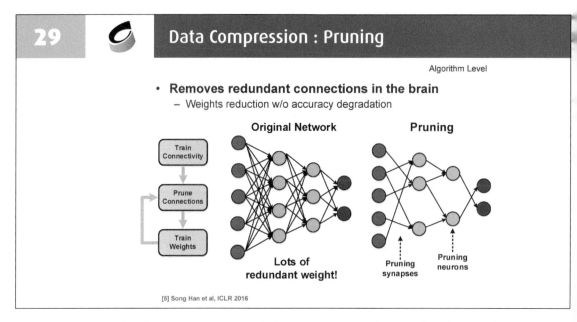

[5] Song Han et al, ICLR 2016

Pruning is a popular data compression method, featuring a much higher accuracy then perforation. The neurons are disabled during training, enabling a higher training speed and a lower energy. This is exactly what happened in brain. Different pruning factors are used for different layers, i.e. 84% for the first layer and 10% for later part.

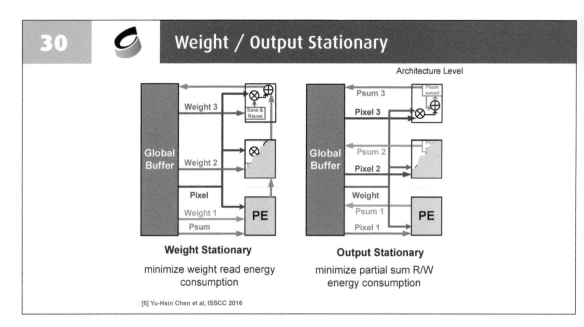

[6] Yu-Hsin Chen et al, ISSCC 2016

The mapping of the processing element is also important. Storage is required for the input, the filter temporary and the final results. For the convolutional operation, processing elements (PEs) with multipliers and adders are needed. Different weights are used for the mapping of the PEs. Weight means the convolutional filter. Convolutional data is stored, while the pixel data is broadcasted, multiplied and sent to the nearest neighbor for the addition process.

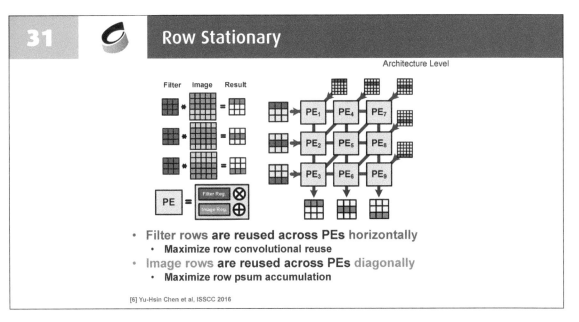

In this case, the pixel data is stored inside the processing element, denoted as weight stationary, because filter data or weight data is stored inside the processing element. Sometimes the weight changes every time, as well as the pixel data. All the operation is done inside with final result popped out. This is denoted as output stationary. Row stationery is used to optimize the operation.

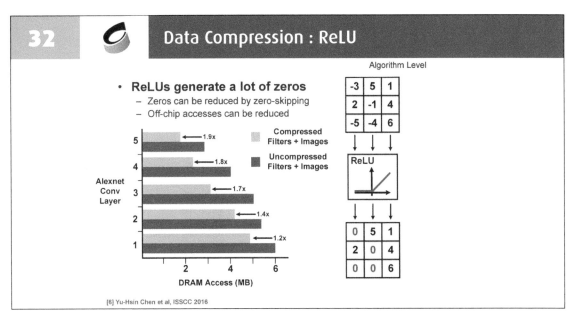

Nonlinear operation, such as the sigmoid operation, is required after convolutional operation. Recently ReLU is commonly used, which set negative parameters all to zero. Operations with zero can be skipped as explained earlier. Thus, a lower power consumption can be expected by using ReLU.

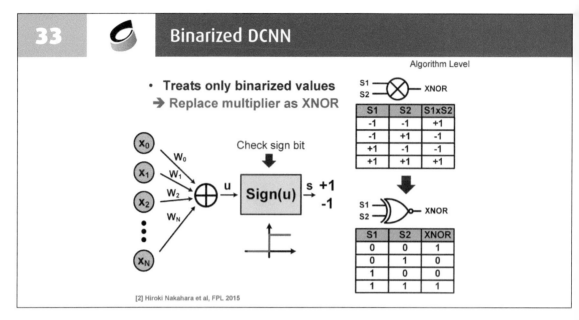

[2] Hiroki Nakahara et al, FPL 2015

Binarized DCNN is another method, which uses only 0 and 1, or 1 and -1, instead of the complicated number system. It is quite useful for many applications. The multiplication and addition can be done by simple logic circuits, i.e. NOR or XNOR. A higher speed, and a lower power consumption can be expected.

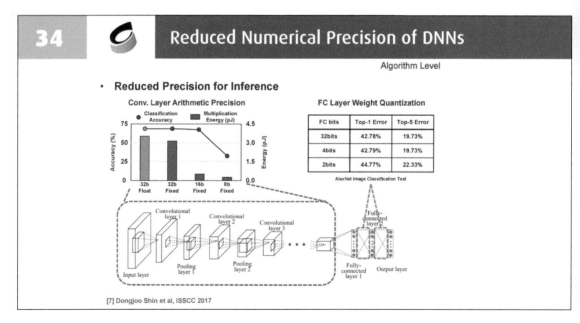

[7] Dongjoo Shin et al, ISSCC 2017

And another interesting idea is reduced number system. Even though the input data employs 32 bit floating point number system, the accuracy of DNN can be reduced to 32-bit fixed, or 16-bit fixed, or 8-bit fixed, or 6-bit fixed or even 2-bit. In addition, the accuracy of different layers can be dynamically changed.

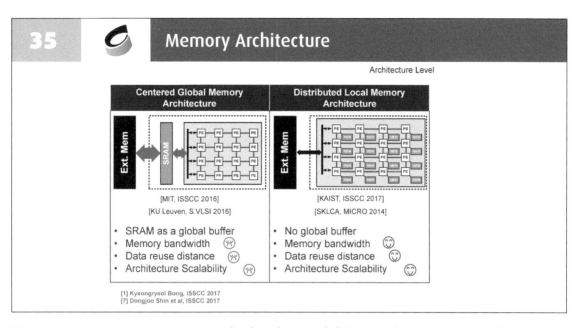

Memory Architecture

Architecture Level

Extern memory, or intern memory, or distributed memory is usually used. Distributed memory system requires a large memory bandwidth, since the high memory access frequency. This issue can be avoided by using local memory. In addition, under the multiple chip integration situation, the memory is used shared by chips.

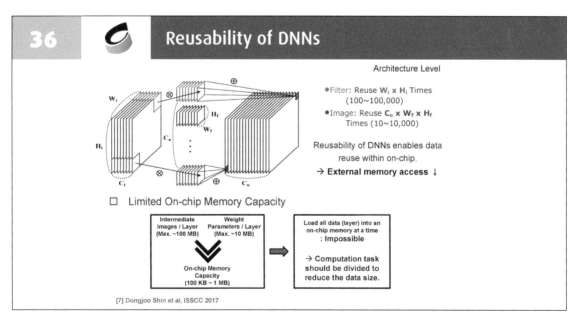

Reusability of DNNs

Due to the limitation of chip memory size, DNN features are usually reused. Image division is more effective in early stage, while channel division is more efficient to reduce the complexity in later stages. A mixed division of image division and channel division features a better quality of the system.

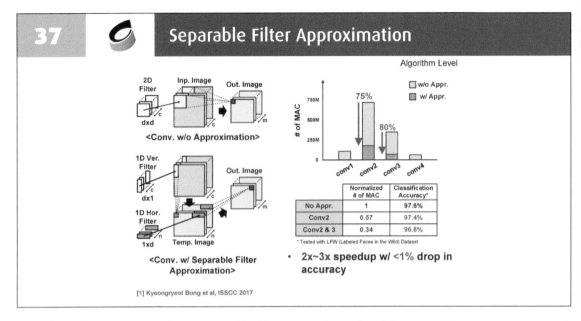

37 **Separable Filter Approximation**

<Conv. w/o Approximation>

<Conv. w/ Separable Filter Approximation>

Algorithm Level

	Normalized # of MAC	Classification Accuracy*
No Appr.	1	97.6%
Conv2	0.57	97.4%
Conv2 & 3	0.34	96.8%

* Tested with LFW (Labeled Faces in the Wild) Dataset

- **2x~3x speedup w/ <1% drop in accuracy**

[1] Kyeongryeol Bong et al, ISSCC 2017

And another interesting idea is separable filter. Two dimensional convolution can be realized using two steps of one dimension convolution. With the separable filter approximation, the accuracy is decreased by less than 1%. However, the hardware can be reduced for more than 75% or 80%. Both vertical and horizontal direction pixels accesses are needed. A special SRAM, i.e. the transposed SRAM is required.

2.3. EXAMPLES OF M/E-DNN

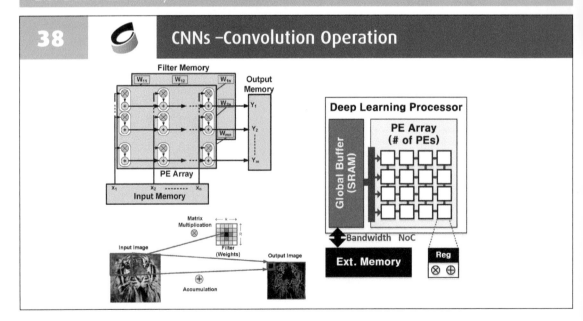

38 **CNNs –Convolution Operation**

This is an image of tiger. Filter and convolutional operation are required. One simplest implementation is CNN. CNN is a two-dimensional array of neurons. The calculation of each pixel requires a summation of the nearest neighbors with proper weight, as well as itself, and the bias. Multiplier and adder are needed.

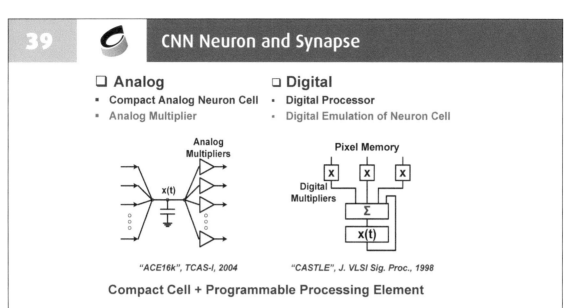

Addition and multiplier can be easily implemented using analog circuit. However, it is a little complicated by using the digital circuit.

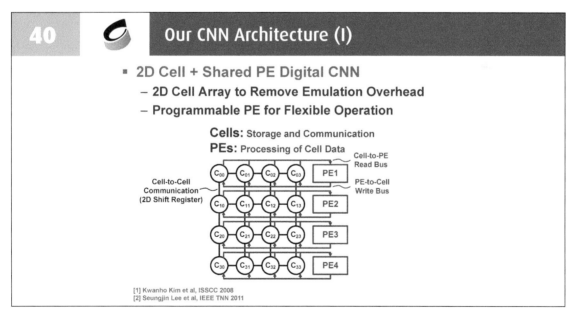

This is the architecture of the CNN, integrating multipliers and adders.

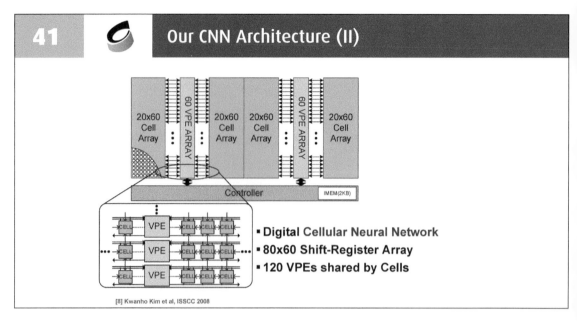

This is a 20x60 cell array, with 60 Pes.

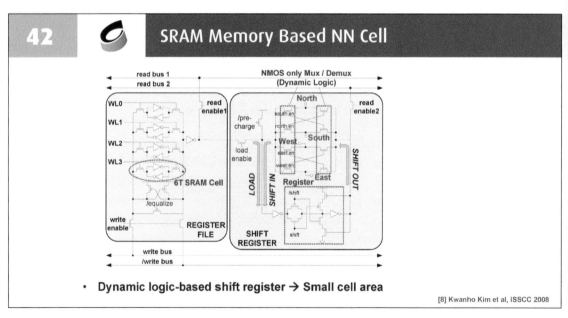

Each cell consists of six transistor SRAM and a shift register. The specially designed shift register is used to connect the cell data to four directions, up, down, left and right, which enables a very fast convolutional operation.

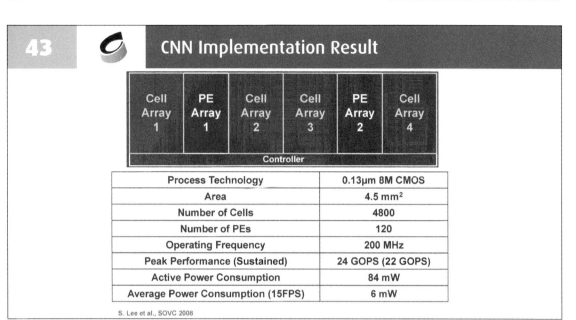

43 — CNN Implementation Result

Cell Array 1	PE Array 1	Cell Array 2	Cell Array 3	PE Array 2	Cell Array 4

Controller

Process Technology	0.13µm 8M CMOS
Area	4.5 mm²
Number of Cells	4800
Number of PEs	120
Operating Frequency	200 MHz
Peak Performance (Sustained)	24 GOPS (22 GOPS)
Active Power Consumption	84 mW
Average Power Consumption (15FPS)	6 mW

S. Lee et al., SOVC 2008

This is the photograph of the fabricated chip [17]. It integrated cell array and processing array. The cell array consists of SRAM cell.

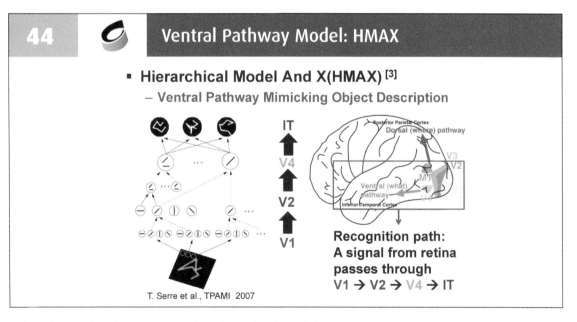

44 — Ventral Pathway Model: HMAX

- **Hierarchical Model And X(HMAX)** [3]
  - **Ventral Pathway Mimicking Object Description**

IT

V4

V2

V1

**Recognition path:
A signal from retina
passes through
V1 → V2 → V4 → IT**

T. Serre et al., TPAMI 2007

This shows the design of another hierarchical model, called HMAX [18]. It mimics the human vision system. The procedure begins from a simple convolutional cell, and then a complex cell to record pooling layer. Thus, simple cell try to extract features, while complex cell try to combine all the data inside as a simple one. And repeat and repeat again. The implementation requires a simple cell array and a complex cell array. The average recognition accuracy rate is 92%.

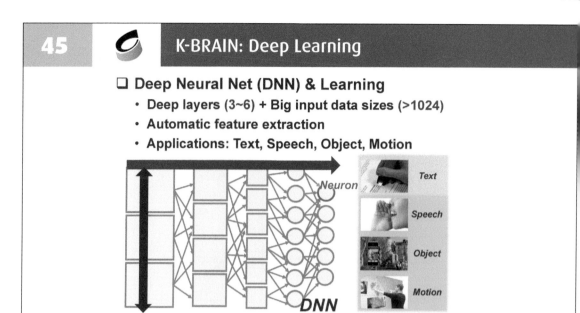

45 · **K-BRAIN: Deep Learning**

❑ **Deep Neural Net (DNN) & Learning**
- **Deep layers (3~6) + Big input data sizes (>1024)**
- **Automatic feature extraction**
- **Applications: Text, Speech, Object, Motion**

K-Brain is another embedded deep neural network [19]. As explained earlier, DNN may consist of more than 100 layers, with more than 1MGB input data. However, the number of layers in an embedded DNN is limited to roughly 6, with an input data size of about thousands. It can be used for text data extraction, recognition of speech, object and motion on mobile devices.

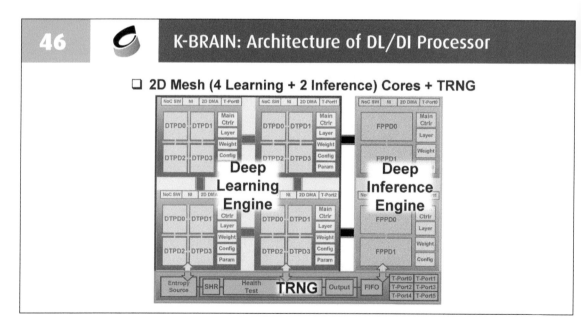

46 · **K-BRAIN: Architecture of DL/DI Processor**

❑ **2D Mesh (4 Learning + 2 Inference) Cores + TRNG**

This is the architecture of the entire design. A learning engine is implemented on chip, as well as a deep inference engine. The training of the network is realized in the learning engine, while recognition is configured in the inference engine. The internal of the DTPD consists of systolic inference, FSM and dynamic memory. Systolic inference is the convolutional multiplication processor.

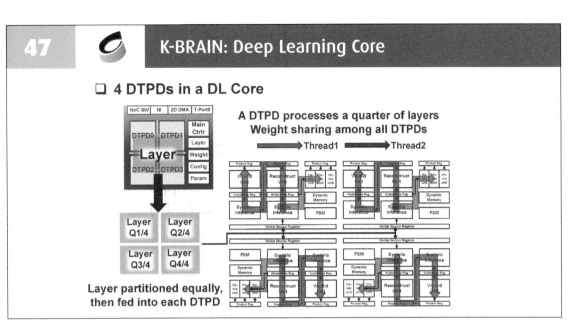

A multi-thread is used in K-Brain. The first thread and the second thread are organized as a pipeline system.

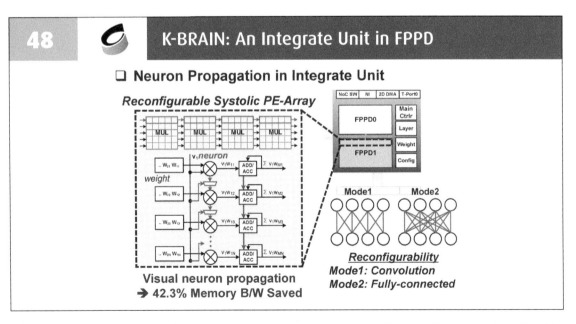

This is the input image and layer 1 to 3. As explained earlier, the output of the first layer is feed into the second layer, while the first layer is ready to receive the operation results from the third layer. The procedure is circulated continuously for higher accurate calculation.

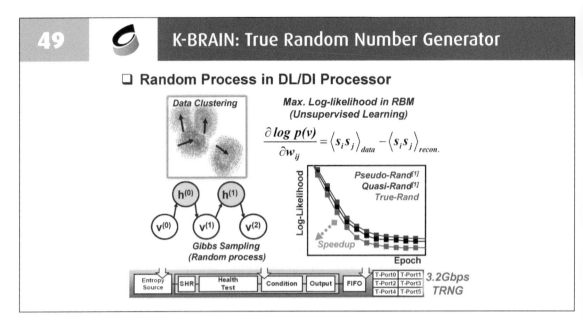

P runing is used for learning. A true random number generator is required for randomly neuron disable operation.

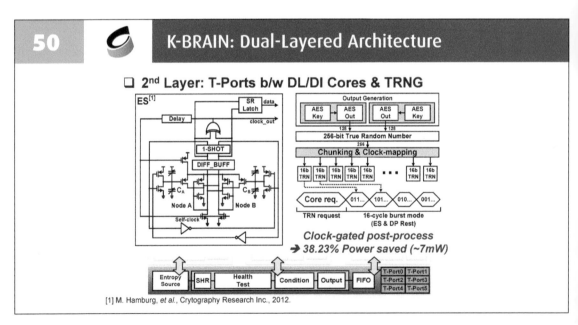

F lip-flop is used as shown in this slide. Two inverters are connected together. The randomness of the TRNG is improved by this proposed architecture, and was published on JSSC [20], [21]. This is a very important analog circuit for the accurate DNN.

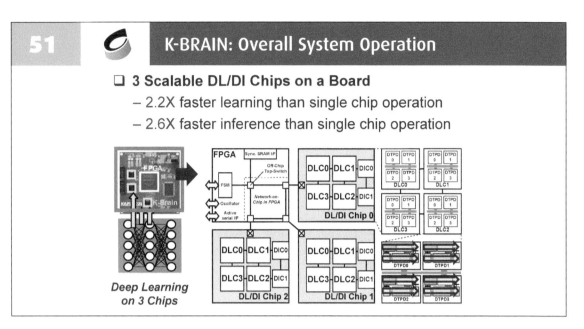

51   **K-BRAIN: Overall System Operation**

❑ **3 Scalable DL/DI Chips on a Board**
  – 2.2X faster learning than single chip operation
  – 2.6X faster inference than single chip operation

Deep Learning on 3 Chips

This is the system operation we integrated on a board. And if we put more chips a faster processing rate can be achieved.

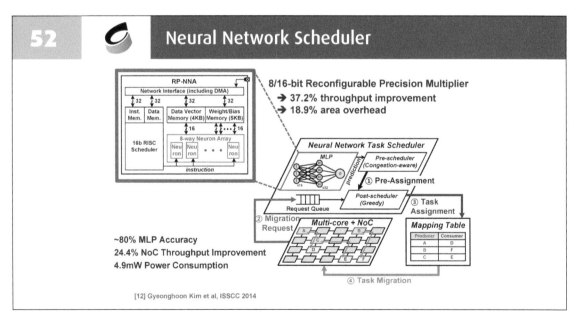

52   **Neural Network Scheduler**

8/16-bit Reconfigurable Precision Multiplier
➔ 37.2% throughput improvement
➔ 18.9% area overhead

~80% MLP Accuracy
24.4% NoC Throughput Improvement
4.9mW Power Consumption

[12] Gyeonghoon Kim et al, ISSCC 2014

A neural network scheduler is used to distribute the workload evenly to all the CPUs by DNN. DNN makes decision according to the status of the CPU and the historical record. In an 8-way neuron array, it monitors the status of each CPU through a queuing buffer. A task mapping table is used for the schedule of the CPUs.

53

M/E Deep Neural Networks

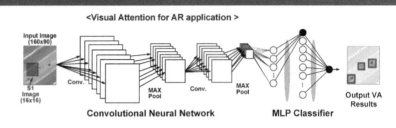

<Visual Attention for AR application >

Input Image (160x90)

S1 Image (16x16)

Conv.

MAX Pool

Conv.

MAX Pool

Convolutional Neural Network

MLP Classifier

Output VA Results

- ~4 CONV & POOL layers + 1 Fully-Connected Layer
- 50x100x1 Fully-Connected Layer
- Low Input Resolution: 160 x 90 / Fixed filter size: 5x5
- Trained with 113 Target Objects
- Latency = 3ms / image

[11] Injoon Hong et al, ASSCC 2015

The designed embedded DNN consisting of 4 convolutional layers and 1 fully connected layer. The resolution of the input image is 160x90. The filter size is 5x5. There are eight 5x5 PE arrays, each of which integrated 25 multipliers. Both local and global adders are needed for convolution operation. It is simple, but very useful for the enhancement of the total function of the chip.

2.4. Ultra-Low Power M/E-DNN

54

Low-power Face Recognition System

- **Conventional FR System**

Ext. CIS

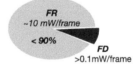

FR ~10 mW/frame

< 90%

FD >0.1mW/frame

- **Functional CIS + FR System**

Functional CIS

Face Detection

Transfer & Process **only detected face images**

With Ultra-low Power CNN Processor

[1] Kyeongryeol Bong et al, ISSCC 2017

In traditional face recognition system, external CIS and vision processor are used with very high power consumption. If the face detection algorithm can be integrated on the CIS with very low power consumption, and the face recognition integrated on ultra low power consumption CNN processor, a low power consumption face recognition system can be realized. In order to realize this concept, distributed SRAM and PE are used with tensor decomposition.

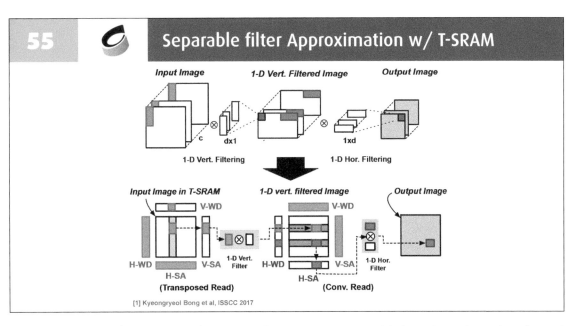

As discussed earlier, 2D convolution can be realized by using 2-step one dimensional filter using a transposed SRAM. In each SRAM cell, seven transistors are added. NTV is used to reduce the total power consumption [12].

2.5. General Purpose DNPU

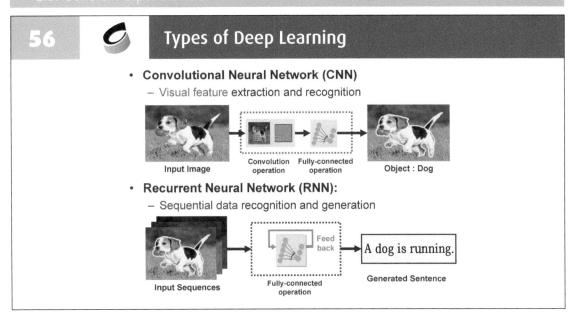

According to literature, CNN is very useful for image recognition on still image, while RNN is good for the sequential images. It can be imaged that a combination of CNN and RNN will be very useful for action recognition.

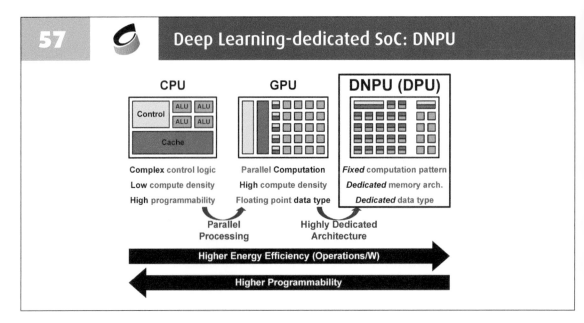

G eneral purpose GPU is very powerful, while NPU is a neural processor unit. We believe DNPU or DPU will be the most important chip set for the AI, because of the programmability.

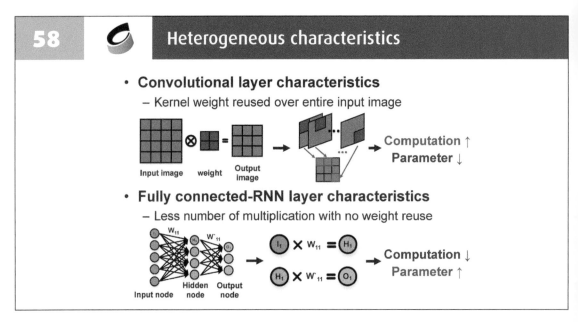

R NN is very complicated due to the required feed-back. However the similarity between the convolutional layer characteristics and fully connected-RNN layer characteristics makes it possible to share the hardware from each other to simplify the design. And the image data can be reused in the dynamic numerical system. The numerical accuracy can be reduced while the number of layers increases.

59 **Convolution Processor**

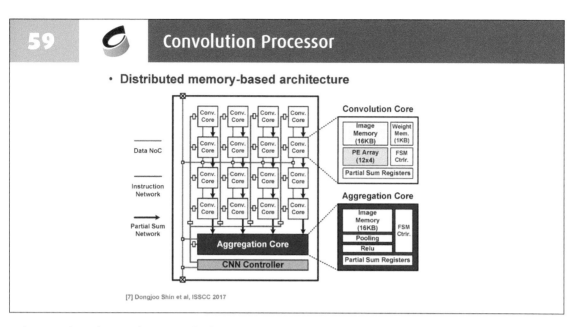

- **Distributed memory-based architecture**

[7] Dongjoo Shin et al, ISSCC 2017

This is the chip architecture [15], consisting convolution cores, aggregation core and CNN controller. The RNN processor and fully connected layer processor are merged to reduce the hardware complexity. The numerical system can be reconfigured layer by layer by changing the number of integer bits and fractional bits, according to different data distribution in layers. Higher precision is required by layers close to input.

60 **LUT-based Reconfigurable Multiplier**

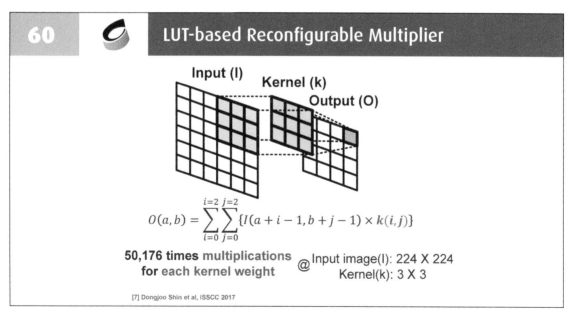

$$O(a,b) = \sum_{i=0}^{i=2}\sum_{j=0}^{j=2}\{I(a+i-1, b+j-1) \times k(i,j)\}$$

50,176 times multiplications for each kernel weight @ Input image(I): 224 X 224
Kernel(k): 3 X 3

[7] Dongjoo Shin et al, ISSCC 2017

LUT-based reconfigurable multiplier is used to reduce the complexity of the multiplication. Long number multiplication can be divided into sum of small number multiplications.

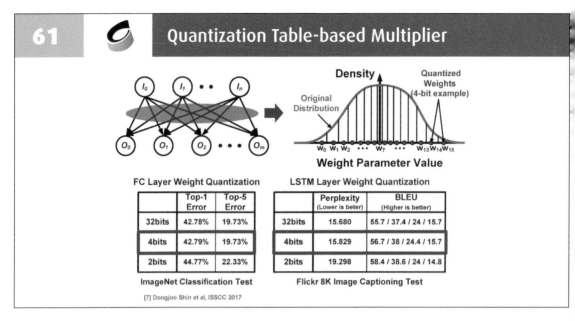

61 — Quantization Table-based Multiplier

	Top-1 Error	Top-5 Error
32bits	42.78%	19.73%
4bits	42.79%	19.73%
2bits	44.77%	22.33%

FC Layer Weight Quantization
ImageNet Classification Test

	Perplexity (Lower is beter)	BLEU (Higher is better)
32bits	15.680	55.7 / 37.4 / 24 / 15.7
4bits	15.829	56.7 / 38 / 24.4 / 15.7
2bits	19.298	58.4 / 38.6 / 24 / 14.8

LSTM Layer Weight Quantization
Flickr 8K Image Captioning Test

[7] Dongjoo Shin et al, ISSCC 2017

The weight parameter can be quantized and parameterized according to its distribution. Parameterized means a clear number of layers and numbers of steps. The multiplication table can be pre-prepared according to prediction, resulting a reduction of the complexity of the multiplication table.

62 — DNPU: Chip implementation

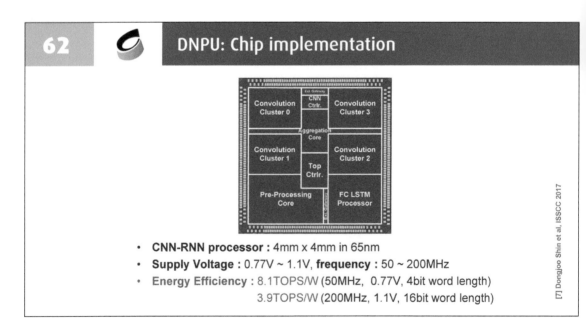

- **CNN-RNN processor :** 4mm x 4mm in 65nm
- **Supply Voltage :** 0.77V ~ 1.1V, **frequency :** 50 ~ 200MHz
- **Energy Efficiency :** 8.1TOPS/W (50MHz, 0.77V, 4bit word length)
  3.9TOPS/W (200MHz, 1.1V, 16bit word length)

[7] Dongjoo Shin et al, ISSCC 2017

The proposed work is fabricated in 65nm CMOS technology, occupying an area of 4mm x 4mm, with a supply voltage of 0.77V to 1.1V and a clock frequency of 50-200MHz. The proposed chip was implemented on a robot for object recognition [22,37].

3. ATTENTION BASED AI PROCESSOR

3.1. SALLENCY BASED ATTENTION

63

There are different methods to integrated a DNN on Von Neumann architecture, i.e. serially, or parallel, or embed one into another. The DNN processor can access the external environment through the Von Neumann or with a separated interface. Object recognition is one of the widely used application. For human, it consists of comparison, matching with acknowledge inside of the brain.

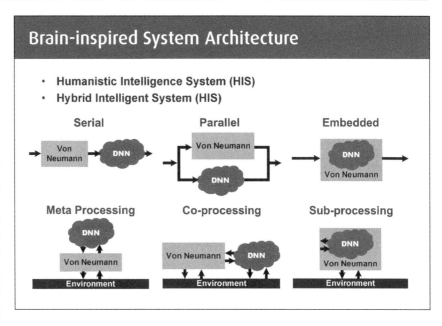

Brain-inspired System Architecture

- **Humanistic Intelligence System (HIS)**
- **Hybrid Intelligent System (HIS)**

Serial | Parallel | Embedded

Meta Processing | Co-processing | Sub-processing

64

In object recognition, the feature of the input is compared with what in the database. Even the feature is clear, the procedure of feature extraction requires heavy computation. It takes about 0.5s for a 2.3GHz CPU to extract one feature with high power consumption. It is even worst, if the feature is not clear, since DNN is applied. Feature extraction algorithm, i.e. Gaussian space generation, is widely used.

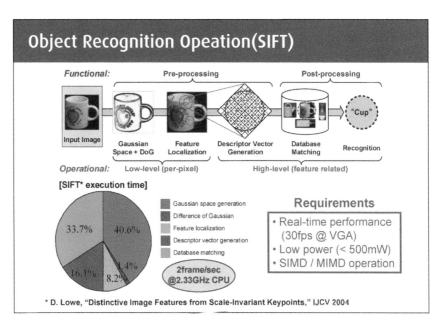

Object Recognition Opeation(SIFT)

Functional: Pre-processing / Post-processing

Input Image | Gaussian Space + DoG | Feature Localization | Descriptor Vector Generation | Database Matching | Recognition | "Cup"

Operational: Low-level (per-pixel) | High-level (feature related)

[SIFT* execution time]

- Gaussian space generation
- Difference of Gaussian
- Feature localization
- Descriptor vector generation
- Database matching

40.6% 33.7% 16.1% 8.2% 1.4%

2frame/sec @2.33GHz CPU

Requirements
- Real-time performance (30fps @ VGA)
- Low power (< 500mW)
- SIMD / MIMD operation

* D. Lowe, "Distinctive Image Features from Scale-Invariant Keypoints," IJCV 2004

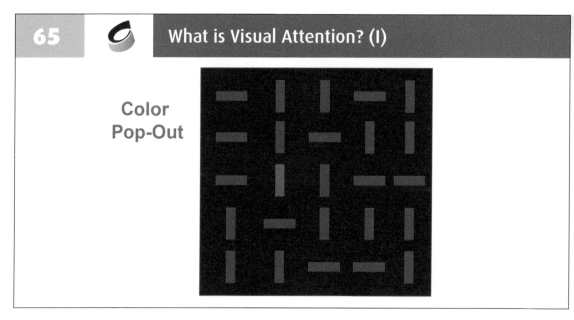

What is the visual attention. For example, whenever you watch this slide, your attention is given to "why color is different".

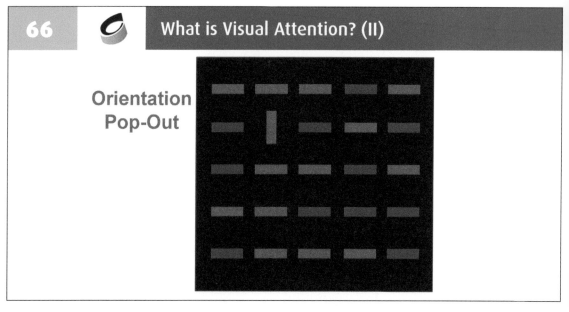

And in this slide, your attention is given to the vertical oriented bar as "why orientation is different".

67   What is Visual Attention? (III)

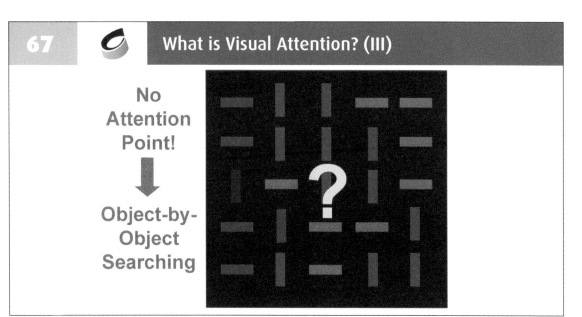

How about this? You have to check one by one. Finally you realize, this one is different. So our human brain. If any object is different from the surrounding object, our attention automatically goes to that object. But computers can not do that. Thus, there is visual attention in human beings.

68   BONE-V2: Visual Attention

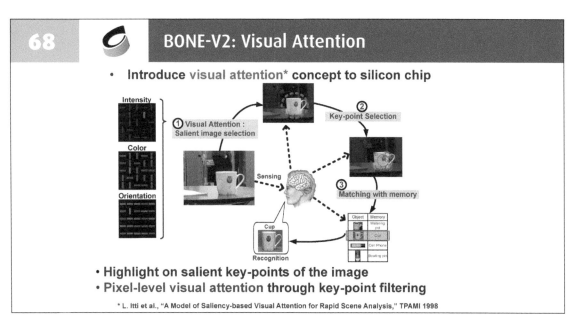

According to visual attention, the most important object, or to say apparent object is extracted first in recognition to speed up the recognition process, as well as to reduce power consumption. For example, for human being, the cup, the salient object is the most important in the picture. However, it is challenge for machine to recognize this. The attention stage is inserted right after the reading of the input image, before the stage of filtering, feature extraction, feature matching and the pop out of the results.

69 · BONE-V2: SIFT + Visual Attention

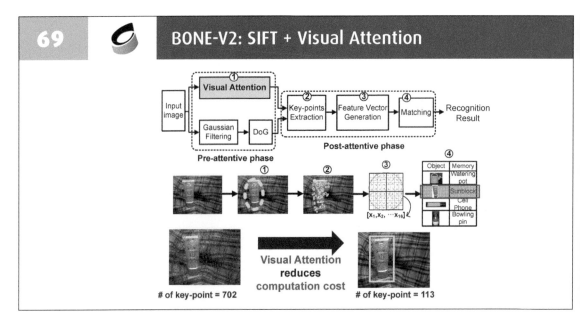

The procedure is divided into 3 steps: multi-scale feature extraction, global attention map, and attention selection. A DNN is used as a global filter in the second step to extract the most salient object. The number of key-point is greatly reduced by applying visual attention. A higher speed and a lower power can be expected.

70 · BONE-V2: Generic Architecture Model

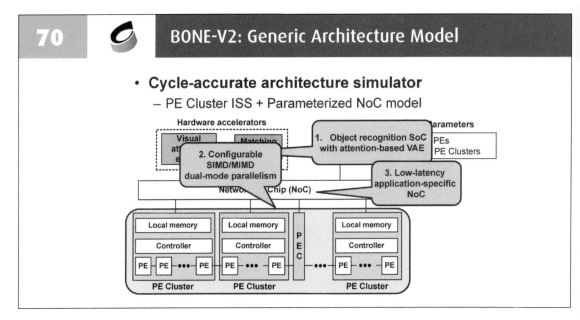

The proposed chip consists of 1) a visual attention engine, 2) a matching acceleration which is to calculate the distance between the current image and image saved inside our database, 3) a many-core PE cluster to perform key-point extraction and vectorization. The object recognition is realized with attention base VAE, a configurable SIMD/MIMD. SIMD is dummy CPUs, receiving the same instruction and operating in the same way, while each CPU in MIMD receives different instructions and operates differently. The arrangement of NoCs is same as the CPUs.

71

In pattern recognition, matching performed in the specified processor, visual attention performed in separate blocks and Gaussian scale space calculation performed in the many-core processor consume the most power. and attention is operated done by separate blocks. The bottleneck can be solved by separate the blocks. For the visual attention, a salience map is form according to features, i.e. color intensity or orientation. CNN is used to calculate the extracted features of color, orientation, intensity, as well as to combine the saliency. SIMD is useful at the beginning stage since same operation is applied to all the pixels, while MIMD is better at the later stage for pattern recognition.

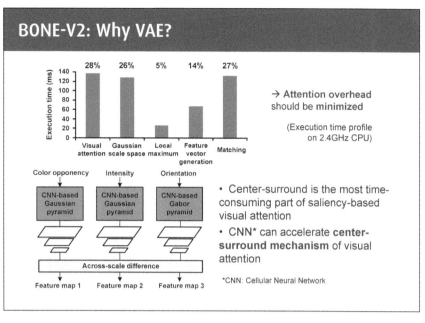

BONE-V2: Why VAE?

→ **Attention overhead** should be **minimized**

(Execution time profile on 2.4GHz CPU)

- Center-surround is the most time-consuming part of saliency-based visual attention
- CNN* can accelerate **center-surround mechanism** of visual attention

*CNN: Cellular Neural Network

72

Reconfigurability between SIMD and MIMD processors is required, where NoC plays an important role, since NoC enables an easy implementation of broadcasting network or unicasting. There are many considerations in the implementation of a NoC protocol, such as switching scheme, routing scheme, queuing scheme. According to the analysis on requirement of the proposed application, wormhole switching, deterministic routing and input queuing are used due to the size efficiency.

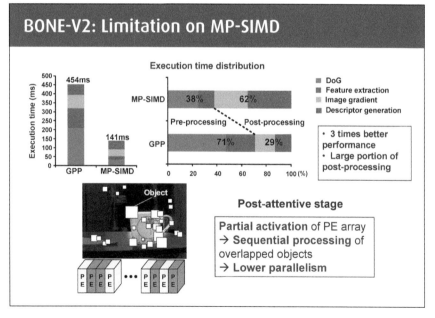

BONE-V2: Limitation on MP-SIMD

- 3 times better performance
- Large portion of post-processing

Post-attentive stage

Partial activation of PE array
→ **Sequential processing** of overlapped objects
→ **Lower parallelism**

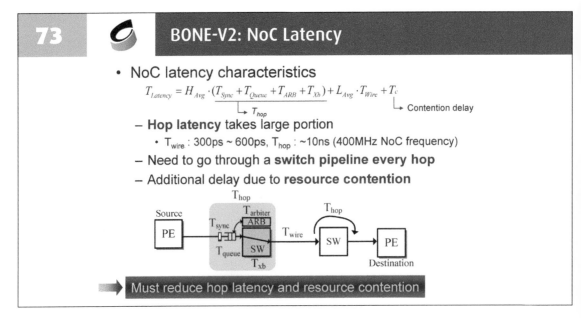

op latency takes most of the operation time. Each return packet features a burst length of about 5, which is the most time hungry. An express channel is introduced to reduce the hop latency. A simple bus is added parallel to the system bus with adaptive circuit and packet switching modules. Switching is applied when a return packet is detected.

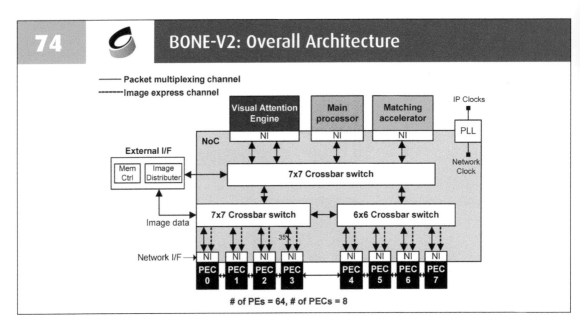

This is the architecture of the proposed chip [23]. It consists of a PE cluster of 8 PEs and crossbar switch. It was fabricated in 0.13um 8M CMOS technology, occupying 6mm x 6mm, with 2M logic gates, consuming 583mW. It was integrated in a robot for target recognition.

3.2. MULTI-OBJECT ATTENTION

There are lots of information inside of human brain, but no attention is paid to. The information will pop out when it is triggered. It likes a spotlight. Human being can give attention to three object simultaneously, but there is no such limitation on a silicon chip. However, for multiple attentions, attention shape maybe different. Traditional method perform the object serially.

75

BONE-V3: Motivation

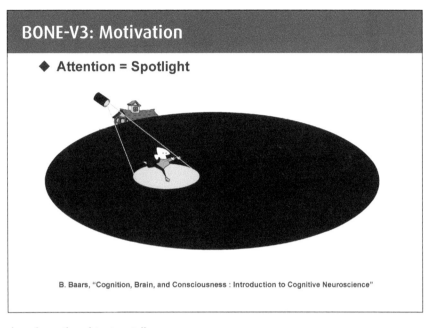

◆ **Attention = Spotlight**

B. Baars, "Cognition, Brain, and Consciousness : Introduction to Cognitive Neuroscience"

In order to realize a parallel recognition, a precise boundary detection is required, which is easy for human beings, but challenge to computer especially there is an overlap. The number of cores, process modules are carefully studied for the NoC. Task scheduling and IP communication protocol is customized. The PEs are grouped according to SIMD or MIMD ways. An optimization is achieved using 8 SIMD/MIMD ways x 16 cores.

76

BONE-V3: Object Detection Based Approach

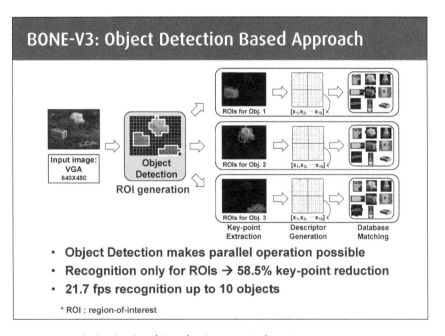

- **Object Detection makes parallel operation possible**
- **Recognition only for ROIs → 58.5% key-point reduction**
- **21.7 fps recognition up to 10 objects**

* ROI : region-of-interest

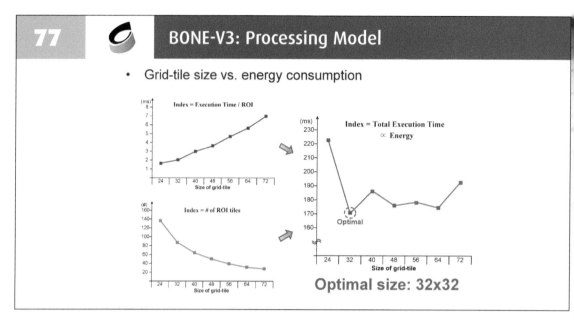

In addition to pixel wise calculation, grid-tile wise calculation is also used. An optimization tile size of 32 x 32 is derived.

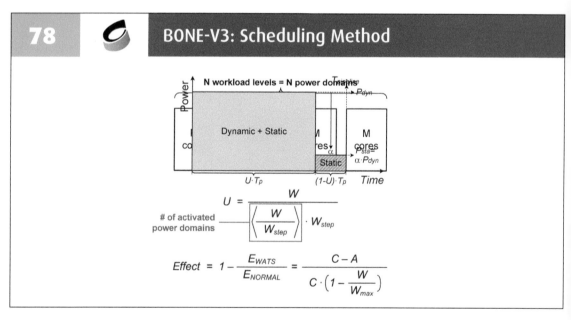

The scheduling method is adaptive. The number of activation is relative to W/W_step, which is the number of activated PEs. A higher effect can be expected, if more PE is activated.

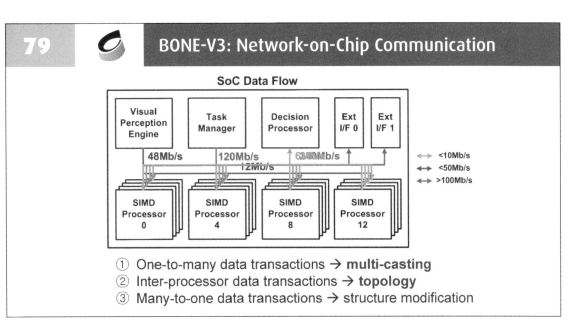

79 · **BONE-V3: Network-on-Chip Communication**

SoC Data Flow

① One-to-many data transactions → **multi-casting**
② Inter-processor data transactions → **topology**
③ Many-to-one data transactions → structure modification

The communication of the design is very complicated. An one-to-N communication is required, since the output of the attention engine is send to all the remaining SIMD. In addition, N-to-N and N-to-one communication are also needed. The h-star ring is the best solution to the requirement. Since this is a NoC, the header address and data from different packet are well synchronized. Thus, Synchronize multicasting is possible.

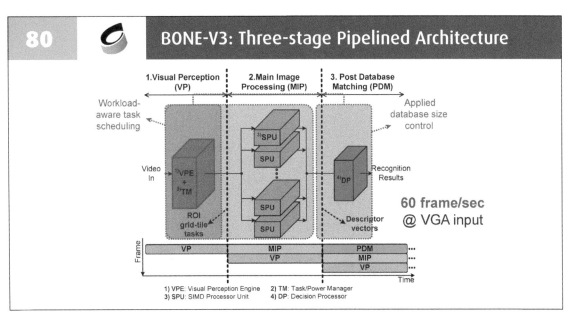

80 · **BONE-V3: Three-stage Pipelined Architecture**

1) VPE: Visual Perception Engine   2) TM: Task/Power Manager
3) SPU: SIMD Processor Unit   4) DP: Decision Processor

Pipeline is also possible due to fact that the time consumed by VPE, TM and MIP are almost the same. So we can make pipeline. The number of SPUs can be changed according to different workload. The balance of the pipeline can be adjusted by changing the frequency of the SPUs.

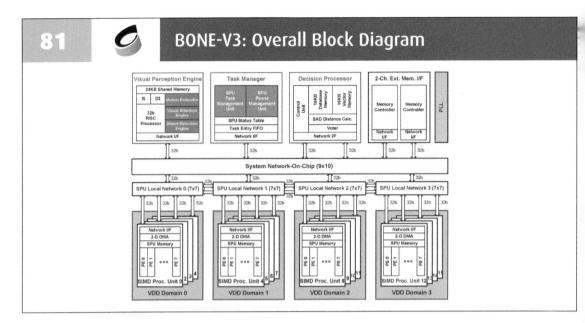

This is the system architecture? The VPE consists of motion estimator, visual attention engine and object detection engine. The chip was fabricated in 0.13um 8M CMOS technology, occupying 7mm x 7mm [24]. It integrated almost 4M gate, bigger than the previous one. The proposed chip was integrated in a Hyundai car for sign recognition about 10 years ago.

3.3. FAMILIARITY BASED ATTENTION

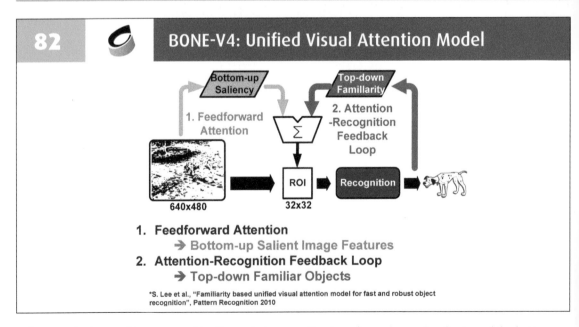

Human brain enables a combination of two attentions for object recognition: bottom-up attention and top-down attention. Top-down attention depends on familiarity, while bottom-up attention is related to saliency attention.

83　BONE-V4: Processing Element Design

- **Function-specific IPs**

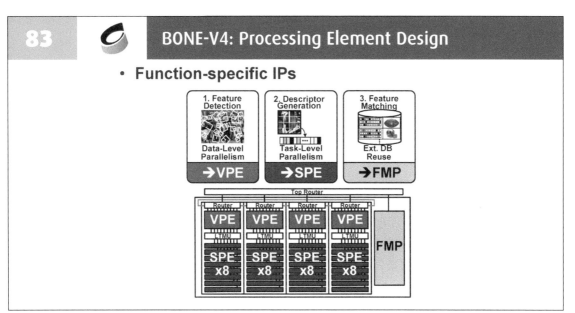

This slide shows the design of the PE. It consists of 1) the VPE, which is an attention engine, 2) the SPE, which is the description generation, and 3) the FMP, which is the feature matching. The three parts are organized in a pipeline.

84　BONE-V4: Chip Block Diagram

- **Cognitive Control Layer: Attention, Power, Scheduling**
- **Parallel Processing Layer: Recognition**
- **51 IPs Connected by Hierarchical Star + Ring NoC**

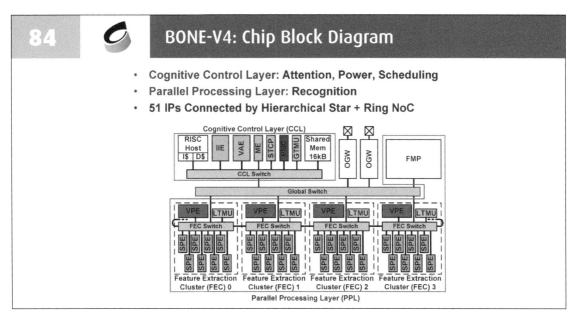

A new architecture is proposed in this design [25,26]. There are two layers. The top layer is the cognitive control layer, including the processor, the inference engine, the attention engine, the power management and the global switch unit. The bottom layer is a parallel processing layer. The bottom layer is a slave layer controlled by the top intelligent layer.

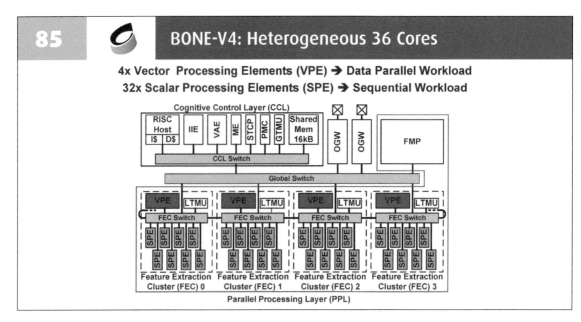

85 BONE-V4: Heterogeneous 36 Cores

4x Vector Processing Elements (VPE) ➔ Data Parallel Workload
32x Scalar Processing Elements (SPE) ➔ Sequential Workload

4 vector parallel processors and 32 scalar processing engines are integrated together. The familiarity decision is very difficult to quantized, but must be turned into quantitative terms in computers. Fuzzy logic is used as the intelligent inference engine.

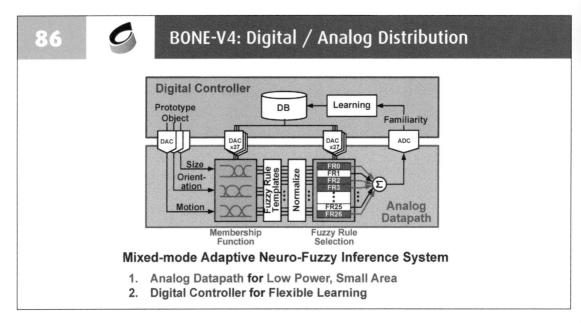

86 BONE-V4: Digital / Analog Distribution

Mixed-mode Adaptive Neuro-Fuzzy Inference System

1. **Analog Datapath for Low Power, Small Area**
2. **Digital Controller for Flexible Learning**

The fuzzy logic is implemented in mixed-mode circuits. Digital controller is used for learning, while the fuzzy module is analog. The membership function is sort of Gaussian function integrated using two differential analog circuits. The fuzzy rule selection is also analog with current mode modification. ADC and DAC connect the digital and analog circuits.

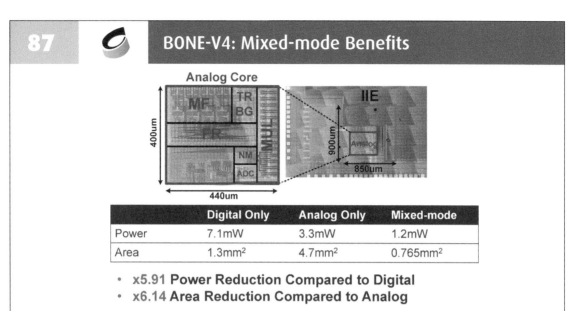

87 — BONE-V4: Mixed-mode Benefits

	Digital Only	Analog Only	Mixed-mode
Power	7.1mW	3.3mW	1.2mW
Area	1.3mm²	4.7mm²	0.765mm²

- **x5.91 Power Reduction Compared to Digital**
- **x6.14 Area Reduction Compared to Analog**

A comparison is performed among the digital only mode, the analog only mode and the mixed-mode, a lowest power consumption is observed in the mixed-mode circuits as well as the silicon area consumption.

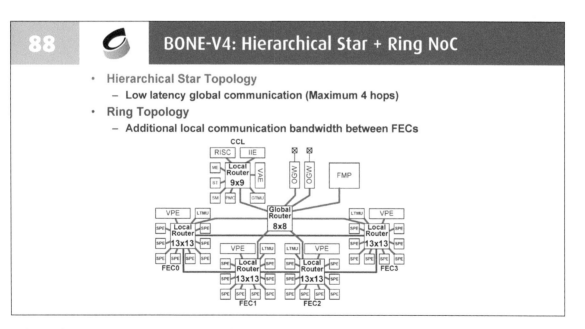

88 — BONE-V4: Hierarchical Star + Ring NoC

- **Hierarchical Star Topology**
  - **Low latency global communication (Maximum 4 hops)**
- **Ring Topology**
  - **Additional local communication bandwidth between FECs**

This is the star topology of the NoC architecture. It's almost the same as the earlier work, but the global router and the local router are different. There are 5 local routers, one of which is different from earlier design. A tradeoff is performed between synchronize slot and pipeline structure.

89 — BONE-V4: Execution Time Equalization

$$t_{FeatureExtraction} = t_{Computation} + t_{Communication}$$

$$\underset{\alpha+N_F\beta}{} \qquad \underset{\gamma(\delta+N_F\varepsilon)}{}$$

1. **Predict Number of Features** N_F
   - **Using heuristics on input data**
   - **Neuro-Fuzzy logic inference**

32x32 Pixel Tile

2. **Regulate Congestion Rate** γ
   - **Control links between FECs and OGWs**
   - **Weighted Round-Robin Arbitration**

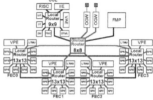

The synchronization of the threads time is realized by changing the Nf, the number of features, and Gamma, the congest rate. The number of the workload can be predicted. Neuro-fuzzy logic can predict the parameters precisely.

90 — BONE-V4: Workload Prediction Heuristics

- **Number of Features** N_F **depends on tile content**
  Complex textures → High N_F
  Smooth surfaces → Low N_F
- **3 Heuristics used as inputs for neuro-fuzzy inference**

32x32 Pixel Tile

1. Intensity Variance
$$\sum(I-E)^2$$

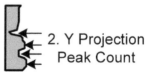

2. Y Projection Peak Count

3. X Projection Peak Count

How we can make prediction? First, the variance of the image is calculated. The variance of each pixel is compared to the average. The slope of the input image is calculated for both X and Y directions. The number of peaks are counted. Based on the above estimations, the workload of each pixel tile can be predicted in the fuzzy logic. The predicted workload well matched the real case.

91 BONE-V4: Weighted Round Robin Packet Arbitration

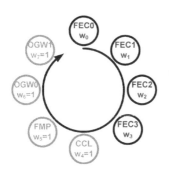

Ownership of output port granted to next ready input port

Number of packets granted per turn = w_n (0~7)

- **Satisfies fairness criterion**
- **Ports with higher weight are granted higher bandwidth ➔ reduced congestion rate γ for those ports**

For the routing algorithm, instead of the round robin arbitration, the weighted round robin arbitration is used to balance the packet throughput.

92 BONE-V4: Chip Implementation

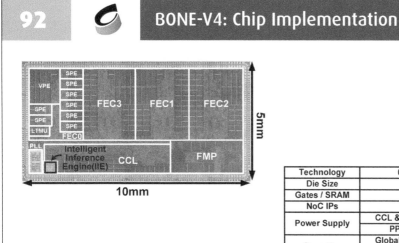

Technology	0.13um 1P8M Logic CMOS	
Die Size	50mm² 10.0mm x 5.0mm	
Gates / SRAM	2.92M Gates / 612 kB	
NoC IPs	51	
Power Supply	CCL & NoC	1.2 V
	PPL	0.65 ~ 1.2 V
Operating Frequency	Global NoC	400MHz (45FO4)
	CCL	200MHz (90FO4)
	PPL	50 ~ 200MHz (90FO4)

The chip was fabricated in 0.13um 1P8M logic CMOS technology. It is a mixed-mode circuit with intelligent inference engine. The proposed chip was assembled as an augmented reality HMD with a power consumption of less than 600mW. It was published at ISSCC 2010 [25].In a later version, the power consumption is reduced to 270mW after training.

3.4. TEMPORAL FAMILIARITY

 93 **BONE-V5: Dynamic Object Recognition**

◆ **Unmanned Aerial Vehicle (UAV) System**

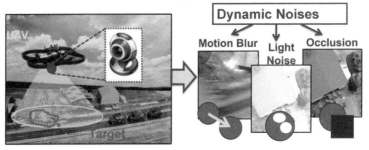

Dynamic Noises in Captured Image
- ☹ Degradation of Recognition Accuracy
- ☹ Degradation of Processing Speed

Drone is very popular recently. It is actually vibrating severely. The picture taken from drone is usually blurred. Or the target object is covered by other objects. An object recognition robust to blur, noise or coverage is expected. Human eyes are sensitive to temporal differences. Temporal familiarity rather than object familiarity can be incorporated for attention.

 94 **BONE-V5: Top Architecture**

◆ **High Performance & Energy-Efficient Processor**

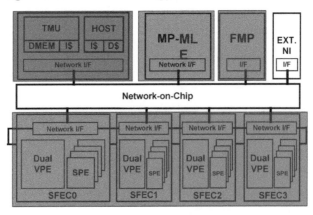

In temporal attention, according to hysteretic results and current result, a prediction can be performed. The difference between the prediction and reality is calculated and converted into temporal familiarity. This is the architecture of the proposed chip [27,28], consisting of vector PE, SPE, task management unit, motion prediction and mapping matching processor.

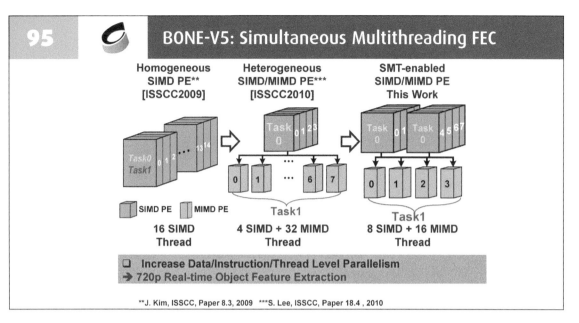

**J. Kim, ISSCC, Paper 8.3, 2009   ***S. Lee, ISSCC, Paper 18.4 , 2010

Compare to previous BONE chips introduced earlier, there are less MIMDs, but more SIMDs. Pipeline is applied for the dual thread vector PE (DVPE), enabling Gaussian filtering, the difference-of-Gaussian, local MIN/MAX extraction, feature extraction and feature matching. There are 3 stages for DVPE with one extra stage for SPE, and one extra stage for FPM.

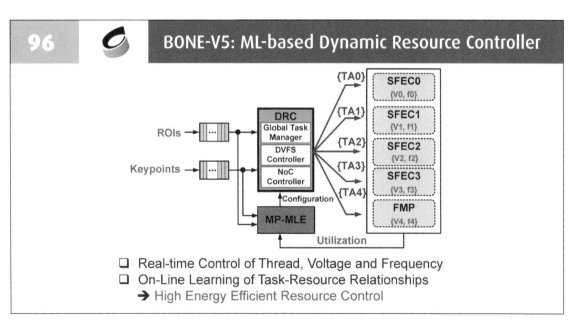

A dynamic resource controller is used to monitor parameters for all the IPs, i.e. the voltage, the frequency. The voltage and frequency can be changed based on different tasks performed in each thread or each IP, according to preset thresholds. More power is required while the workload is increasing. The utilization depends on the threads control. A temporal familiarity based thread control is applied.

97 BONE-V5: Chip Implementation

The chip was fabricated in 0.13um technology [27], occupying 4mm x 8mm. According to experimental results, the power consumption during operation with on-chip control or pre-programmed control is similar. An average power consumption of 320mW and 9.6 mJ/frame is achieved. The proposed chip was

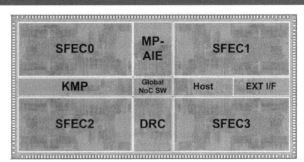

SFEC0	MP-AIE	SFEC1	
KMP	Global NoC SW	Host	EXT I/F
SFEC2	DRC	SFEC3	

- ❏ *4mm x 8mm* by 0.13mm CMOS Technology
- ❏ *342GOPS* Peak Performance
- ❏ *320mW* Average/ *560mW* Peak Power Consumption
- ❏ *9.6mJ/frame, 10.5nJ/pixel Energy Efficiency* for 720p

implemented in an airplane, realizing deblurring, tracking and recognition of overlaid object.

4. AUGMENTED REALITY AI SOC

4.1. MARKERLESS AUGMENTED REALITY

98 BONE-AR: Augmented Reality

AR and VR are popular nowadays. It is a combination between real world environment and 3-D computer graphic. But most of the AR applications requires customized SoCs, which is unfortunately not available in the market today. The implementation methods of AR can be divided into two categories: marked or markerless. Marked AR, i.e. GPS, Pokeman Go, implements object recognition based on the marker or bar code. It is not good at the recognition of orientation. On the other hand, markerless AR is a real AR, robust to rotation or changing of view point. Pattern recognition, 3D graphics and view point information are required for AR implementation.

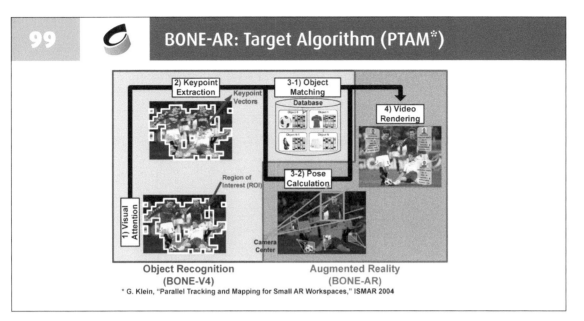

We proposed a target algorithm consisting of object key point extraction, object recognition, 3D rendering, and pose estimation. Pose estimation is an angle detection of the camera. The object recognition processor is integrated with the AR processor. Vision attention is applied to detect the most salient object. Then the key points are extracted for object matching. The camera angle is calculated for the generation of the computer graphics based on available information of the target after object recognition.

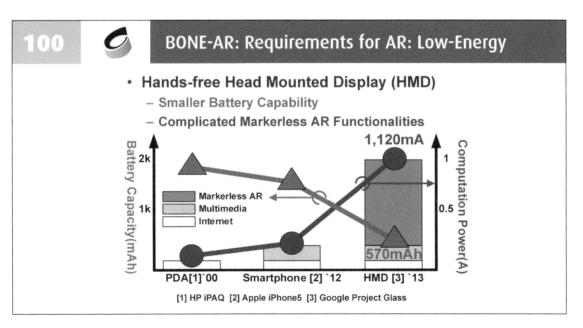

Different from marked AR, markerless AR requires natural feature extraction with much higher computational complexity than marked AR, which lays a great challenge for hardware implementation on a single chip. In addition, consider the application of AR, i.e. Google glasses, the power budget is very limited.

101　　BONE-AR: Scale-space Cluster

❑ 256-way Massively Parallel for Fast Image Filtering

To 2D-mesh NoC Grid
38b x2

Network Interface w/ 2D DMA

32b

Host Controller

| 3-stage μ-controller |
| IMEM |
| DMEM |
| Config. Reg. |

Wide System Bus

128b
128b
128b

Convolution Accelerator

256-way Massively Parallel PE Array

2D Pixel Cache

| Pre-fetcher | 2D Pixel Memory |

→ 61.25 GMAC/sec sustained @ 250 MHz
→ 2.5x speedup compared to BONE-V6 (2013)

This is the architecture of our chip [29,30]. It consists of 1) the visual attention engine to detect the most salient part, 2) the scale space based processor to deal with pictures with different sizes and angles,

3) the key point detector and matching for object recognition and pose estimation for 3D graphic generation, 4) descriptor generation or vectorization.

102　　BONE-AR: Descriptor Generation Cluster

❑ 16-way 16b Vector ALU with 256b SIMD Memory

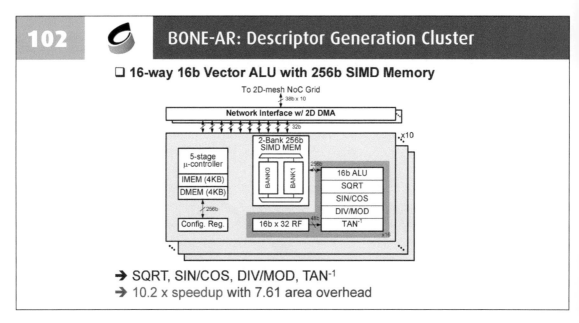

To 2D-mesh NoC Grid
38b x 10

Network Interface w/ 2D DMA

32b

x10

| 5-stage μ-controller |
| IMEM (4KB) |
| DMEM (4KB) |
| Config. Reg. |

2-Bank 256b SIMD MEM

BANK0　BANK1

256b

| 16b ALU |
| SQRT |
| SIN/COS |
| DIV/MOD |
| TAN⁻¹ |

256b

16b x 32 RF

48b

x16

→ SQRT, SIN/COS, DIV/MOD, TAN^{-1}
→ 10.2 x speedup with 7.61 area overhead

The scale space cluster consists of a convolution accelerator, an instruction memory and data memory. It is a 16-way, 16-bit vector ALU with 256b massively parallel architecture, and 256-way

massively parallel PLA. It features 61.25 GMAC/sec sustained operation at 250MHz. The processing is sped up by 2.5 times compared to earlier BONE versions.

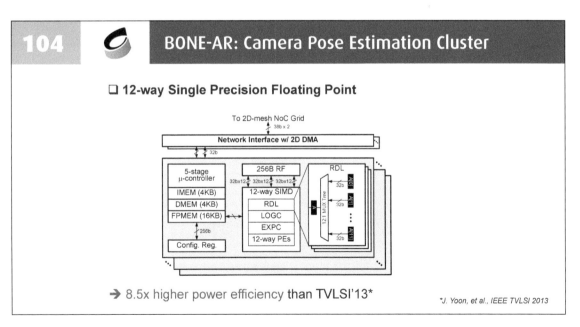

103 BONE-AR: Keypoint Detection Cluster

❑ 32-way 8b Vector ALU with 256b SIMD Memory

→ 108 msec for one image tile (32 x 32) detection
→ 1.7x speedup compared to BONE-V6 (2013)

According to trade-off analysis, 4 bank 256 bit SIMD memory is used for the 32 way 8 bit vector ALU. It takes 108ms to perform the process on a single image tile, which is 1.7 times faster than earlier version. Operations, i.e. square root, sine, cosine, division and modulus, are performed in customized modules. Vocabulary forest with 4 vocabulary trees is used for matching. 4 tree searching are performed simultaneously organized by a tree controller. The proposed chip features a 1.26 higher throughput and a 74% energy reduction.

104 BONE-AR: Camera Pose Estimation Cluster

❑ 12-way Single Precision Floating Point

→ 8.5x higher power efficiency than TVLSI'13*

*J. Yoon, et al., IEEE TVLSI 2013

Pose estimation requires log or exponential operations, which is performed in customized single precision floating point circuit in the proposed design. It features an 8.5 times higher power efficiency than ever reported.

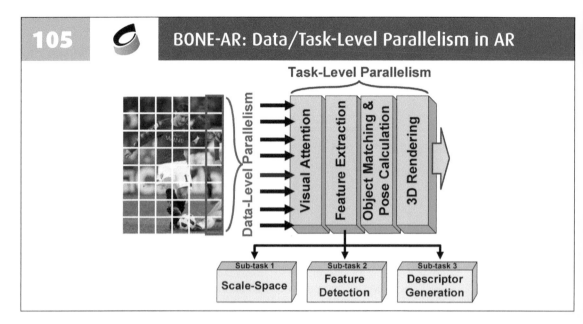

In the design, visual attention, feature extraction, object matching, pose calculation and 3D rendering are performed independently to each pixel in the pipeline stage. It is a data level parallelism. In addition, all the tasks are divided into three sub tasks, scale space formation, feature detection and descriptor generation, which realize a pipeline of task level parallelism. By doing this, the pipeline utilization is improved by 2.75 times.

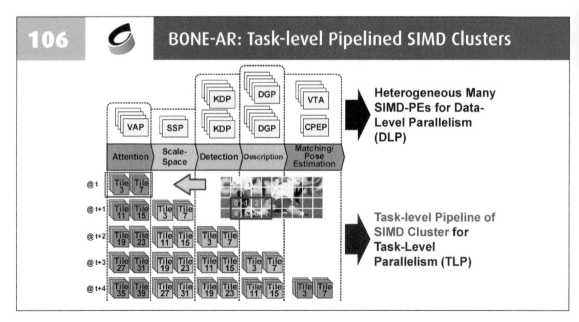

For example, tile 3 and 7 first were fed into the attention engine and then the scale space. When tile 11 and 15 were read in, another pipeline was used, while tile 3 and 7 enter the key point detector. The size of the pipeline is designed as eight by eight by eight. Thus, six tiles, tile 3, 7, 11, 15, 19, 23 can follow the same pipeline stage. This scheme is denoted as task level pipeline.

107 BONE-AR: Chip Micrograph

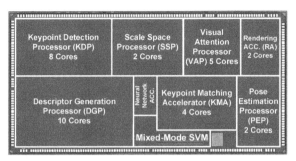

- **4mm x 8mm by 65nm CMOS Technology**
- **1.22*TOPS* Peak Performance**
- **381*mW* Average / 778*mW* Peak Power Consumption**
- **1.57*TOPS/W* Energy Efficiency for 720p Video**

This is the chip photo of the proposed design [29]. The design was fabricated in 65nm technology, occupying an area of 4mm x 8mm, featuring a peak performance of 1.22TOPS, an average power of 381mW, a 1.57TOPS/W for the 720p video input. The proposed chip was assembled in a K-Glass, demonstrating smart shopping in bookstore or Lego shop.

4.2. AUGMENTED REALITY WITH EYE MOUSE

108 BONE-V8: Proposed "EyeClick" Smart Glass

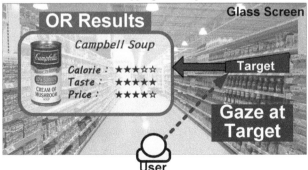

Smart glass, i.e. Google Glass, require voice comments to trigger a process, i.e. take a picture. Everyone surround you can hear you, which is sometime not very convenient. Gesture recognition has been used as an input, which is better than voice, but still way to go. A new intention-concealed UI rather than the speech and gesture UI is needed very much. Gaze UI is one option, which automatically pop out information on the target object the user starred at.

109 — BONE-V8: Gaze User Interface

Step1. "Point" the Cursor by "Gaze"

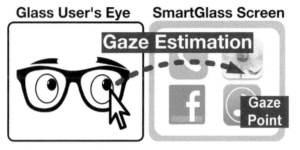

Glass User's Eye   **SmartGlass Screen**

Gaze Estimation

Gaze Point

→ **Low-power(<50mW [1]) Always-On Gaze UI**
- **Previous Work [2] : 100mW**

[1] Power Consumption of Wireless Optical Mouse Reported in "Mouse Battery Life Comparison Test Report", Microsoft Corporation, 2004
[2] D. Kim, et al., "A 5000S/s Single-chip Smart Eye-tracking Sensor", ISSCC 2008

An eye mouse is required for the implementation of Gaze UI, which can trigger and control the movement of a cursor. Click is realized by winking eyes. The privacy is better for sure. But the power consumption will be higher, since the camera is always on. In addition, the error on eyeball detection must be less than a pixel. This is a challenge to the designer.

110 — BONE-V8: Overall System Architecture

1. 320x240 Gaze Image Sensor (GIS)
2. Multi-core OR Processor (ORP)

Gaze Image Sensor

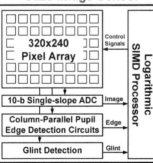

320x240 Pixel Array

Control Signals

Logarithmic SIMD Processor

10-b Single-slope ADC — Image

Column-Parallel Pupil Edge Detection Circuits — Edge

Glint Detection — Glint

Object Recognition Processor

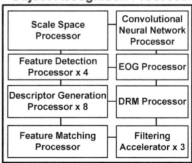

Scale Space Processor	Convolutional Neural Network Processor
Feature Detection Processor x 4	EOG Processor
Descriptor Generation Processor x 8	DRM Processor
Feature Matching Processor	Filtering Accelerator x 3

Two chips were designed for BONE-V8. One is object recognition processor, consisting of scale space processor, feature detection descriptor generation, feature matching and convolutional neural network processor. It can recognize the difference between unintentional blinks and intentional winks. The other chip is a gaze image sensor. It is CMOS image sensor with computational analog circuits, enabling pupil boundary detection and orientation detection of the gaze.

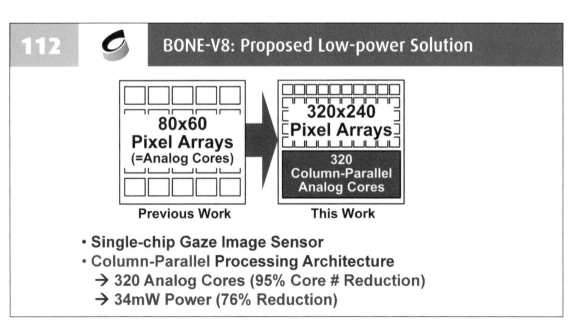

111   **BONE-V8: Pupil Edge Algorithm : XY-PD**

Iris

Pupil

Find Pupil Edge Along X-Axis

Find Pupil Edge Along Y-Axis
<Proposed XY-PD Algorithm>

· **Intensity of Pupil (I_p) is Darker than Iris (I_r)**
→ I_p-I_r < Threshold, I_p is Pupil Edge

· **XY-PD = Find Pupil Edge along X/Y Axis**
→ **Efficient Processing along X/Y axis of Image Sensor**

The boundary of the iris is first detected according to the vertical direction and horizontal abrupt intensity change. It is calculated on the CMOS image sensor at pixel level with an analog pupil edge detection circuit (PEDC). The PEDC enable edge detection for different directions.

112   **BONE-V8: Proposed Low-power Solution**

80x60 Pixel Arrays (=Analog Cores)

320x240 Pixel Arrays

320 Column-Parallel Analog Cores

Previous Work   **This Work**

• **Single-chip Gaze Image Sensor**
• **Column-Parallel Processing Architecture**
    → **320 Analog Cores (95% Core # Reduction)**
    → **34mW Power (76% Reduction)**

An 80x60 pixel array is used with pixel level analog cores has proposed in literature [31] with high power consumption and low fill factor. We proposed a column parallel analog core with few extra transistors added in each pixel, and low power consumption.

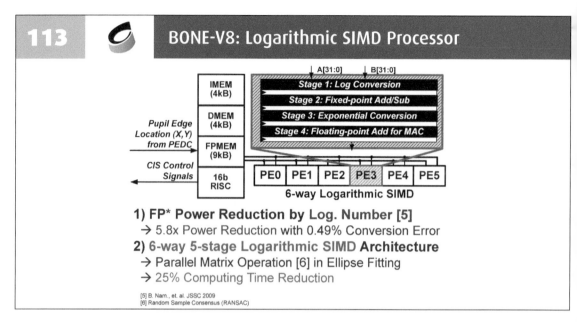

113 — **BONE-V8: Logarithmic SIMD Processor**

1) FP* Power Reduction by Log. Number [5]
→ 5.8x Power Reduction with 0.49% Conversion Error
2) 6-way 5-stage Logarithmic SIMD Architecture
→ Parallel Matrix Operation [6] in Ellipse Fitting
→ 25% Computing Time Reduction

[5] B. Nam., et. al. JSSC 2009
[6] Random Sample Consensus (RANSAC)

Logarithmic SIMD processor is used for pupil calculation. The required multiplication and division operations can be easily realized with low hardware expense and power consumption.

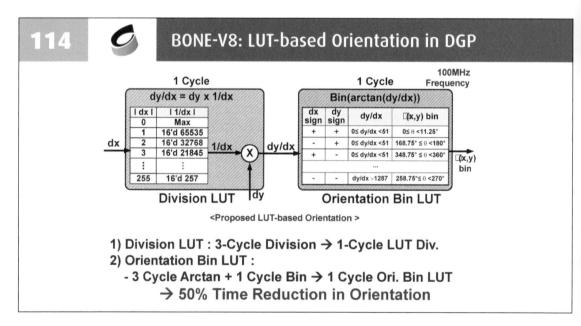

114 — **BONE-V8: LUT-based Orientation in DGP**

1) Division LUT : 3-Cycle Division → 1-Cycle LUT Div.
2) Orientation Bin LUT :
 - 3 Cycle Arctan + 1 Cycle Bin → 1 Cycle Ori. Bin LUT
 → 50% Time Reduction in Orientation

Descriptor generation processor, DGP, is used to detect the orientation. The orientation detection takes up 50% process time, while histogram takes 33% and normalization takes 17%. Lookup table (LUT) is used to reduce the time consumption. Only selected angles are calculated. Division can be reduced from 3 cycles to 1 cycle. Orientation operation expense can be reduced from 4 cycles to 1 cycle. A LUT based pipeline was designed, featuring 4x higher throughput and an 88.4% pipeline utilization.

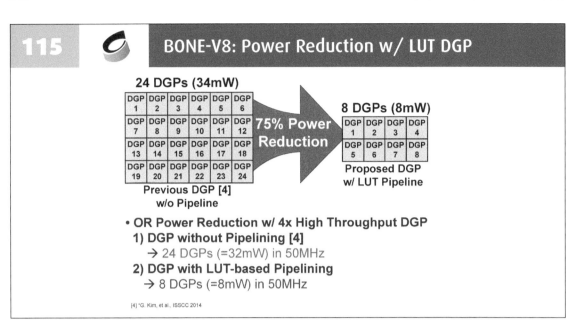

The number of DGPs can be reduced to 8 with 8mW power consumption by applying pipeline. A 75 % power reduction is achieved.

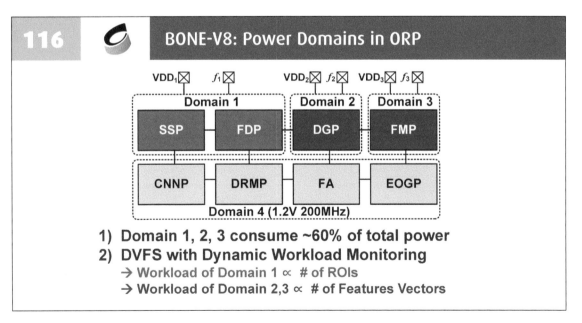

The chip is divided into 4 different power domains with different voltage and frequency, i.e. 0.7V at 20MHz, 0.8V at 40MHz, 0.9V at 60MHz, 1.2V at 100MHz, based on different workload. Domain 1, 2, and 3 consume 60% of the total power. The workload of domain 1 depends on the number of ROIs (Region of interest), while 2 and 3, proportionally to the feature vectors since the distance of the input vectors store inside the memory.

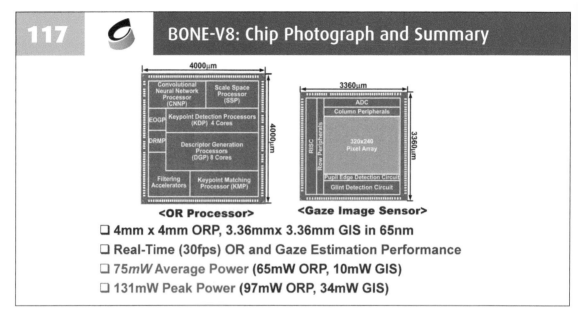

117 **BONE-V8: Chip Photograph and Summary**

<OR Processor>   <Gaze Image Sensor>

❑ 4mm x 4mm ORP, 3.36mmx 3.36mm GIS in 65nm
❑ Real-Time (30fps) OR and Gaze Estimation Performance
❑ 75mW Average Power (65mW ORP, 10mW GIS)
❑ 131mW Peak Power (97mW ORP, 34mW GIS)

The proposed chip was fabricated in 65nm technology integrated a 320x240 pixel array. Pupil edge detection circuits, glint detection circuits, real time object recognition and gaze estimation are implemented on chip. An average power consumption of 75mW and 131mW at peak is achieved.

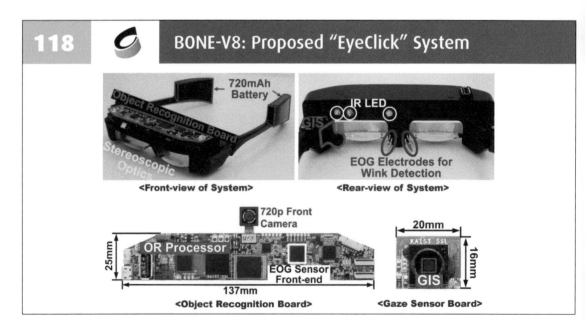

118 **BONE-V8: Proposed "EyeClick" System**

<Front-view of System>   <Rear-view of System>

<Object Recognition Board>   <Gaze Sensor Board>

The proposed chip was assembled on a smart glass with battery, stereoscope camera, IR LED and a HP grade 720p front camera. A blink and wink can be tell correctly using the EOG sensor, which detects the changes of the muscle while winking. The cursor follows the eye movement. A maintenance menu embedded in the k-glass can be selected with eye blinking.

119   Multi-Modal UI/UX

After BONE-V8, the next generation integrated both speech and gesture recognition as the user interaction/expedience (UI/UX). The voice recognition system can detect whether there is a voice or not. A stereo vision is also integrated, enabling distance detection between the glass and the sound source. A finger motion detector is

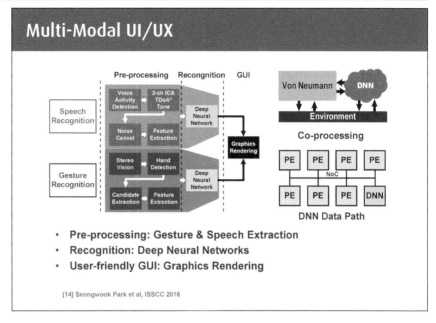

- **Pre-processing: Gesture & Speech Extraction**
- **Recognition: Deep Neural Networks**
- **User-friendly GUI: Graphics Rendering**

[14] Seongwook Park et al, ISSCC 2016

also integrated. The UI/UX system consumes higher than 600mW. A 250mAh, 3.7V battery can only support 1 hour operation. It must be optimized.

120   Overall System Architecture

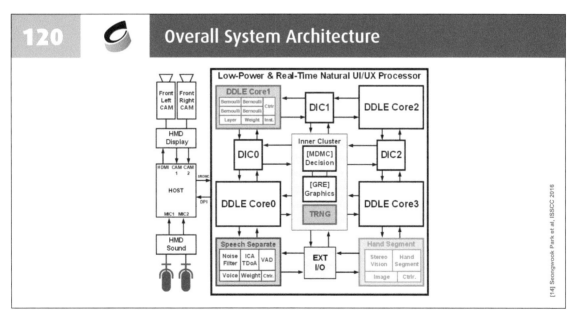

[14] Seongwook Park et al, ISSCC 2016

A multi-modal deep neural network (DNN) is applied for the processing of audio signal and hand image, respectively. The proposed processor consists of a TRNG based deep learning engine, preprocessors, a deep inference core. Both voice and hand gesture recognition are performed in the inference core. The inference core enables the recognition display on the screen as well.

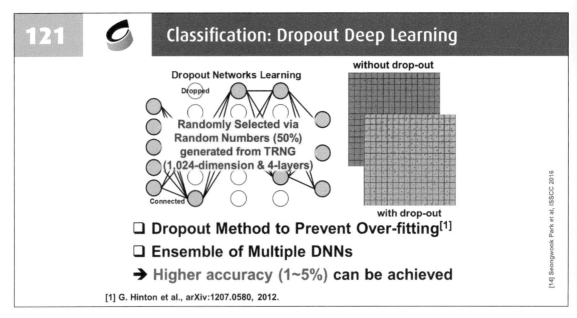

[1] G. Hinton et al., arXiv:1207.0580, 2012.

The drop-out procedure is very important for a precise learning, since it prevents the over-fitting and also ensemble of multiple DNNs. It can improve the accuracy by 1-5%. The neurons should be disabled randomly based on the random numbers.

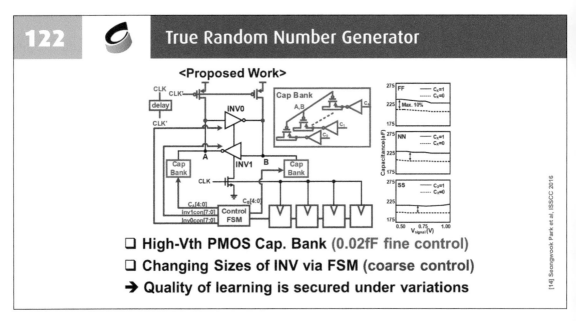

This is the schematic of a true random number generator. A precise bi-stability is obtained, realizing a high randomness of the stability. The standard test suite from national institute of standard (NIST) is used for the test. The proposed designed demonstrated a very good performance with 15 test packages passed. It can be used in quantum computing and quantum encryption as well.

123 UI/UX Specialized SoC–K-Glass 3

UI/UX Processor

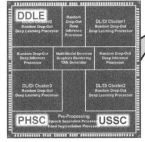

K-Glass 3 System

[14] Seongwook Park et al. ISSCC 2016

- **4mm x 4mm by 65nm CMOS Technology**
- **126mW Deep Learning Processor**
- **1.80TOPS/W Smart Glasses Processor**

The proposed design was fabricated in 65nm technology, implementing 4.81 million gates. It features 1.8 TOPS/W and 36.5 GOPS/mm2. It was assembled into K-Glass 3, similar to Microsoft HoloLens. Almost looks the same. But HoloLens doesn't integrate such natural UI/UX processor. Demonstration shows a successful control of mailbox and key pad.

4.4. ULTRA-LOW POWER AI SOC

124 NT-FMP: Power Breakdown of MAV

Intelligence operation, i.e. CNN is power hungry. The power consumption of a GPU is 5W. MicroGlider, as an example, is a micro robot, requires a capability of intelligence operation, i.e. object recognition. The battery is small and light weighted with limited capability. According to the power analysis, the computer vision module consumes the most power. It must be optimized sacrificing some accuracy as a trade-off. An as low as 1mW power consumption with a higher than 90% accuracy is expected.

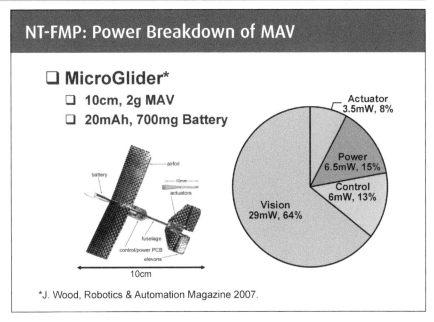

❑ **MicroGlider***
- ❑ **10cm, 2g MAV**
- ❑ **20mAh, 700mg Battery**

*J. Wood, Robotics & Automation Magazine 2007.

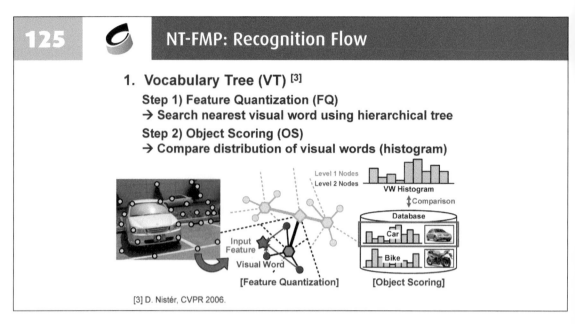

125 — **NT-FMP: Recognition Flow**

1. **Vocabulary Tree (VT)** [3]

Step 1) Feature Quantization (FQ)
→ **Search nearest visual word using hierarchical tree**

Step 2) Object Scoring (OS)
→ **Compare distribution of visual words (histogram)**

Level 1 Nodes
Level 2 Nodes
VW Histogram
↕ Comparison
Database
Car
Bike

Input Feature
Visual Word

[Feature Quantization]   [Object Scoring]

[3] D. Nistér, CVPR 2006.

A vocabulary tree is used with feature quantization. All the object images are stored in the database. Each histogram, each word is compared for the object scoring.

Generalized Hough transform (GHT) was used to remove the background. A large memory and high computation are required. The proposed processor consists of a distance calculation, a single-stage GHT and a 4-stage vocabulary tree database. There are 600 cycles in each pipeline.

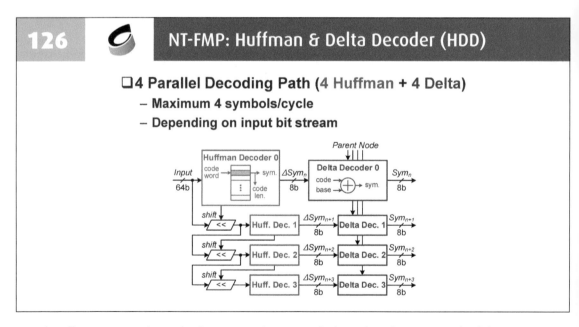

126 — **NT-FMP: Huffman & Delta Decoder (HDD)**

❏ **4 Parallel Decoding Path (4 Huffman + 4 Delta)**
  – **Maximum 4 symbols/cycle**
  – **Depending on input bit stream**

Parent Node

Huffman Decoder 0
Input 64b
code word → sym.
code len.
ΔSym_n 8b

Delta Decoder 0
code → ⊕ → sym.
base
Sym_n 8b

shift << → Huff. Dec. 1 → ΔSym_{n+1} 8b → Delta Dec. 1 → Sym_{n+1} 8b

shift << → Huff. Dec. 2 → ΔSym_{n+2} 8b → Delta Dec. 2 → Sym_{n+2} 8b

shift << → Huff. Dec. 3 → ΔSym_{n+3} 8b → Delta Dec. 3 → Sym_{n+3} 8b

B oth Huffman compression and Delta compression are applied to reduce the memory. The delta compression, which is applied first, can improve the sharpness of the distribution of the data.

127 — NT-FMP: Proposed 10T Level Shifter (10T-LS)

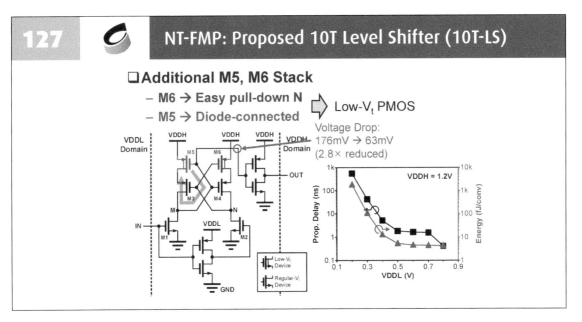

Near threshold circuit was applied, reducing the supply voltage to 0.5V with good analogy efficiency. A process variation detection circuits, i.e. low VT or high VT, are designed to reduce the effect due to process variation. Down NMOS and diode connection are used to improve the stability of the latch circuit.

128 — NT-FMP: Chip Implementation

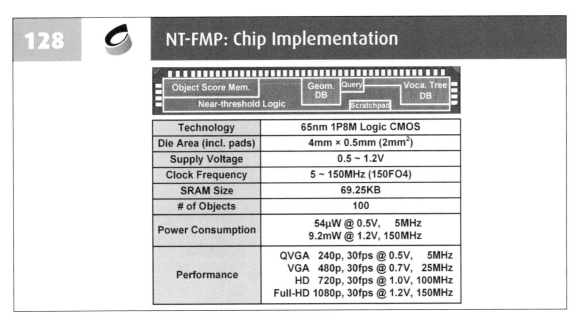

Technology	65nm 1P8M Logic CMOS
Die Area (incl. pads)	4mm × 0.5mm (2mm^2)
Supply Voltage	0.5 ~ 1.2V
Clock Frequency	5 ~ 150MHz (150FO4)
SRAM Size	69.25KB
# of Objects	100
Power Consumption	54µW @ 0.5V, 5MHz 9.2mW @ 1.2V, 150MHz
Performance	QVGA 240p, 30fps @ 0.5V, 5MHz VGA 480p, 30fps @ 0.7V, 25MHz HD 720p, 30fps @ 1.0V, 100MHz Full-HD 1080p, 30fps @ 1.2V, 150MHz

The proposed design was fabricated in 65nm technology, occupying 4mm x 0.5mm. It consists of object score memory near threshold logic, Geom database for the vocabulary tree and scratchpad memory. It consumes 54uW at 0.5V / 5MHz, with QVGA image quality.

4.5. ULTRA-LOW POWER ALWAYS-ON FACE RECOGNITION

129 **Hybrid CIS: Motivation**

• **Smart Face Recognition as a New UI**

Always-on face recognition system

❑ **"Intelligence of Things"**
- **Things recognizing users**
- **More interactive with users**

❑ **Requirements**
- **Ultra-low power consumption**
  ➔ Hybrid CIS
- **Interactive intelligence**
  ➔ CNNP (Next topic)

Always-on face recognition will be an important role in new user interface (UI) design, if the power consumption is less than 1mW. It will enable an "IoT", not internet of thing, but "Intelligence of Thing". Ultra-low power consumption and interactive intelligence. A two-chip implementation is needed, one for hybrid CMOS image sensor, and the other for convolutional neural network processor.

130 **Hybrid CIS: Principle –Digital Unit**

• **Haar-like Filtering w/ Integral Image Generation**
  - **Simple haar-like filtering computation** ☺
    • **Integral image by intensity accumulation**
    • **Block summation with only 3 ADD/SUB**
  - **Inefficient for early-rejected windows** ☹
    • **Reusability ↓ ➔ Preprocessing burden ↑**

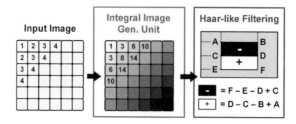

Haar-like filters are used for the implementation of the hybrid CMOS image sensor. It is designed using sequentially simple filters to get more accurate features. It only requires summation and subtraction if the input image is carefully rearranged. However, the re-arrange of the input image data is not efficient. An adapt dynamic analog memory is applied to realize Haar-like filter with integral image.

131 Hybrid CIS: Concept

- **AHFC for 1st Stage**
  - **Static current flowing at only 1 stage**

- **DHFU for Remaining Stages**
  - **Integral image gen. for** only passed windows

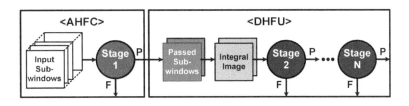

Analog Haar-like filter (AHFC) is used for earlier stage, while digital Haar-like filter (DHFC) is used for later stage to achieve a high power efficiency. The AHFC, as the first stage, can tell whether there is a face in the image. When a face is detected, DHFC is triggered for pattern recognition and face recognition. The total power consumption can be reduced.

132 Chip & System implementation

Always-on face detector (FD)

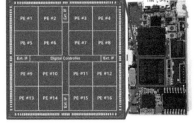

Ultra low-power CNN processor

[1] Kyeongryeol Bong et al, ISSCC 2017

- **Always-on FD: 3.3mm x 3.36mm CIS in 65nm**
  - 320 x 240 pixel array, Haar-like filtering circuit & analog MEM
- **CNN processor: 4mm x 4mm CNNP in 65nm**
  - 4 x 4 PE array with local T-SRAM

The proposed design is fabricated in 65nm technology, integrated a 320x240 pixel array, occupying 3.3mm x 3.36mm. For the hybrid processing, the power of the analog consumption can be reduced by more than 30%. After the first 3 stages or after the analog filtering stage, more than 60% unnecessary windows, with no face detected, can be discarded. An ultra-low power CNN chip is also fabricated in 65nm occupying 4mm x 4mm.

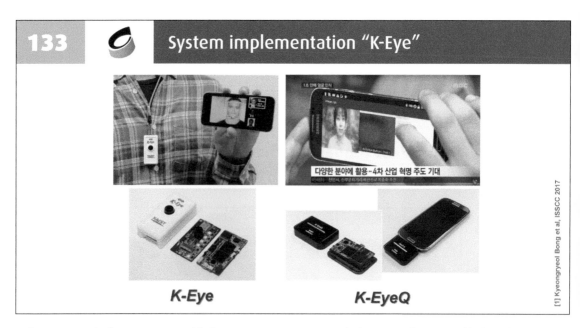

The proposed chip was assembled in a smart necklace. Detected face will be displayed on the smartphone screen. The power consumption is only 0.62mW, 10,000 times less than GPU. The proposed chip was also assembled in the k-eye or k-eyeQ, which perform face recognition triggered by a touch on the smartphone with very low power consumption.

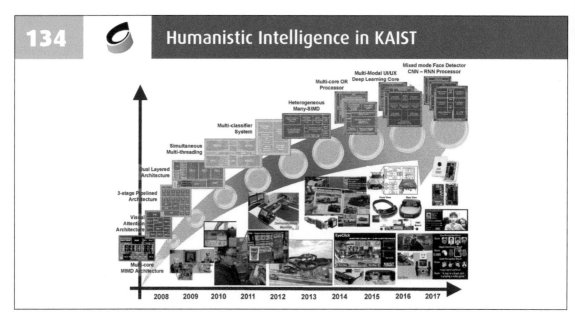

We published a paper at 2008 ISSCC [23] and every year we publish 1 or 2 and sometimes 3 intelligence chips at ISSCC [12, 15, 19, 24, 25, 27, 29, 31-37]. I think our group published more papers than any other groups in the world about the AI chips. We have longest history of course on the AI chips. We also published mobile 3D graphics SoC book [38] and Low-Power NoC book [39].

135 · Humanistic Engineering in SSL

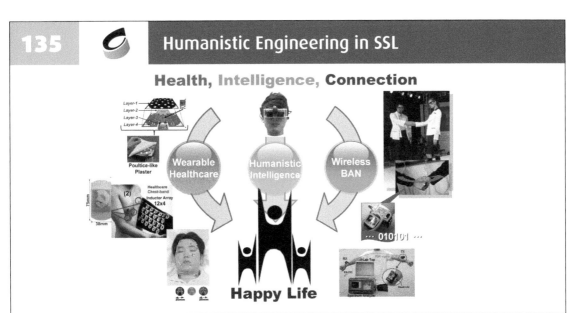

The theme of my research group is making happy life. Happiness is very difficult to define. As an engineer, I'd like to categorize happiness.

If you are healthy, you are happy. If you are intelligent, smart, then you are happy. If you are more social, if you are better connected with other people, then you are happy. Those three important factors are indispensable for happy life. For the healthy life, I'm studying wearable healthcare system. We develop special wearable ECG monitoring system and EEG monitoring system. Also for the intelligence, we are developing the AI chips. For the better connection, we are developing body channel communication. Your data can be shared by your friends by shaking hands. Or your MP3 music flows through your body and you can hear through your earlobe receiver. This kind of body channel communication, wireless BAN, body area network. All the three parts are my research areas.

136 · Reference

1. Kangmin Lee, "A 51mW 1.6GHz on-chip network for low-power heterogeneous SoC platform", *2004 IEEE Int. Solid-State Circuits Conf. (ISSCC) Dig. Tech. Papers*, pp. 152-518, 2004

2. Yong-Ha Park, Seon-Ho Han, Jung-Hwan Lee, and Hoi-Jun Yoo, "A 7.1-GB/s low-power rendering engine in 2-D array-embedded memory logic CMOS for portable multimedia system", *IEEE J. Solid-State Circuits*, vol. 36, no. 6, pp. 944–955, Jun. 2001.

3. NVIDIA, "NVIDIA Launches the World's First Graphics Processing Unit: GeForce 256", 1999.

4. P. Merolla, J. Arthur, F. Akopyan, N. Imam, R. Manohar, and D. S. Modha, "A digital neurosynaptic core using embedded crossbar memory with 45pJ per spike in 45nm", *Proc. Cust. Integr. Circuits Conf.*, pp. 1–4, 2011.

5. B. V. Benjamin *et al.*, "Neurogrid: A mixed-analog-digital multichip system for large-scale neural simulations", *Proc. IEEE*, vol. 102, no. 5, pp. 699–716, 2014.

6. P. H. Pham, D. Jelaca, C. Farabet, B. Martini, Y. LeCun, and E. Culurciello, "NeuFlow: Dataflow vision processing system-on-a-chip", *Midwest Symp. Circuits Syst.*, pp. 1044–1047, 2012.

7. W. S. McCulloch and W. Pitts, "A Logical Calculus of the Idea Immanent in Nervous Activity", *Bull. Math. Biophys.*, vol. 5, pp. 115–133, 1943.

8. F. Rosenblatt, "The perceptron: A probabilistic model for information storage and organization in the brain", *Psychol. Rev.*, vol. 65, no. 6, pp. 386–408, 1958.

9. M. Minsky and S. A. Papert, *Perceptrons: An Introduction to Computational Geometry*. The MIT Press, 1969.

10. D. E. Rumelhart, G. E. Hinton, and R. J. Williams, "Learning representations by back-propagating errors", *Nature*, vol. 323, no. 6088, pp. 533–536, 1986.

11. G. E. Hinton, S. Osindero, and Y.-W. Teh, "A Fast Learning Algorithm for Deep Belief Nets", *Neural Comput.*, vol. 18, no. 7, pp. 1527–1554, 2006.

12. K. Bong, "A 0.62mW Ultra-low-power Convolutional Neural Network Face Recognition Processor and a CIS Integrated with Always-on Haar-like Face Detector", *2017 IEEE Int. Solid-State Circuits Conf. (ISSCC) Dig. Tech. Papers*, pp. 248-249, 2017.

13. S. Han, H. Mao, and W. J. Dally, "Deep Compression: Compressing Deep Neural Networks with Pruning, Trained Quantization and Huffman Coding", *Int. Conf. Learn. Represent.*, pp. 1–14, Oct. 2016.

14. C. Chen, T. Krishna, J. Emer, and V. Sze, "Eyeriss: An Energy-Efficient Reconfigurable Accelerator for Deep Convolutional Neural Networks", *Deep Convolutional Neural Networks*, vol. 52, no. 1, pp. 127–138, 2017.

15. D. Shin, "DNPU: An 8.1 TOPS/W reconfigurable CNN-RNN processor for general-purpose deep neural networks", *2017 IEEE Int. Solid-State Circuits Conf. (ISSCC) Dig. Tech. Papers*, pp. 240-241, 2017

16. H. Nakahara and T. Sasao, "A deep convolutional neural network based on nested residue number system", *2015 25th Int. Conf. F. Program. Log. Appl.*, pp. 1–6, 2015.

17. S. Lee, K. Kim, M. Kim, J. Y. Kim, and H. J. Yoo, "The brain mimicking visual attention engine: An 80x60 digital cellular neural network for rapid global feature extraction", *IEEE Symp. VLSI Circuits, Dig. Tech. Pap.*, pp. 26–27, 2008.

18. T. Serre, L. Wolf, S. Bileschi, M. Riesenhuber, and T. Poggio, "Robust object recognition with cortex-like mechanisms", *IEEE Trans. Pattern Anal. Mach. Intell.*, vol. 29, no. 3, pp. 411–426, 2007.

19. S. Park, K. Bong, D. Shin, J. Lee, S. Choi, and H. J. Yoo, "4.6 A1.93TOPS/W scalable deep learning/ inference processor with tetra-parallel MIMD architecture for big-data applications", *2015 IEEE Int. Solid-State Circuits Conf. (ISSCC) Dig. Tech. Papers*, pp. 80-81, 2015.

20. M. Hamburg, P. Kocher, and M. E. Marson, "Analysis of Intel's Ivy Bridge digital random number generator", pp. 1–22, 2012.

21. S. G. Bae, Y. Kim, Y. Park, and C. Kim, "3-Gb/s High-Speed True Random Number Generator Using Common-Mode Operating Comparator and Sampling Uncertainty of D Flip-Flop", *IEEE J. Solid-State Circuits*, vol. 52, no. 2, pp. 605–610, 2017.

22. Y. Kim, D. Shin, J. Lee, Y. Lee, and H.-J. Yoo, "A 0.55 V 1.1 mW Artificial Intelligence Processor With On-Chip PVT Compensation for Autonomous Mobile Robots", *IEEE Trans. Circuits Syst. I Regul. Pap.*, vol. 65, no. 2, pp. 567–580, Feb. 2018.

23. K. Kim *et al.*, "A 125 GOPS 583mW network-on-chip based parallel processor with bio-inspired visual-attention engine", *2008 IEEE Int. Solid-State Circuits Conf. (ISSCC) Dig. Tech. Papers*, pp. 308-310, 2008.

24. J. Y. Kim, M. Kim, S. Lee, J. Oh, K. Kim, and H. J. Yoo, "A 201.4 GOPS 496 mW real-time multi-object recognition processor with bio-inspired neural perception engine", *IEEE J. Solid-State Circuits*, vol. 45, no. 1, pp. 32–45, 2010.

25. S. Lee, "A 345 mW heterogeneous many-core processor with an intelligent inference engine for robust object recognition", *2010 IEEE Int. Solid-State Circuits Conf. (ISSCC) Dig. Tech. Papers*, pp. 332-333, 2010.

26. S. Lee *et al*, "A 345 mW heterogeneous many-core processor with an intelligent inference engine for robust object recognition," *IEEE J. Solid-State Circuits*, vol. 46, no. 1, pp. 42–51, 2011.

27. J. Oh, "A 320 mW 342GOPS real-time moving object recognition processor for HD 720p video streams", *2012 IEEE Int. Solid-State Circuits Conf. (ISSCC) Dig. Tech. Papers*, pp. 220-224, 2012.

28. J. Oh *et al.*, "A 320 mW 342 GOPS real-time dynamic object recognition processor for HD 720p video streams," *IEEE J. Solid-State Circuits*, vol. 48, no. 1, pp. 33–45, 2013.

29. G. Kim, "A 1.22 TOPS and 1.52 mW/MHz augmented reality multi-core processor with neural network NoC for HMD applications", *2014 IEEE Int. Solid-State Circuits Conf. (ISSCC) Dig. Tech. Papers*, pp. 182-183, 2014-Feb.

30. G. Kim *et al.*, "A 1.22 TOPS and 1.52 mW/MHz augmented reality multicore processor with neural network NoC for HMD applications," *IEEE J. Solid-State Circuits*, vol. 50, no. 1, pp. 113–124, 2015.

31. I. Hong et al., "A 2.71nJ/Pixel 3D-stacked Gaze-Activated Object Recognition system for Low-power Mobile HMD Applications", *2015 IEEE Int. Solid-State Circuits Conf. (ISSCC) Dig. Tech. Papers*, pp. 182-183, 2015.

32. J. Oh, "A 57 mW embedded mixed-mode neuro-fuzzy accelerator for intelligent multi-core processor", *2011 IEEE Int. Solid-State Circuits Conf. (ISSCC) Dig. Tech. Papers*, pp. 130-131, 2011.

33. J. Park, "A 646 GOPS/W multi-classifier many-core processor with cortex-like architecture for super-resolution recognition", *2013 IEEE Int. Solid-State Circuits Conf. (ISSCC) Dig. Tech. Papers*, pp. 168-169, 2013.

34. Y. Kim, "A 0.5V 54µW Ultra-Low-Power Recognition Processor with 93.5% Accuracy Geometric Vocabulary Tree and 47.5% Database Compression", *2015 IEEE Int. Solid-State Circuits Conf. (ISSCC) Dig. Tech. Papers*, pp. 1-3, 2015.

35. S. Park, "A 126.1mW real-time natural UI/UX processor with embedded deep-learning core for low-power smart glasses", *2016 IEEE Int. Solid-State Circuits Conf. (ISSCC) Dig. Tech. Papers*, pp. 254-255, 2016.

36. K. Lee, " A 502 GOPS and 0.984mW Dual-Mode ADAS SoC with RNN-FIS Engine for Intention Prediction in Automotive Black-Box System", *2016 IEEE Int. Solid-State Circuits Conf. (ISSCC) Dig. Tech. Papers*, pp. 256-257, 2016.

37. Y. Kim, " A 0.55V 1.1mW Artificial-Intelligence Processor with PVT Compensation for Micro Robots", *2016 IEEE Int. Solid-State Circuits Conf. (ISSCC) Dig. Tech. Papers*, pp. 258-259, 2016.

38. H.-J. Yoo, J.-H. Woo, J.-H. Sohn, and B.-G. Nam, *Mobile 3D Graphics SoC: From Algorithm to Chip*. John Wiley & Sons, 2010.

39. H.-J. Yoo, K. Lee, and J. K. Kim, *Low-Power NoC for High-Performance SoC Design*. CRC Press, 2008.

CHAPTER 05

Ultimate High-Speed Wireless Link

Challenge to Terahertz CMOS Transceivers

Minoru Fujishima

School of Advanced Sciences of Matter
Hiroshima University

C MOS transceivers are used not only in personal mobile terminals but in all sensors and actuators in the Internet-of-Thing (IoT) era. In this seminar, among the many CMOS transceivers ever reported, we will discuss two ultimate transceivers using ultra-wideband. The first topic is ultrahigh-speed wireless communication. We will introduce terahertz CMOS communication which realizes ultrahigh-speed communication by utilizing ultra-wideband. The second topic is low-power wireless communication. We introduce UWB pulse communication which realizes ultimate low-power consumption. Through these two wireless communications, we discuss what is the ultimate wireless system.

1. INTRODUCTION TO TERAHERTZ COMMUNICATION

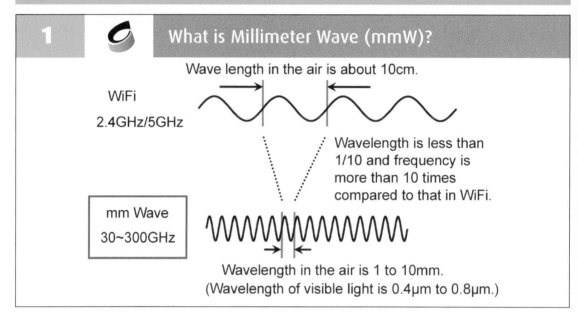

1 — What is Millimeter Wave (mmW)?

Wave length in the air is about 10cm.

WiFi
2.4GHz/5GHz

Wavelength is less than 1/10 and frequency is more than 10 times compared to that in WiFi.

mm Wave
30~300GHz

Wavelength in the air is 1 to 10mm.
(Wavelength of visible light is 0.4µm to 0.8µm.)

What is microwave? Wi-Fi is one of the microwave, using 2.4ghz and 5ghz, wave length in the air of microwave is ten centimeters. In millimeter wave, millimeter wave frequency range is from 30Ghz to 300 Ghz. The wave length of millimeter wave is less than 1/10, and frequency is more than ten times compared to that in Wi-Fi.

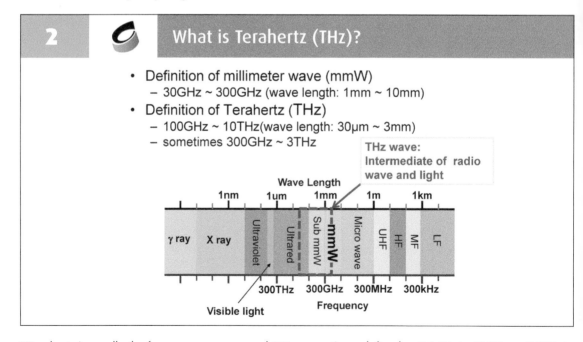

2 — What is Terahertz (THz)?

- Definition of millimeter wave (mmW)
  - 30GHz ~ 300GHz (wave length: 1mm ~ 10mm)
- Definition of Terahertz (THz)
  - 100GHz ~ 10THz(wave length: 30µm ~ 3mm)
  - sometimes 300GHz ~ 3THz

THz wave:
Intermediate of radio wave and light

Wave Length

1nm   1um   1mm   1m   1km

γ ray | X ray | Ultraviolet | Ultrared | Sub mmW | mmW | Micro wave | UHF | HF | MF | LF

Visible light

300THz   300GHz   300MHz   300kHz
Frequency

Terahertz is usually the frequency range around 1THz, sometimes defined as 0.1 THz to 10 THz, or 0.3THz to 3THz, since 3THz is the upper limit of radio wave.

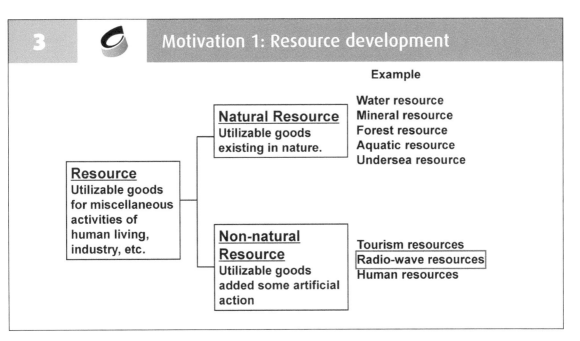

Why we are studying terahertz of millimeter wave? One motivation is to explore resource development, i.e. human life and the industries. The resource has two categories, one is natural resource using material existing nature, for example, water resource, mineral resource, forest resource, aquatic resource, undersea resource. The other one is non-natural resource used with substance added some artificial action, tourism resources.

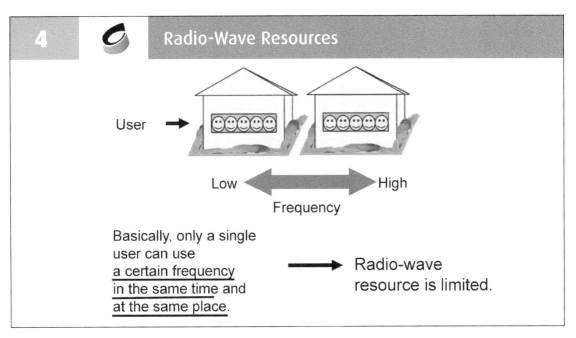

Why radio wave is resources? Because it can only use a certain frequency in the same time and at the same place basically. And also upper limit of the frequency is determined by the technology. We have to concern radio wave is limited resources, it's not unlimited resource, radio wave is limited resource.

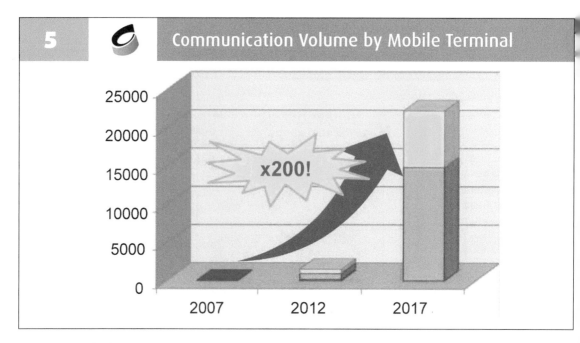

Unfortunately, the number of mobile terminal is increasing exponentially, this slide shows the communication volume of mobile terminal. From ten years ago to this year, the communication volume is improving about 200 times, so this number is increasing exponentially in future. Every terminal or every person has to occupy one frequency at the same time. But frequency resource is limited one, so we have to think about it.

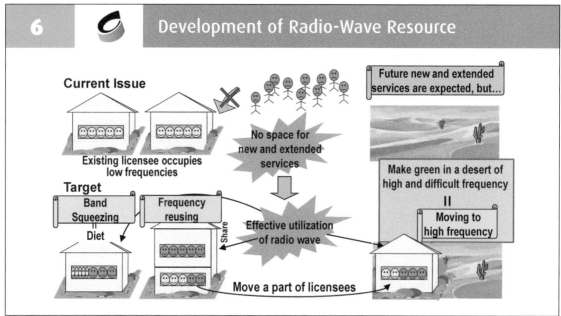

There is no space left for new and expanding extended services because existing licenses occupies low frequencies that is a problem. To solve it, we have to think effective use of radio wave. The study includes band squeezing, frequency reusing and moving to high frequency.

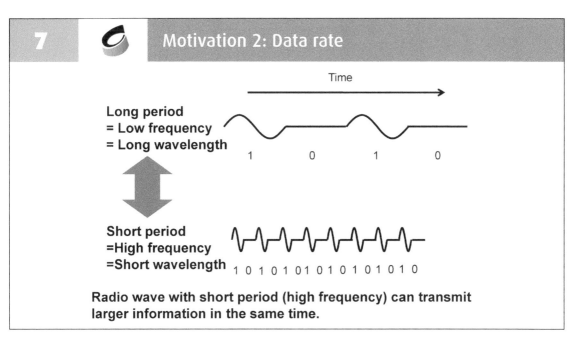

The second motivation is data rate. If you use a high frequency, data rate becomes high. Larger information can be transmitted in the same time, to this is the second motivation and this is a very big motivation.

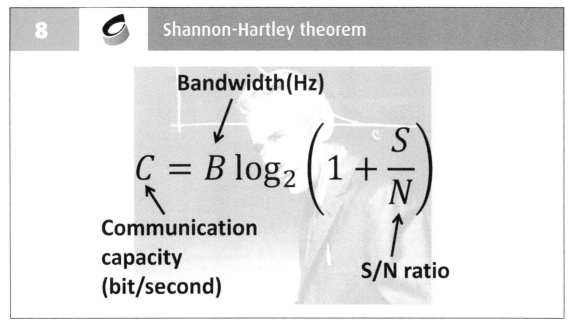

This is Shannon-Hartley theorem, according to Shannon-Hartley theorem, communication capacity is equal to bandwidth times log one plus signal-noise ratio. Communication capacity that bit per second is proportion to bandwidth, if the bandwidth is large, communication capacity that is data rate can be improved.

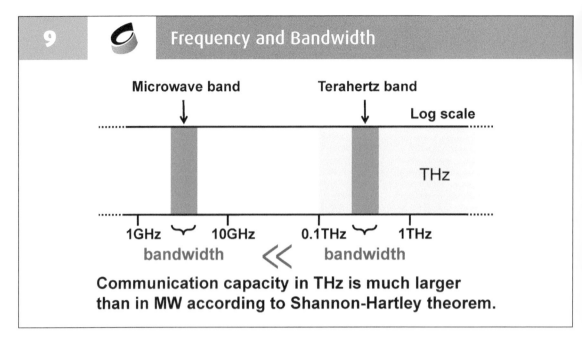

9 **Frequency and Bandwidth**

Communication capacity in THz is much larger than in MW according to Shannon-Hartley theorem.

If you are located bandwidth in the microwave region, compare to terahertz, the shape is the same, but this is brought in log scale. Bandwidth of microwave is much smaller than bandwidth of the terahertz. The communication capacity in terahertz is much larger than in microwave according Shannon-Hartley theorem.

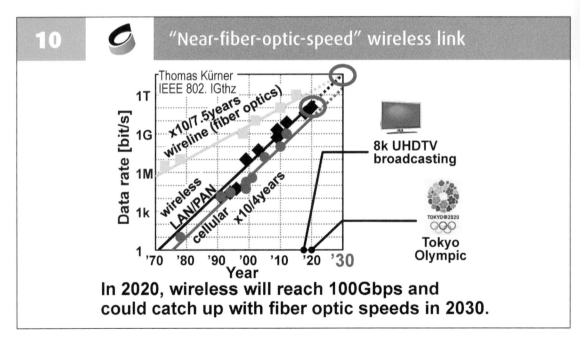

10 **"Near-fiber-optic-speed" wireless link**

In 2020, wireless will reach 100Gbps and could catch up with fiber optic speeds in 2030.

That's why data rate is increasing year by year. This slide shows that data rate as a function of the year. Black line shows indoor communication and red line shows the outdoor communication. For the wireline, data rate is increasing ten times every seven half years so far. On the other hand, the data rate of wireless communication is increasing ten times every four years to the increasing speed of wireless communication is much higher than the increasing speed of the wireline.

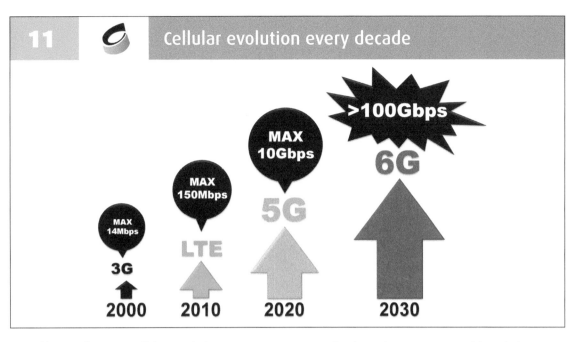

In addition, there is cellular evolution every ten years since 2000. Year to 2020, 3 years from now, 5G will start, 5G maximum data rate 10Gbps. How do we realize how do you per second by wireless? One answer is using terahertz.

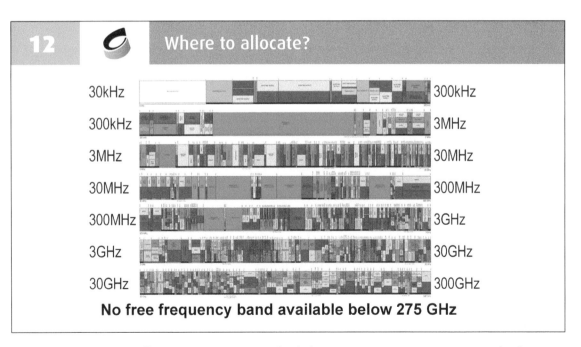

This is a frequency allocation map. In United States, many applications are already allocated in frequencies as no free frequency band is available below 275 GHz. Now over 275GHz is under discussion and in near future around 300GHz are available for wireless communication.

2. WHAT IS 300GHZ BAND REALIZING 100GBPS

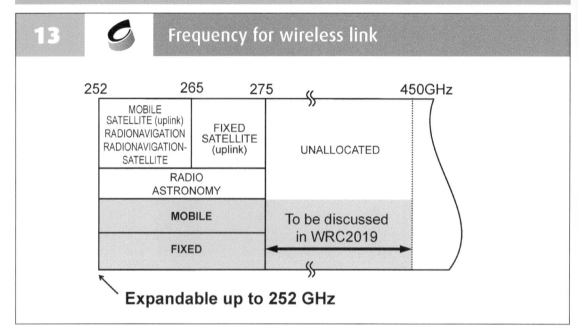

13 Frequency for wireless link

Expandable up to 252 GHz

From 250 to 275 wireless communication is already allocated for mobile use and fixed use and over 275GHz is currently under discussion but maybe this frequency range is used for wireless communication too. The main target in this frequency.

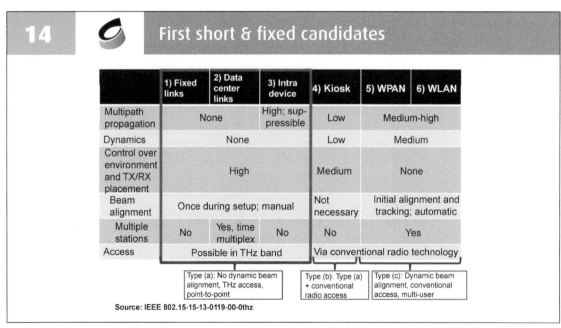

14 First short & fixed candidates

	1) Fixed links	2) Data center links	3) Intra device	4) Kiosk	5) WPAN	6) WLAN
Multipath propagation	None		High; suppressible	Low	Medium-high	
Dynamics	None			Low	Medium	
Control over environment and TX/RX placement	High			Medium	None	
Beam alignment	Once during setup; manual			Not necessary	Initial alignment and tracking; automatic	
Multiple stations	No	Yes, time multiplex	No	No	Yes	
Access	Possible in THz band			Via conventional radio technology		

Type (a): No dynamic beam alignment, THz access, point-to-point

Type (b): Type (a) + conventional radio access

Type (c): Dynamic beam alignment, conventional access, multi-user

Source: IEEE 802.15-15-13-0119-00-0thz

The potential application of terahertz communication is discussed in several groups. This is a potential operation reported from the IEEE terahertz group. According to their reports, the most applications are fixed links and data center links and intra-device communication.

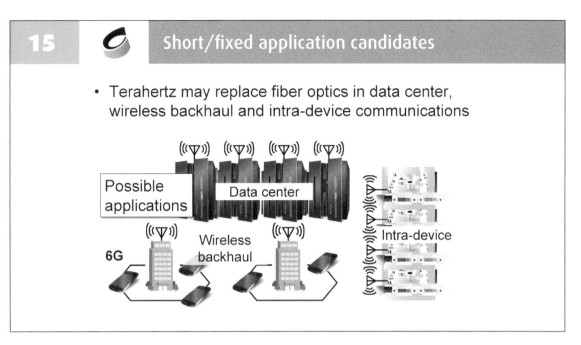

15   Short/fixed application candidates

- Terahertz may replace fiber optics in data center, wireless backhaul and intra-device communications

In the data center, currently many fiber optics are used. The total length of the fiber optics is approaching to four thousand kilometers in the big date center. Very complicated network is built with the fiber optics if that fiber optics can be replaced by wireless communication. Maintainability can be improved and also if we want to enjoy 6G for very high speed commutation.

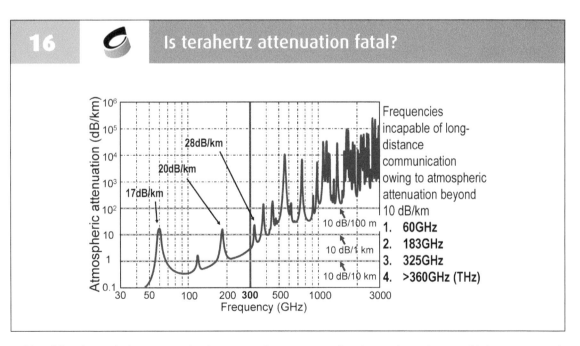

16   Is terahertz attenuation fatal?

This slide showed that atmospheric attenuation as a function of the frequency. If atmospheric attenuation exceeds 10dB. Maybe wireless communication becomes difficult so people think attenuation in terahertz is very high, so we can't expect long distance communication in terahertz. But if we avoid these frequencies long distance communication can be possible.

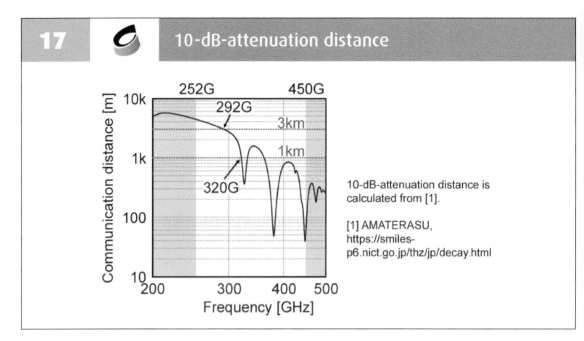

This slide shows the communication distance as a function of frequency. If we limit up to 290GHz 3km communication can be possible, because 10dB attenuation appears at 3km. Thus, if we use a terahertz communication, moderator long communication can be possible up to 292GHz. One kilometer communication is possible up to 320ghz is possible.

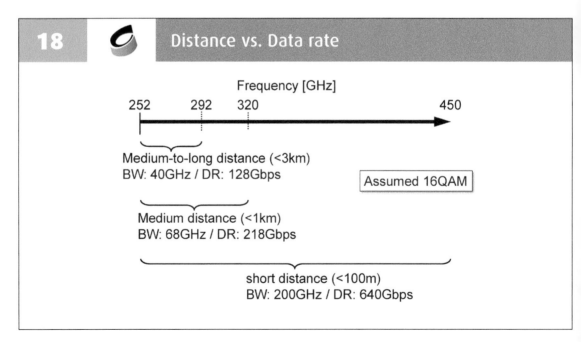

From 275GHz to 292GHz, three kilometer distance communication is expected. Combined these frequencies we can use 40GHz bandwidth here. If we assume 16QAM, a date rate of 100Gbps can be reached with 40ghz bandwidth data rate and 128Gbps so a date. If you expect short distance communication, like indoor communication, data rate is more than 600Gbps approaching to 1Tbps. So according to your expectation of the communication distance. Date rate is changing negative.

19 Propagation loss and wavelength

Friis' transmission equation

Antenna gain

Propagation loss

$$\frac{P_r}{P_t} = G_t G_r \left(\frac{\lambda}{4\pi R} \right)^2$$ wavelength

distance

Antenna gain $G_I = \dfrac{4\pi}{\lambda^2} A_e$ Effective area

Effective area

$$\frac{P_r}{P_t} = A_t A_r \left(\frac{1}{\lambda R} \right)^2$$ The propagation loss decreases at high frequencies if the effective area is constant.

And another reason why terahertz is concerned to be limited in a short and fixed distance if the Friis' transmission equation. Pr is receiving power and Pt is transmission power. According to this equation, since receiving power in terahertz is very small, but antenna gain is proportional to effective antenna area, if we do not reduce the size of antenna, even if the frequency is high, if the antenna size is not reduced, you can obtain higher power at the receiver.

20 Wavelength and antenna size

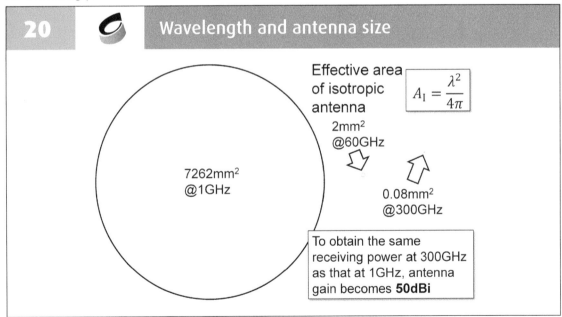

Effective area of isotropic antenna

$$A_I = \frac{\lambda^2}{4\pi}$$

7262mm² @1GHz

2mm² @60GHz

0.08mm² @300GHz

To obtain the same receiving power at 300GHz as that at 1GHz, antenna gain becomes **50dBi**

Effective antenna area of isotropic antenna is expressed by this equation, lambda square over four pi. If you don't reduce the antenna size, antenna gain is increasing according to the increased frequency. To obtain the same receiving power as 300ghz as that at 1ghz, gain becomes as high as 50dbi.

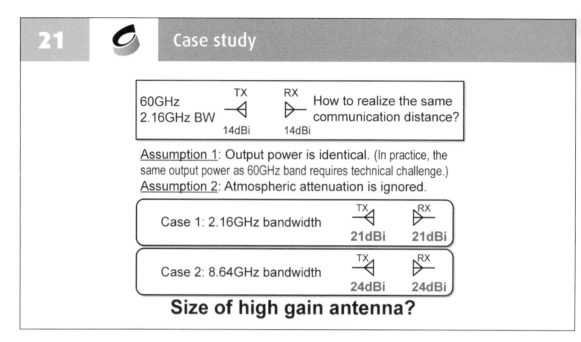

21   Case study

60GHz
2.16GHz BW   TX   RX   How to realize the same
14dBi   14dBi   communication distance?

Assumption 1: Output power is identical. (In practice, the same output power as 60GHz band requires technical challenge.)
Assumption 2: Atmospheric attenuation is ignored.

Case 1: 2.16GHz bandwidth   TX   RX
21dBi   21dBi

Case 2: 8.64GHz bandwidth   TX   RX
24dBi   24dBi

Size of high gain antenna?

Bandwidth of 60ghz communication is 2.16ghz. And typical antenna gain of 60ghz is 14dbi for both transmitter and receiver. How can we realize the same communication distance at 300ghz?

There are two assumptions. One is out of power is identical to the 60ghz and 300ghz. The other is atmospheric attenuation is ignored. That means the communication distance is very short.

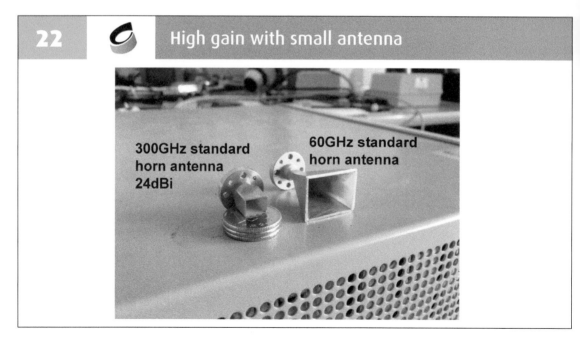

22   High gain with small antenna

300GHz standard horn antenna 24dBi

60GHz standard horn antenna

This is a comparison of antenna for 300ghz and 60ghz. Both are standard 24dbi gain horn antenna. The size of 300GHz antenna is very small. You can't broadcast the information using 300ghz because antenna gain is very high.

23 What does "high speed" mean?

1. "High data rate"
   – Fiber-optics is "faster" than wireless
   – Backbone data rates ≥ 100Gb/s

2. "Low latency"
   – Wireless is 50% "faster"
   – $250m invested to link NY and Chicago in 2011–2013*

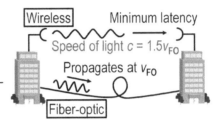

Wireless — Minimum latency
Speed of light $c = 1.5v_{FO}$
Propagates at v_{FO}
Fiber-optic

*C. Cookson, "Time is Money When It Comes to Microwaves," *FT Magazine*, May 2013.

There are two meanings of high speed. One is high data rate, the other is low latency. The promising future application is real time wireless communication. For example, two 250 million US dollar invested to link New York and Chicago in 2011 to 2013. This is a stock exchange. real time trading is very important in stock marketing. So they invested in wireless communication from New York to Chicago for real time trading. So real time application is promising for wireless application.

24 Uncompressed full spec 8K streaming

http://www.ikegami.com/

Ultra HD **8K**

http://www.xbitlabs.com/

72 Gbps max. (Full spec. / 60fps)

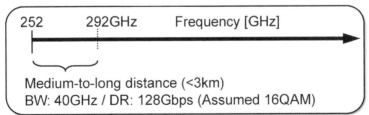

252   292GHz   Frequency [GHz]

Medium-to-long distance (<3km)
BW: 40GHz / DR: 128Gbps (Assumed 16QAM)

One example for real time application in uncompressed full spec 8K streaming. 8K streaming is 72Gbps maximum or 144Gbps maximum. Full spec 72Gbps is in the case of 60fps and 144 case of 120fps. In both cases from 252ghz to 292ghz maximum data rate is 128Gbps so you can enjoy uncompressed full spec 8K streaming by using this frequency.

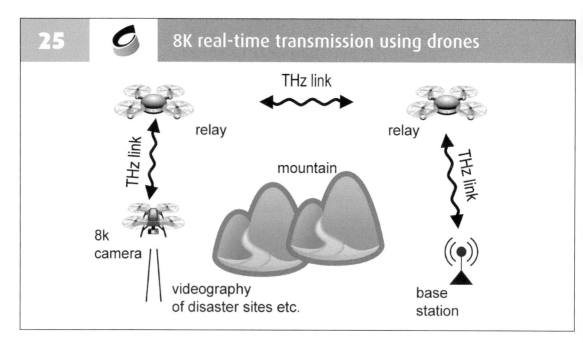

Where should we use this kind of uncompressed full spec 8K stream? If you mount 8K camera on a drone as a monitor the data can be transmitted to the base station in real time. Terahertz is almost line of sight communication because the antenna gain is very high.

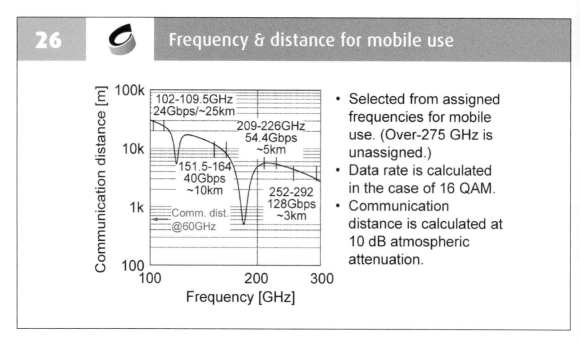

If you decrease the frequency, you can increase the communication distance. If you decrease the frequency bandwidth, the maximum data rate is decreasing. communication at the same time.

3. GENERAL CONSIDERATIONS FOR HIGH-FREQUENCY CMOS DESIGN

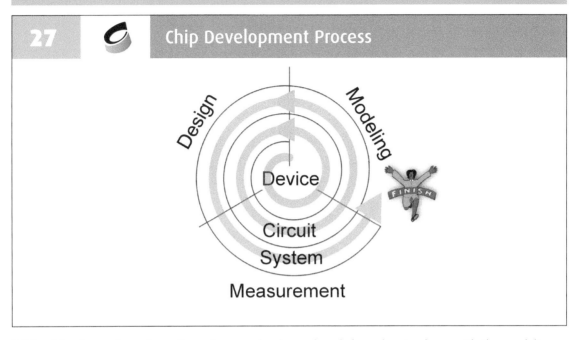

This slide shows the schematic review graph of chip development process. There are three layers, device layer and circuit layer and system layer, and each layer has to down. with the modeling, a measurement modeling and design.

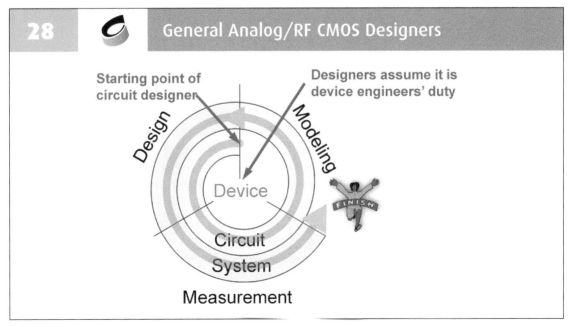

General analog and RF CMOS designers start from here, the model is provided from the foundry, so designer assume it is devises engineers' duty. Don't care about this way and starting point of circuit design is here certainly.

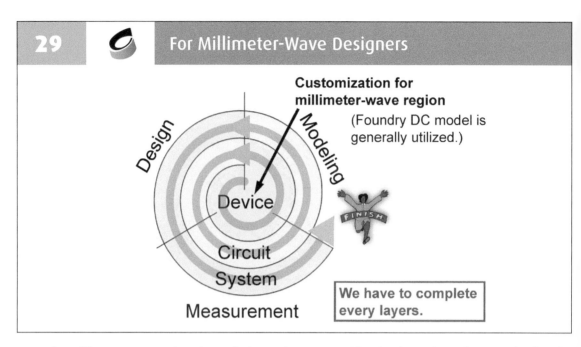

29 · For Millimeter-Wave Designers

Customization for millimeter-wave region

(Foundry DC model is generally utilized.)

Design
Modeling
Device
Circuit
System
Measurement

We have to complete every layers.

But for millimeter wave and terahertz designers have to consider the device layer, because the foundry does not provide high frequency model. this device design is a layout technique.

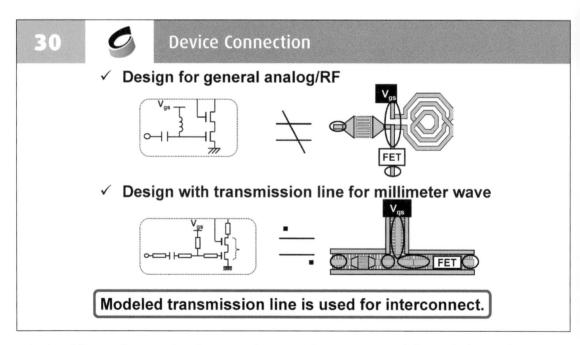

30 · Device Connection

✓ **Design for general analog/RF**

V_{gs}

V_{gs}
FET

✓ **Design with transmission line for millimeter wave**

V_{gs}

V_{gs}
FET

Modeled transmission line is used for interconnect.

The big difference between low frequency design and high frequency design is device connection. In low frequency design firstly each device is placed and connected by wires. It is not easy to obtain parasitic inductance. LPE can't be applied for inductance. To avoid this issue in high frequency, we are using transmission line instead of using this kind of bear connection.

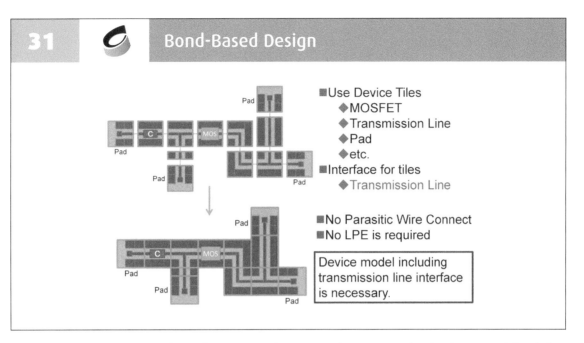

31 Bond-Based Design

- ■Use Device Tiles
  - ◆MOSFET
  - ◆Transmission Line
  - ◆Pad
  - ◆etc.
- ■Interface for tiles
  - ◆Transmission Line

- ■No Parasitic Wire Connect
- ■No LPE is required

Device model including transmission line interface is necessary.

The design technique bond-base design is used. Device tiles are designed and used. No bed parasitic exception is required because no connections there. But we have to consider the device model including transmission line interface.

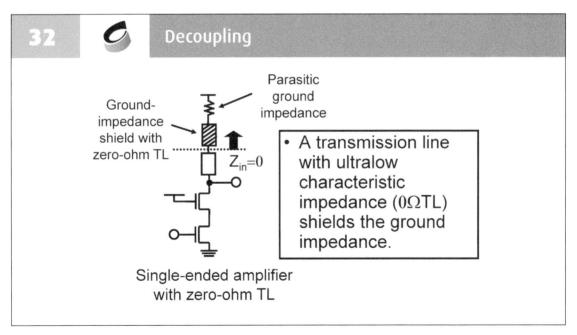

32 Decoupling

- A transmission line with ultralow characteristic impedance (0ΩTL) shields the ground impedance.

Parasitic ground impedance

Ground-impedance shield with zero-ohm TL

$Z_{in}=0$

Single-ended amplifier with zero-ohm TL

And the decoupling is also important. This parasitic ground impedance will affect the circuit performance. In high frequency if you put a large capacitor, it has large inductance too. To solve this issue, you should insert ground impedance shield with zero-ohm transmission line, which is a transmission line, with ultralow characteristic impedance.

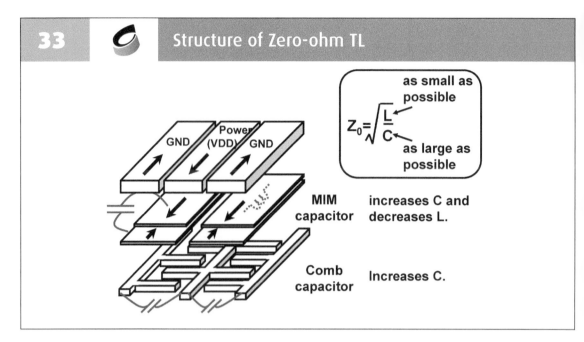

33   Structure of Zero-ohm TL

$$Z_0 = \sqrt{\frac{L}{C}}$$

as small as possible

as large as possible

MIM capacitor   increases C and decreases L.

Comb capacitor   Increases C.

How do we realize very small characteristic impedance transmission line? Characteristic impedance is determined by the root square of L over C. If you want to this subzero, approaching to zero L should be as small as possible, and C should be as large as possible.

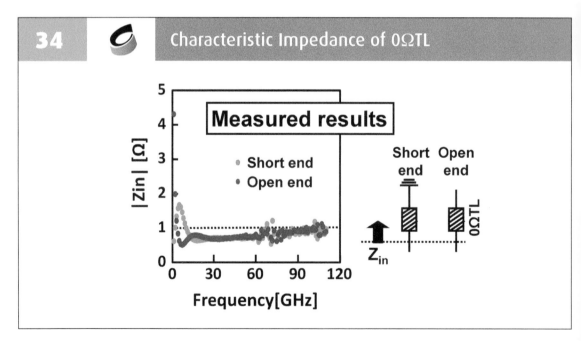

34   Characteristic Impedance of 0ΩTL

Measured results

- Short end
- Open end

$|Zin|$ [Ω]

Frequency[GHz]

Short end   Open end

Z_{in} 0ΩTL

The input impedance changes according to the frequency. If you can carefully layout zeros ohm transmission line, zero impedance can be shield like this. So this kind of decoupling capacitors, capacitors devices are used in high frequency layout.

35   Accurate Probing

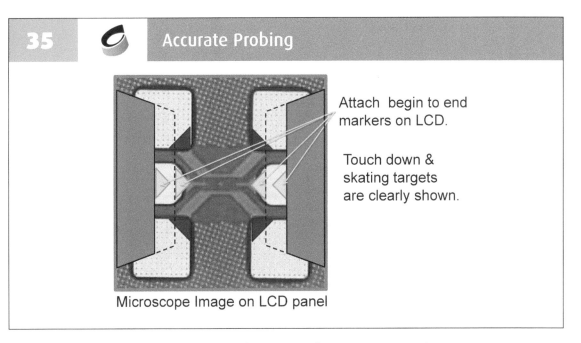

Attach begin to end
markers on LCD.

Touch down &
skating targets
are clearly shown.

Microscope Image on LCD panel

In the measurement of high frequency device, microwave probe, i.e. GSD probe, is used. Potential of probe will affect performance or measurement result. So precise probing is very important. Sometimes engineer made align marker on the chip for alignment.

36   Scratches on Pads

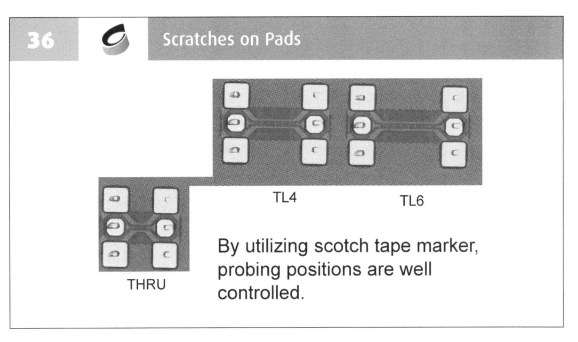

TL4

TL6

THRU

By utilizing scotch tape marker,
probing positions are well
controlled.

After probing, you can monitor this stretch on the pad. This is an example stretches, through and different length of transmission lines, you will see not only the position but also the shape is almost same, even though that left probe and right probe the distance is different.

4. THRU-REFLECTION-LINE DE-EMBEDDING

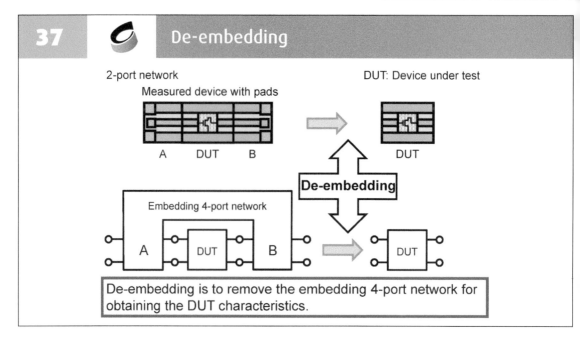

37   De-embedding

2-port network   DUT: Device under test
Measured device with pads

A   DUT   B   DUT

De-embedding

Embedding 4-port network

A   DUT   B   DUT

De-embedding is to remove the embedding 4-port network for obtaining the DUT characteristics.

In the measurement pad is attached to the device itself. The measurement results include left pad and right band which we have to remove to obtain the core device characteristics. Generally, these two pads can be described by embedding 4-port network if we understand the characteristics of embedding 4-port network. This is the de-embedding.

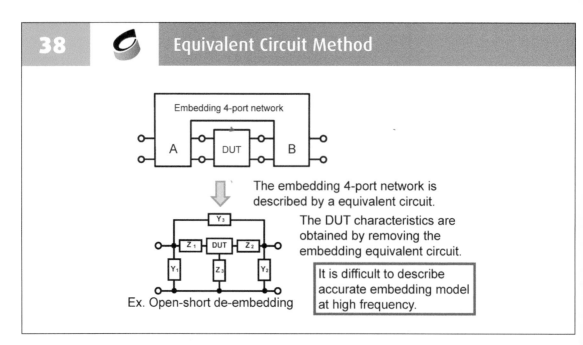

38   Equivalent Circuit Method

Embedding 4-port network

A   DUT   B

The embedding 4-port network is described by a equivalent circuit.

The DUT characteristics are obtained by removing the embedding equivalent circuit.

It is difficult to describe accurate embedding model at high frequency.

Ex. Open-short de-embedding

First method to apply de-embedding is using an equivalent circuit method. One example is open short de-embedding. In the open short de-embedding. It is not so easy to describe accurate embedding model at high frequencies.

39 Cascade Matrix Method

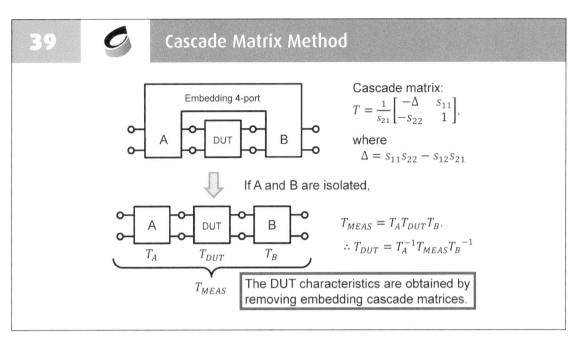

Cascade matrix:
$$T = \frac{1}{s_{21}}\begin{bmatrix} -\Delta & s_{11} \\ -s_{22} & 1 \end{bmatrix},$$
where
$$\Delta = s_{11}s_{22} - s_{12}s_{21}$$

If A and B are isolated,

$$T_{MEAS} = T_A T_{DUT} T_B.$$
$$\therefore T_{DUT} = T_A^{-1} T_{MEAS} T_B^{-1}$$

The DUT characteristics are obtained by removing embedding cascade matrices.

That is cascade matrix method. A pad and B pad are isolated. The interaction of the left pad and right pad is not so large, because the distance of left pad and the right pad is not so closed.

If we use the T matrix, conversed from the S parameter, the measure T parameter can be expressed by the multiplication od T_A, T_{DUT} and T_B. T_A and T_B are T matrix of A and B pad that can be caliber.

40 TRL (Thru-Reflect-Line)

- TRL does not require all standards to be either ideal or fully known, introduced by Engen and Hoer in 1979.

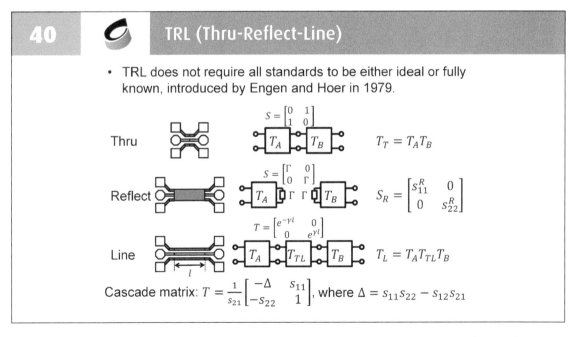

$$S = \begin{bmatrix} 0 & 1 \\ 1 & 0 \end{bmatrix}$$

Thru $\qquad T_T = T_A T_B$

$$S = \begin{bmatrix} \Gamma & 0 \\ 0 & \Gamma \end{bmatrix}$$

Reflect $\qquad S_R = \begin{bmatrix} s_{11}^R & 0 \\ 0 & s_{22}^R \end{bmatrix}$

$$T = \begin{bmatrix} e^{-\gamma l} & 0 \\ 0 & e^{\gamma l} \end{bmatrix}$$

Line $\qquad T_L = T_A T_{TL} T_B$

Cascade matrix: $T = \frac{1}{s_{21}}\begin{bmatrix} -\Delta & s_{11} \\ -s_{22} & 1 \end{bmatrix}$, where $\Delta = s_{11}s_{22} - s_{12}s_{21}$

One very famous method is TRL, Thru-Reflect-Line, introduced by Engen and Hoer in 1979. TRL does not require all standards to be either ideal or full known. TRL using a thru, a reflect, and a line to caliber IA and TB.

41 **Propagation Constant y of TL**

Define as $T_D \equiv T_L T_T^{-1}$ $T_A T_{TL} T_B$ $T_B^{-1} T_A^{-1}$ $T_A T_{TL} T_A^{-1}$

$$T_D \equiv \quad \times \quad = $$

Note that $T_D T_A = \quad = T_A T_{TL} \quad \longrightarrow \quad \therefore \boxed{T_{TL} = T_A^{-1} T_D T_A}$

When reference impedance is identical to the characteristic

impedance of the line, $\quad \overset{T_{TL}}{=} \begin{bmatrix} e^{-\gamma l} & 0 \\ 0 & e^{\gamma l} \end{bmatrix}$ diagonalization!

Thus, $e^{-\gamma l}$ is derived from the eigenvalue λ of T_D,
when scattering matrix transformed from T_D is symmetric.

Finally, we obtain $\gamma = \dfrac{-\ln(\lambda) + j2\pi k}{l}$,
where k is an integer giving correct delay.

From TL and TT we can caliber TD, which is defined by TL times TT inverse. The derivation of the propagation constant of TL is shown in this slide.

42 **Pad Derivation (I)**

$$T_A = r_1 \begin{bmatrix} a_1 & b_1 \\ c_1 & 1 \end{bmatrix}, \; T_B = r_2 \begin{bmatrix} a_2 & b_2 \\ c_2 & 1 \end{bmatrix}$$

$b_1, c_1/a_1$ are obtained by solving the eigenvalue problem of
$T_{TL} = T_A^{-1} T_D T_A$.

From Thru, $T_T = T_A T_B \to r_t \begin{bmatrix} a_t & b_t \\ c_t & 1 \end{bmatrix} = r_1 r_2 \begin{bmatrix} a_1 & b_1 \\ c_1 & 1 \end{bmatrix} \cdot \begin{bmatrix} a_2 & b_2 \\ c_2 & 1 \end{bmatrix}$

$$\therefore c_2 = \frac{c_t - a_t c_1/a_1}{1 - b_t c_1/a_1},$$

$$\frac{b_2}{a_2} = \frac{b_t - b_1}{a_t - b_1 c_t},$$

$$a_1 a_2 = \frac{a_t - b_1 c_t}{1 - b_t c_1/a_1}$$

This slide shows the derivation of the T matrix of the two pads.

43   Pad Derivation (II)

From Reflect,

$$s_{11}^R = \frac{a_1\Gamma + b_1}{c_1\Gamma + 1}$$

$$S_R = \begin{bmatrix} s_{11}^R & 0 \\ 0 & s_{22}^R \end{bmatrix}$$

$$\therefore a_1 = \frac{s_{11}^R - b_1}{\Gamma(1 - s_{11}^R c_1/a_1)} \quad (1)$$

Similarly, $\quad a_2 = \dfrac{s_{22}^R - b_2}{\Gamma(1 - s_{22}^R b_2/a_2)}$

$$\therefore a_1/a_2 = \frac{(s_{11}^R - b_1)(1 + s_{22}^R b_2/a_2)}{(s_{22}^R + c_2)(1 + s_{11}^R c_1/a_1)}$$

$$\therefore a_1 = \pm\sqrt{\frac{(s_{11}^R - b_1)(1 + s_{22}^R b_2/a_2)(a_t - b_1 c_t)}{(s_{22}^R + c_2)(1 - s_{11}^R c_1/a_1)(1 - b_t c_1/a_1)}}, a_2 = \frac{(a_t - b_1 c_t)}{a_1(1 - b_t c_1/a_1)}$$

Sign of a_1 is determined by the consistency with Eq. (1)

U sually, not only one line but also multiple lines are used, because there is always error term in measurement. Propagation constant gamma can be averaged, if lines with moderate insertion loss are used.

---

44   De-embedding by TRL

Now, $\begin{bmatrix} a_1 & b_1 \\ c_1 & 1 \end{bmatrix} = \dfrac{1}{r_1}T_A$ and $\begin{bmatrix} a_2 & b_2 \\ c_2 & 1 \end{bmatrix} = \dfrac{1}{r_2}T_B$ are determined.

However, r_1 and r_2 are not individually determined although its product $r_1 r_2 (= r_t)$ is obtained from Thru.

From $\dfrac{1}{r_1}T_A$, $\dfrac{1}{r_2}T_B$ and $r_1 r_2$, DUT is de-embedded as follows.

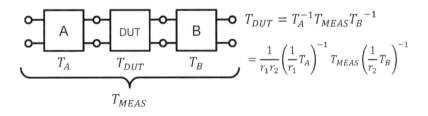

$$T_{DUT} = T_A^{-1} T_{MEAS} T_B^{-1}$$

$$= \frac{1}{r_1 r_2}\left(\frac{1}{r_1}T_A\right)^{-1} T_{MEAS} \left(\frac{1}{r_2}T_B\right)^{-1}$$

I n real situation, propagation constant gamma can be averaged by multiple line, but there are errors in real measurement. Alpha and beta can be precisely extracted even delta L1 and delta L2 are short. If measurement result is fluctuated, the insertion loss is high.

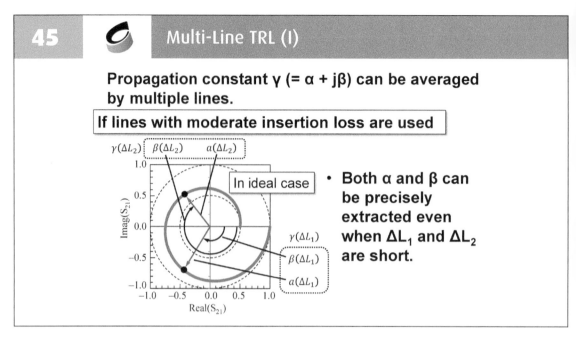

45 **Multi-Line TRL (I)**

Propagation constant γ (= α + jβ) can be averaged by multiple lines.

If lines with moderate insertion loss are used

However, if the insertion loss is not so high, low loss lines will be available. In the low loss of transmission line, trajectory is shown in this slide. Even though beta is changed greatly, it can be precisely extracted from delta L1 and delta L2. The changing of the alpha is not so large, therefore alpha or attenuation constant can't be extracted precisely.

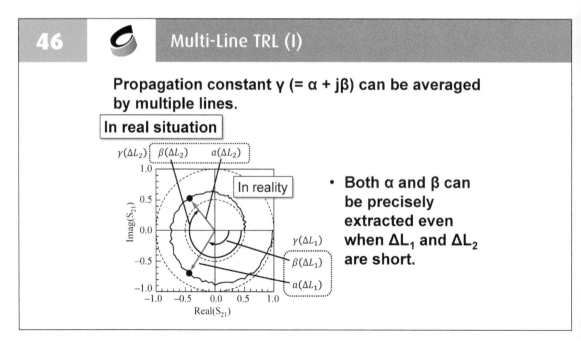

46 **Multi-Line TRL (I)**

Propagation constant γ (= α + jβ) can be averaged by multiple lines.

In real situation

This is one example in the low frequency circuit. alpha is attenuation constant in the low frequency. Under 100ghz, alpha is very simple. Exceeding 100ghz, alpha is changing.

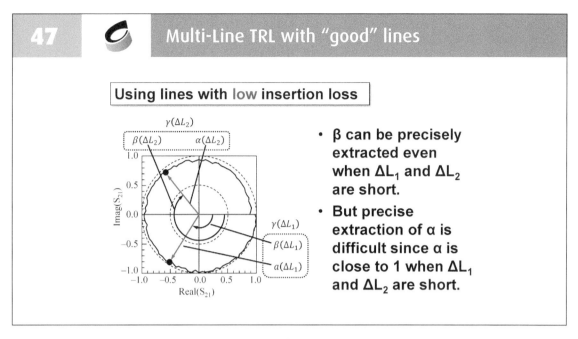

47 — **Multi-Line TRL with "good" lines**

Using lines with low insertion loss

- **β can be precisely extracted even when ΔL₁ and ΔL₂ are short.**
- **But precise extraction of α is difficult since α is close to 1 when ΔL₁ and ΔL₂ are short.**

n addition, alpha bring the length long depends on the line length this.

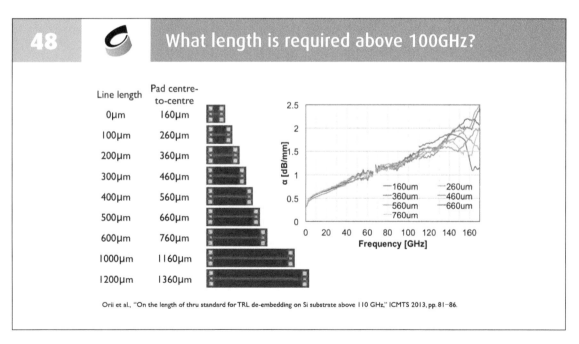

48 — **What length is required above 100GHz?**

Line length	Pad centre-to-centre
0µm	160µm
100µm	260µm
200µm	360µm
300µm	460µm
400µm	560µm
500µm	660µm
600µm	760µm
1000µm	1160µm
1200µm	1360µm

Orii et al., "On the length of thru standard for TRL de-embedding on Si substrate above 110 GHz," ICMTS 2013, pp. 81–86.

f we increase the frequency up to 325ghz, this fluctuation expands and we can't obtain precise alpha. This slides shows the electromagnetic simulation from the 3d electromagnetic simulation. But the measurement results are completely different from simulation. If you change the line length, the measurement result will change.

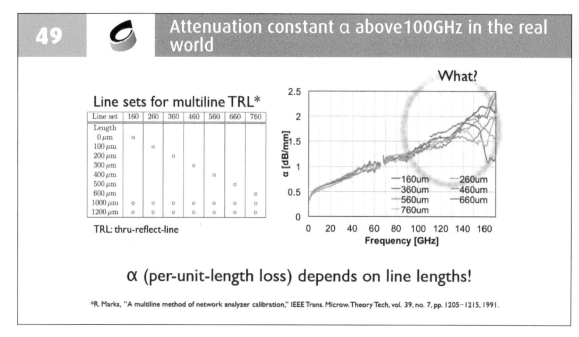

One solution to avoid this issue is to use very long transition line. Attenuation is increasing. The difference of alpha is increasing. We can precisely extract alpha.

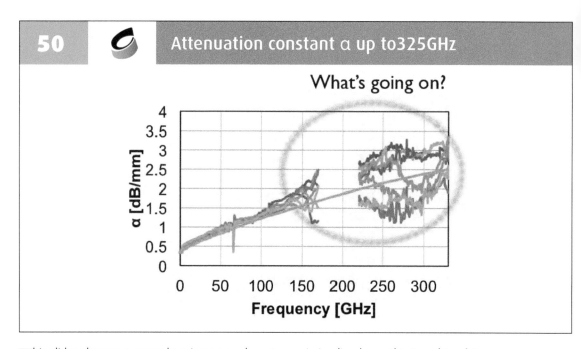

This slides shows an example using a very long transmission line here. This is a chip of 8mm.

51 — Multi-Line TRL with "long" lines

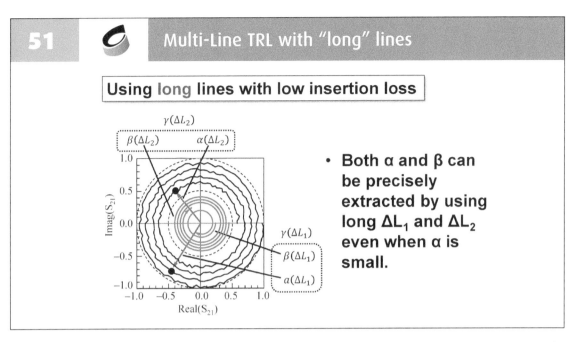

Using long lines with low insertion loss

- **Both α and β can be precisely extracted by using long ΔL₁ and ΔL₂ even when α is small.**

This is alpha function of frequency from zero to 325GHz. If we want to evaluate transmission line precisely, we have to use very long transmission line otherwise measurement result will fluctuate. But it is not in that even though alpha, alpha is fluctuated a short transmission line.

52 — New transmission line chip(65nm CMOS)

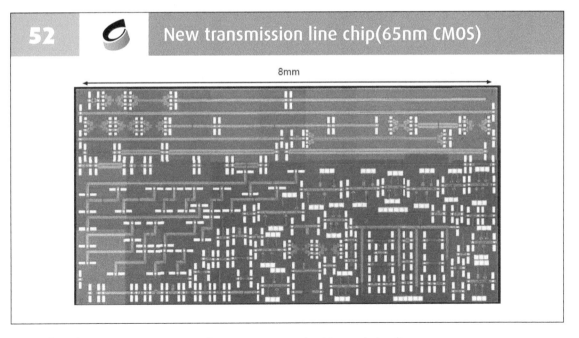

8mm

Beta is a phase constant very smooth even you use a short transmission line.

53 — Summary

- **Characteristics of devices obtained with TRL give the best fit with the simulation results.**
- **TRL method is suitable for high-frequency CMOS circuits.**
- **Long lines are preferable for precise multiline TRL de-embedding.**

This is the short summary, characteristic of the device obtained with TRL give the best fit with the simulation results and the TRL method is suitable for high frequency CMOS circuit and long line are preferable for precise multiline TRL de-embedding.

5. MATERIAL-PARAMETER EXTRACTION FOR ELECTROMAGNETIC FIELD ANALYSIS

Electromagnetic simulation is very important, not only the leading wire but also passive devices including inductance or capacitance in high frequency.

54 — Introduction

- **The characteristics of passive devices are simulated by electromagnetic field (EM) analysis**

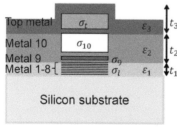

Standard CMOS process

Process parameters are required for EM analysis

σ : conductivity
t : thickness of dielectrics
ε : dielectric constant
$\tan\delta$: dielectric loss tangent

Material parameters should be calibrated to improve EM simulation accuracy in millimeter wave and terahertz.

The transmission line parameters are changed by shape, i.e. per-unit-length inductors and per-unit-length capacitors. Conductors is changed by the shape and material.

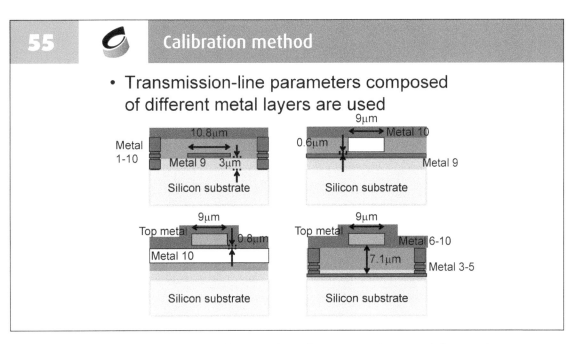

55 **Calibration method**

- Transmission-line parameters composed of different metal layers are used

For example, skin effect of the material or metal is proportional to square one over conductor. And from tangent delta and alpha, alpha is attenuation constant and vp is a phase velocity from these parameters tangent delta is like this, and also inductance is a function of shape and capacitance is function of shape and dielectric constant assuming that permeate constant is one in the normal CMOS process.

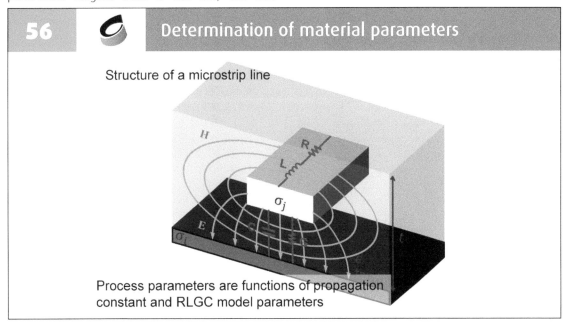

56 **Determination of material parameters**

Structure of a microstrip line

Process parameters are functions of propagation constant and RLGC model parameters

Then we can obtain precise propagation constant including phase constant from the beta and attenuation constant. In the low frequency, we can obtain all RLGC model.

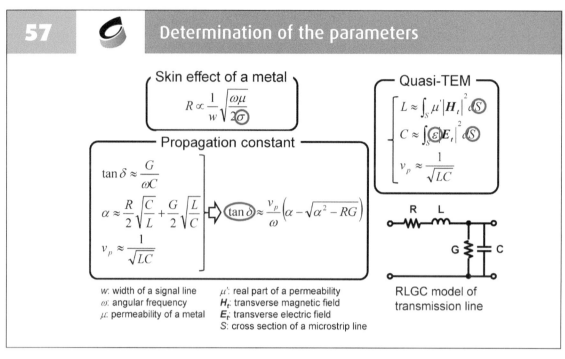

57 — Determination of the parameters

Skin effect of a metal

$$R \propto \frac{1}{w}\sqrt{\frac{\omega\mu}{2\sigma}}$$

Quasi-TEM

$$L \approx \int_S \mu' |H_t|^2 \, dS$$

$$C \approx \int_S \varepsilon |E_t|^2 \, dS$$

$$v_p \approx \frac{1}{\sqrt{LC}}$$

Propagation constant

$$\tan\delta \approx \frac{G}{\omega C}$$

$$\alpha \approx \frac{R}{2}\sqrt{\frac{C}{L}} + \frac{G}{2}\sqrt{\frac{L}{C}} \quad \Rightarrow \quad \tan\delta \approx \frac{v_p}{\omega}\left(\alpha - \sqrt{\alpha^2 - RG}\right)$$

$$v_p \approx \frac{1}{\sqrt{LC}}$$

RLGC model of
transmission line

w: width of a signal line
ω: angular frequency
μ: permeability of a metal
μ': real part of a permeability
H_t: transverse magnetic field
E_t: transverse electric field
S: cross section of a microstrip line

We use two millimeter transmission lines for this experiment with thru from 1ghz to 330ghz.

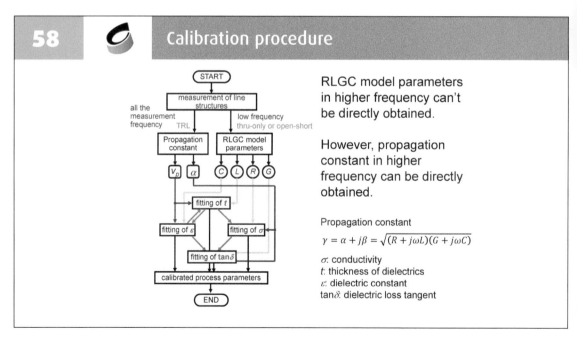

58 — Calibration procedure

START

measurement of line
structures

all the
measurement
frequency   TRL   low frequency
thru-only or open-short

Propagation
constant

RLGC model
parameters

v_p α C L R G

fitting of t

fitting of ε   fitting of σ

fitting of tanδ

calibrated process parameters

END

RLGC model parameters
in higher frequency can't
be directly obtained.

However, propagation
constant in higher
frequency can be directly
obtained.

Propagation constant

$$\gamma = \alpha + j\beta = \sqrt{(R + j\omega L)(G + j\omega C)}$$

σ: conductivity
t: thickness of dielectrics
ε: dielectric constant
tanδ: dielectric loss tangent

Two millimeter transmission lines is not sufficient for precise evaluation. We need eight millimeter transmission lines, but the cost is high.

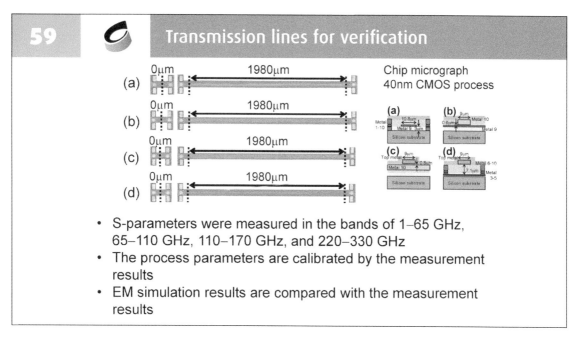

59

Transmission lines for verification

- S-parameters were measured in the bands of 1–65 GHz, 65–110 GHz, 110–170 GHz, and 220–330 GHz
- The process parameters are calibrated by the measurement results
- EM simulation results are compared with the measurement results

This is measurement result and calibration. There are four patterns of transmission line, a, b, c, d.

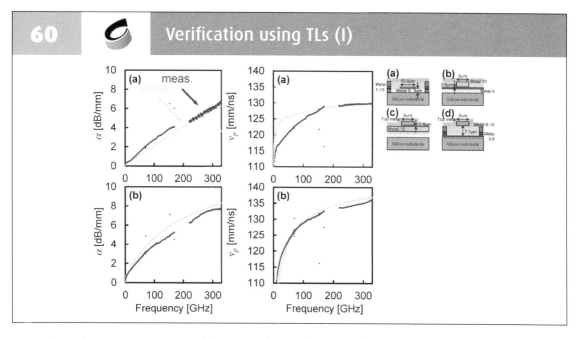

Verification using TLs (I)

We have four measurement result by calibrating material parameters. Even though we can see some discrepancy between the EM simulation after calibration of the material constant measurement results of.

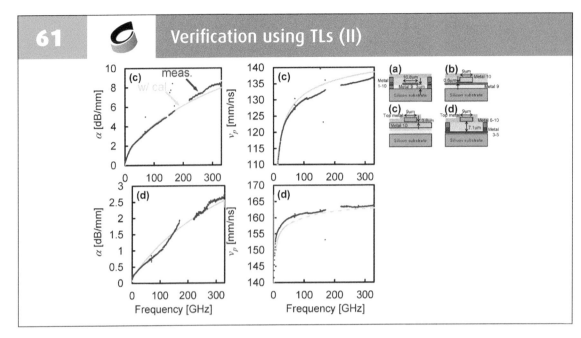

And these are results for transmission line c and d. The measurement result almost fits to the EM simulation. We could calibrate the material parameters by using a transmission line. If we can apply longer transmission line, we can obtain much more good data.

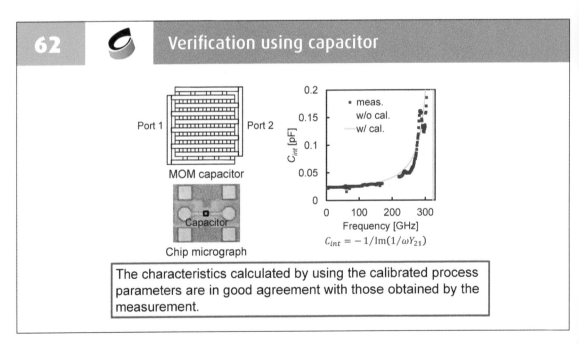

This is another result of the MOM capacitor. After de-embedding, the EM simulation results almost fit to the measurement results.

63 Conclusions

- Process parameters are calibrated for CMOS technology up to 330 GHz.
- Four kinds of transmission lines are used to extract the process parameters of each layer.
- EM simulation results with the calibrated process parameters were in good agreement with the measurement results up to 330 GHz.

The calibration is mostly fit to the measurement result. This is a verification of the parameter extraction. The process parameters like this. Otherwise EM simulation results will not fit the measurement result.

6. LINEAR MODELING EQUIVALENT CIRCUIT EXTRACTION

64 Millimeter-Wave Design

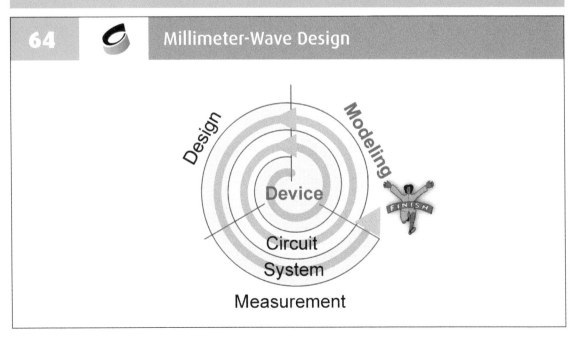

After de-embedding, we can obtain S parameters of the core circuit. The S parameters can be changed to any matrix in this case, i.e. admittance matrix.

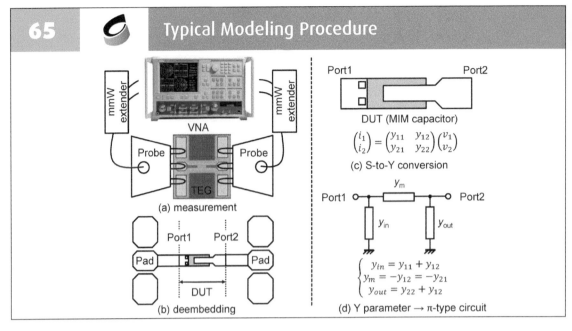

65 — **Typical Modeling Procedure**

Port1   Port2

DUT (MIM capacitor)

$$\begin{pmatrix} i_1 \\ i_2 \end{pmatrix} = \begin{pmatrix} y_{11} & y_{12} \\ y_{21} & y_{22} \end{pmatrix} \begin{pmatrix} v_1 \\ v_2 \end{pmatrix}$$

(c) S-to-Y conversion

(a) measurement

(b) deembedding

y_m

Port1 o———————o Port2

y_{in} y_{out}

$$\begin{cases} y_{in} = y_{11} + y_{12} \\ y_m = -y_{12} = -y_{21} \\ y_{out} = y_{22} + y_{12} \end{cases}$$

(d) Y parameter → π-type circuit

A fter obtaining y_{in} and y_{out}, equivalent circuit can be applied.

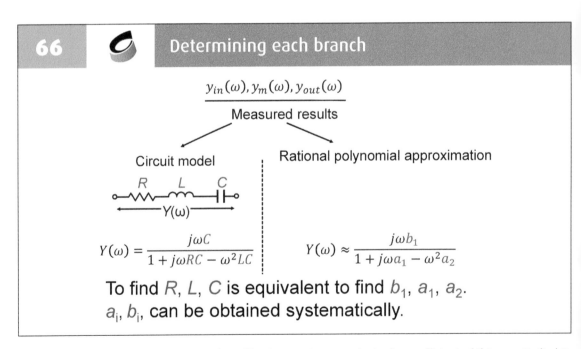

66 — **Determining each branch**

$$y_{in}(\omega), y_m(\omega), y_{out}(\omega)$$

Measured results

Circuit model   Rational polynomial approximation

R L C

———Y(ω)———

$$Y(\omega) = \frac{j\omega C}{1 + j\omega RC - \omega^2 LC}$$

$$Y(\omega) \approx \frac{j\omega b_1}{1 + j\omega a_1 - \omega^2 a_2}$$

To find R, L, C is equivalent to find b_1, a_1, a_2.
a_i, b_i, can be obtained systematically.

I f we want to obtain the parameters of RLC like this, we have to obtain the coefficient of this one. To find R, L, C equivalent to find b1, a1, a2. And general ai, bi can be obtained systematically.

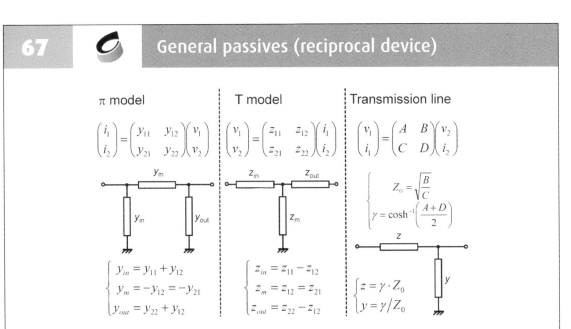

This slides shows different types of equivalent circuit and the relationship between admittance matrix component

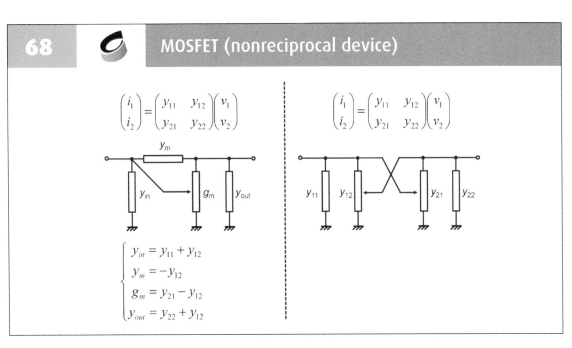

MOSFET is nonreciprocal device, which usually is evaluated by admittance matrix. Parameters of each branch is related to these components by this equation. And sometimes admittance matrix can be directly written by the equivalent circuit.

69 Least squares approximation

$$\underset{\substack{\text{Measured}\\\text{results}}}{\underline{Y(\omega)}} = \underset{\substack{\text{Real}\\\text{part}}}{\underline{G(\omega)}} + \underset{\substack{\text{Imaginary}\\\text{part}}}{\underline{jB(\omega)}} \approx \underset{\substack{\text{Rational polynomial}}}{\frac{j\omega b_1}{1 + j\omega a_1 - \omega^2 a_2}}$$

$$G(\omega) - \omega B(\omega)a_1 - \omega^2 B(\omega)a_2 +$$
$$j\{B(\omega) + \omega G(\omega)a_1 - \omega^2 B(\omega)a_2 - \omega b_1\} \approx 0$$

$$\begin{cases} \frac{\partial J}{\partial a_1} = 2\omega^2\big(G^2(\omega) + B^2(\omega)\big)a_1 - 2\omega^2 G(\omega)b_1 = 0 \\[2mm] \frac{\partial J}{\partial a_2} = 2\omega^4\big(G^2(\omega) + B^2(\omega)\big)a_2 + \\[1mm] \qquad\quad 2\omega^3 B(\omega)b_1 - 2\omega^2\big(G^2(\omega) + B^2(\omega)\big) = 0 \\[2mm] \frac{\partial J}{\partial b_1} = -2\omega^2 G(\omega)a_2 + 2\omega^3 B(\omega)a_2 + 2\omega^2 b_1 - 2\omega B(\omega) = 0 \end{cases}$$

This slide shows how to determine these coefficients. Squares approximation is used to find these parameters. We already obtain measurement results, measurement results have real part and imaginary part.

70 Matrix expression

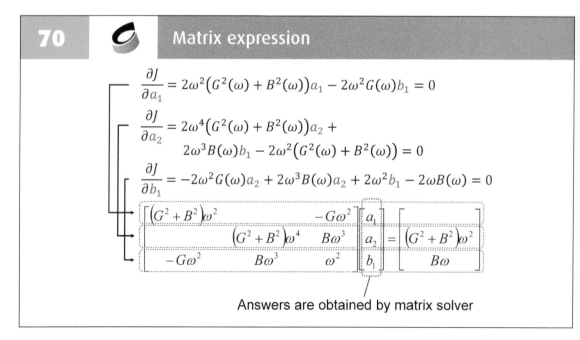

$$\frac{\partial J}{\partial a_1} = 2\omega^2\big(G^2(\omega) + B^2(\omega)\big)a_1 - 2\omega^2 G(\omega)b_1 = 0$$

$$\frac{\partial J}{\partial a_2} = 2\omega^4\big(G^2(\omega) + B^2(\omega)\big)a_2 +$$
$$\qquad\quad 2\omega^3 B(\omega)b_1 - 2\omega^2\big(G^2(\omega) + B^2(\omega)\big) = 0$$

$$\frac{\partial J}{\partial b_1} = -2\omega^2 G(\omega)a_2 + 2\omega^3 B(\omega)a_2 + 2\omega^2 b_1 - 2\omega B(\omega) = 0$$

$$\begin{bmatrix} \big(G^2 + B^2\big)\omega^2 & & -G\omega^2 \\ & \big(G^2 + B^2\big)\omega^4 & B\omega^3 \\ -G\omega^2 & B\omega^3 & \omega^2 \end{bmatrix} \begin{bmatrix} a_1 \\ a_2 \\ b_1 \end{bmatrix} = \begin{bmatrix} \big(G^2 + B^2\big)\omega^2 \\ \\ B\omega \end{bmatrix}$$

Answers are obtained by matrix solver

If we can write this matrix, then easily we can obtain these coefficients. Then finally we obtain R, C, L models.

71   Circuits and rational-polynomial expr.

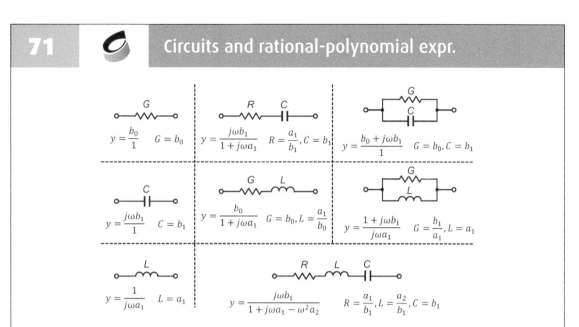

There are many models can be used as the equivalent circuit.

72   Conclusions

- Systematic equivalent-circuit generation is introduced for high frequency device modeling
  - Rational polynomial approximation
  - Smoothing and extrapolation with high accuracy
- Approximation results can be converted to equivalent circuit
- Approximation can be extended to rational polynomial with arbitrary dimensions

This is systematic approach and we already have a program for arbitrary order polynomial, rational polynomial equation and fitting. So we can build any type of equivalent circuit by this kind of modelling.

7. PARASITIC RESISTANCES

73 — Parasitic resistances of MOSFET

- **Extraction of parasitic resistances are important for modeling of MOSFETs**

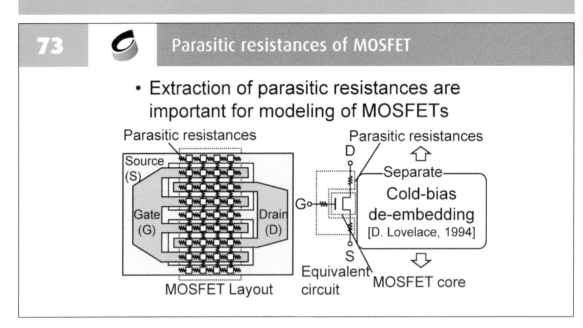

xtraction of parasitic resistance is important to modeling of MOSFET. So usually there are many parasitic resistances in the MOSFET. After extraction of the parasitic resistances, MOSFET core can be extracted like de-embedding. Extraction is important to precise model of the core MOSFET. Somehow we have to separate these parasitic resistances and MOSFET core. One method to separate parasitic resistance and MOSFET core is cold bias de-embedding.

74 — "RF" parasitic resistances

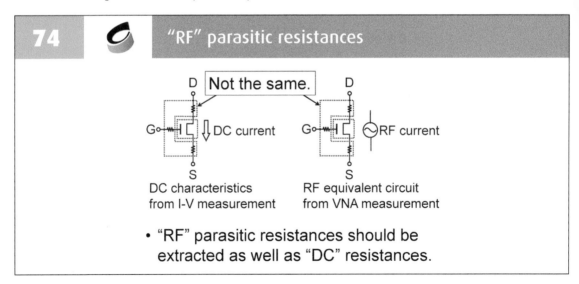

- **"RF" parasitic resistances should be extracted as well as "DC" resistances.**

F parasitic resistance are not the same as the DC parasitic resistances. Parasitic resistance can be measured in DC, but sometimes RF parasitic resistance is different from DC parasitic resistance. Resistance is the function of temperature. In DC, temperature is time invariant. Even though the temperature is changed, temperature time constant is very long. But in RF, due to the time constant, temperature is always changing because the current is always changing. RF resistance is not always equal to DC resistance.

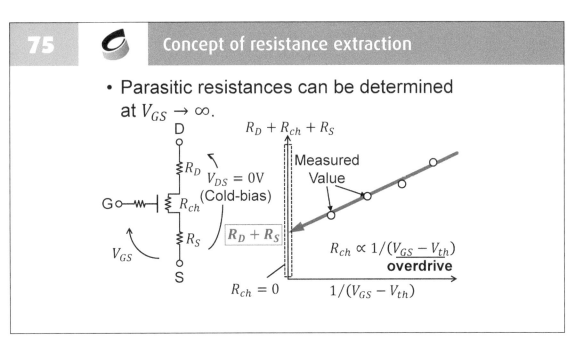

75 — Concept of resistance extraction

- Parasitic resistances can be determined at $V_{GS} \to \infty$.

$R_D + R_{ch} + R_S$

Measured Value

$R_D + R_S$

$R_{ch} \propto 1/(V_{GS} - V_{th})$
overdrive

$R_{ch} = 0$ $1/(V_{GS} - V_{th})$

Parasitic resistance can be determined when gate source voltage is infinite. If the gate source voltage is infinite of course is a real MOSFET apply infinite gate source voltage, MOSFET is destroyed. But theoretically, we can obtain these parasitic resistances using cold-bias de-embedding method.

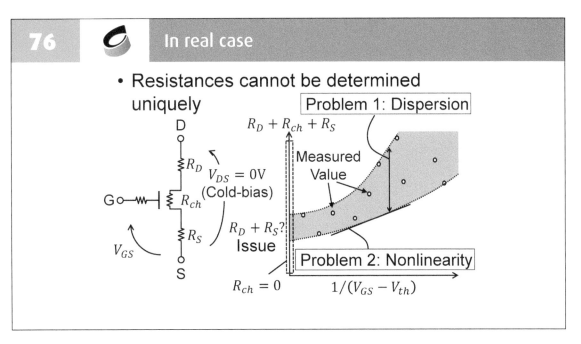

76 — In real case

- Resistances cannot be determined uniquely

Problem 1: Dispersion

$R_D + R_{ch} + R_S$

Measured Value

$R_D + R_S$?
Issue

Problem 2: Nonlinearity

$R_{ch} = 0$ $1/(V_{GS} - V_{th})$

But in real case in art, resistance cannot be determined uniquely. There is fluctuated in the measured value. The second is not linear. To solve these two issues, one way is to extract the resistance in both low and high frequency version.

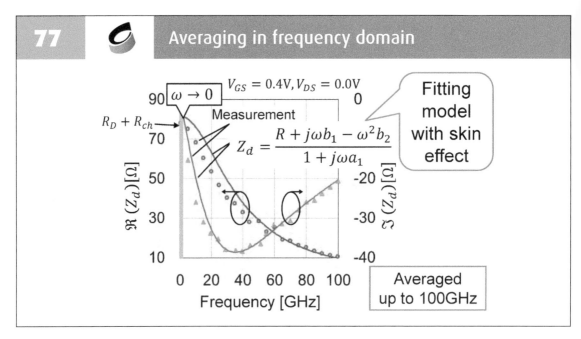

This is one example. We want to obtain a zero frequency. However, there are always fluctuations in the measurement. So, we apply equivalent circuit instruction using many frequencies.

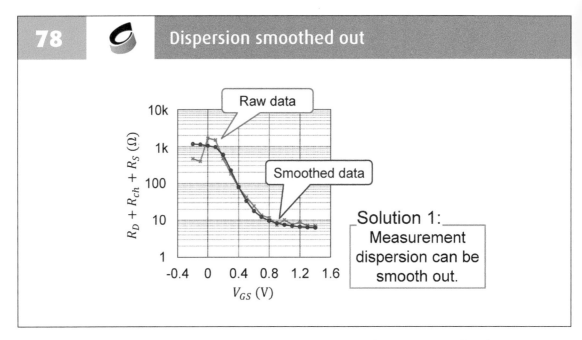

By applying this kind of equivalent circuit model, these measure data is smooth out. then by using the equivalent circuit if we substitute omega is zero, then these values are removed and only zero hertz risk as existed. By using high frequency information, fluctuation can be smooth out.

79   Linearity vs threshold Voltage

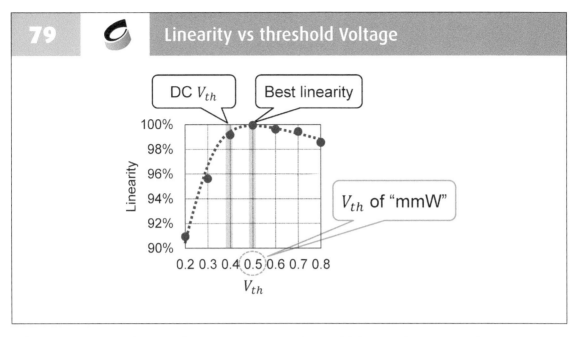

The next issue is non-linearity. The resistance is not proportional to one over V_{GS} minus V_{th}. Channel resistance dependent should be linear according to this model, but in real case it is not linear. We assume DC threshold voltage is not equal to millimeter wave threshold voltage.

80   "mmW V_{th}" linearizes R_{ch}

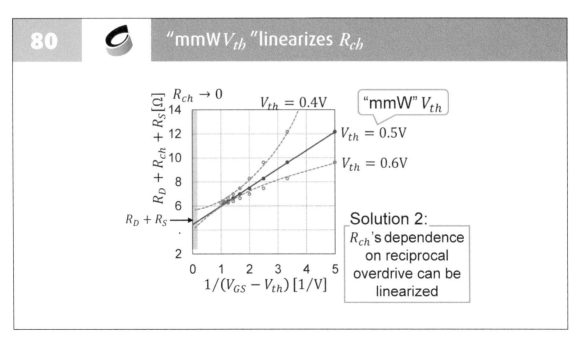

This is one result if we change the threshold. The virtual threshold voltage, DC resistance voltage is zero point for volt, but we can change threshold voltage in calibration. Then this curve shows linearity. If we change threshold voltage from zero point four volt to zero point five volt, we can obtain almost best linearity.

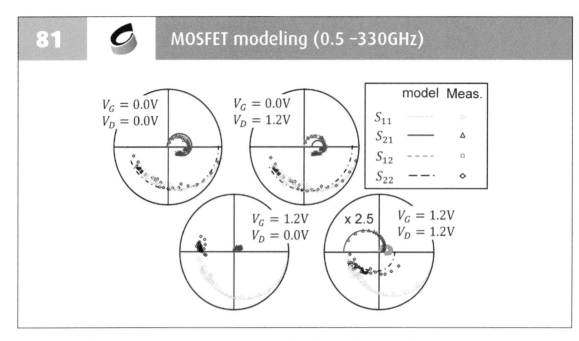

This is a result from 0.5Hz to 300ghz. Measure data fits the simulation results.

82   Conclusion

- Improved cold-bias de-embedding is introduced
    - Bias-independent parasitic resistances can be extracted reliably by
        - Smoothing out measurement dispersion
        - Introducing "mmW V_{th}," which linearizes dependence of channel resistance on reciprocal overdrive voltage
    - Bias-dependent core MOSFET model becomes simple
- 65nm MOSFET up to 330 GHz is successfully modeled

To conclude, the precise parasitic extraction is very important for MOSFET. It can be done by applying the method shown in this slide.

8. CIRCUIT DESIGN IN HIGH FREQUENCY

83   **Millimeter-Wave Design**

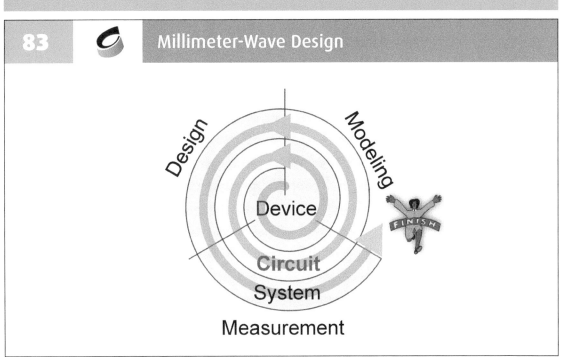

84   **Numerical Optimization**

- **Multi-stage mmW amplifiers are difficult to optimize due to non-negligible reverse isolation (S_{12}) of MOSFETs.**
  → **Apply computational optimization**

Ex. Four-stage E-band amplifier

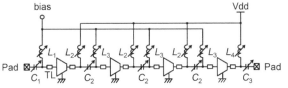

- C_1, C_2, C_3, L_1, L_2, L_3 and L_4 are optimized computationally.

First of all, we have to know how to design amplifier. This slide shows a typical amplifier. It is a very complicated process to increase the gain and make flat of the gain. Generally a numerical optimization instead of manual optimization is used.

85 Transfer Function

- **Parallel elements (stub / load *L*) are converted into *Z* matrixes.**
- **Series elements (TL / cascode stage / *C*) are converted into *T* matrixes.**

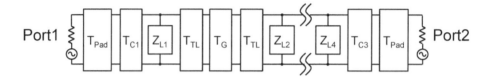

ow do we apply numerical optimization? We are using matrix component, if all the components can be written by T matrix or the T matrix can be multiplied, total T matrix can be calibrating by multiplying all the T matrix. Each component has parameters and if we apply an initial value. Initial T components amplifier is obtained.

86 Numerical Optimization

- **Conjugate gradient (CG) engine maximizes the goal function by changing vector x.**

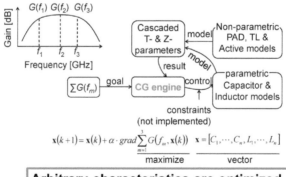

Arbitrary characteristics are optimized by redefining goal function.

f we change a little bit of all this parameters after we have an initial value, we can obtain a slope of the goal function.

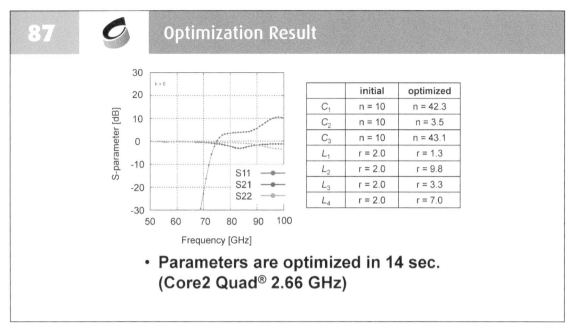

	initial	optimized
C_1	n = 10	n = 42.3
C_2	n = 10	n = 3.5
C_3	n = 10	n = 43.1
L_1	r = 2.0	r = 1.3
L_2	r = 2.0	r = 9.8
L_3	r = 2.0	r = 3.3
L_4	r = 2.0	r = 7.0

- **Parameters are optimized in 14 sec. (Core2 Quad® 2.66 GHz)**

This is one example for a 4-stage amplifier, embedding amplifier. The gain and return coefficient will be changed by applying a numerical optimization. Even though the coding of the program takes some time but calibration is very fast.

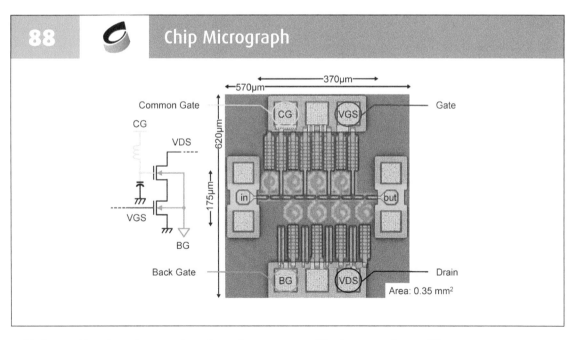

This is our chip micro photography using a four stage amplifier is a cascade amplifier.

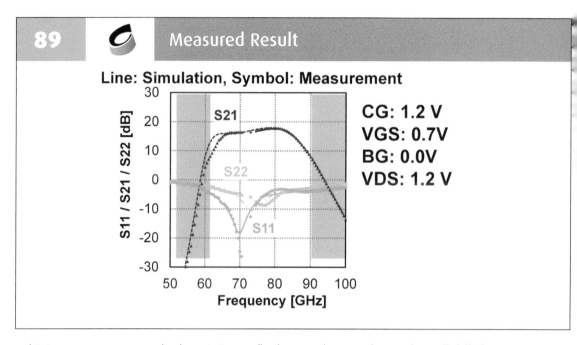

This is a measurement result. The gain is very flat from 60ghz to 80ghz. Maybe well skilled engineer cannot realize this kind of flat results without using numerical optimization.

9. 300GHZ CMOS TRANSMITTER – REALIZE IMPOSSIBILITY

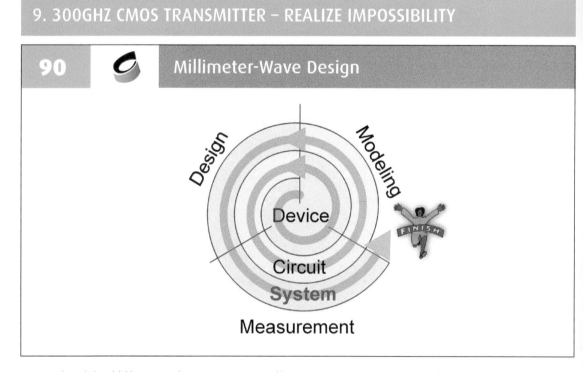

From this slide, I'd like to explain transceiver itself, my main topic is transceiver but for the moment, up to the previous slide, all explanation is preparation.

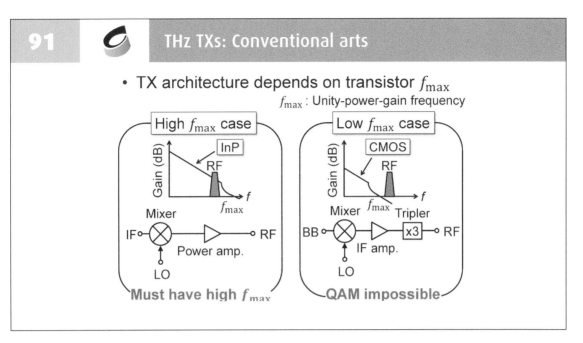

91 THz TXs: Conventional arts

- TX architecture depends on transistor f_{max}

f_{max} : Unity-power-gain frequency

Conventional transmitter depend on the unity power gain frequency of transistor, fmax. If fmax is sufficiently high, normal architecture with mixer and power amplifier is applied. If fmax is low, one conventional solution is to use QAM, Quadrature Amplitude Modulation. However the tripler used in QAM is non-linear when the frequency is very high, i.e. higher than 300GHz.

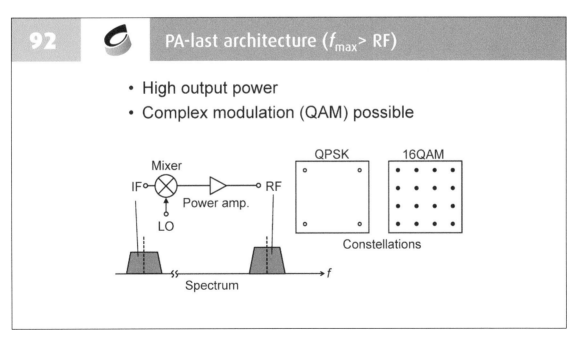

92 PA-last architecture (f_{max}> RF)

- High output power
- Complex modulation (QAM) possible

This is a case of PA-last architecture, which can be applied to a sufficiently high transistor. High output power can be obtained by a power amplifier and also complex modulation, i.e. QAM.

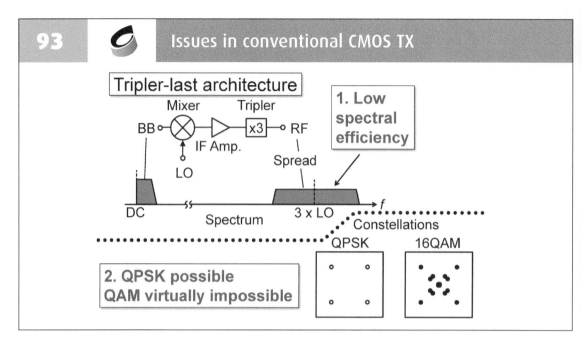

93 **Issues in conventional CMOS TX**

Tripler-last architecture

1. Low spectral efficiency

2. QPSK possible
QAM virtually impossible

But if we use a tripler. Both the frequency and the bandwidth are tripled by the tripler. The efficiency will be very low, because of the non-linearity of the tripler. QPSK is fine for tripler, by comparing symbols for the constellations of QPSK is preserved. However due to the non-linearity of the tripler if we apply 16QAM here, output of the tripler constellation will change to her so 16QAM constellation is destroyed by tripler, this also reduces the spectral efficiency.

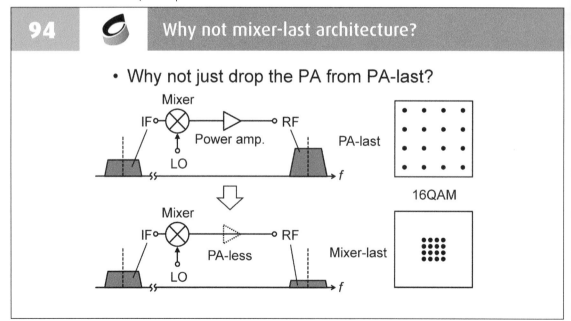

94 **Why not mixer-last architecture?**

• Why not just drop the PA from PA-last?

One solution is to remove power amplifier in the CMOS, since fmax is not so sufficient. Of course, in this case we can generate 16QAM. However output power will be very small.

95   **Issues in "mixer last"**

- f_{max} < RF means very low output power
- Massive power combining necessary but layout would be extremely complicated

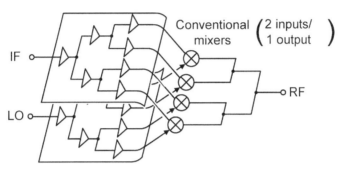

Conventional mixers $\left(\begin{array}{l}\text{2 inputs/}\\\text{1 output}\end{array}\right)$

Mixer is arranged in parallel. Mixer has two input, one is modulated IF signal one is moderate LO signal. So not only IF signal is spread for all the mixers but also the LO driver. But these frequencies are already very high, so this kind of cross section transmission line will degree the performance.

96   **Technical challenges**

- 300-GHz PA-less TX in CMOS operating beyond f_{max}
  - QAM is preferable for high data-rates
    - Thus, no frequency multiplier should be used
  - Massive power combining is inevitable without undue layout complication

    | How can we reconcile these conflicting requirements? |

This slide summarizes the technical challenges; we want to realize 300ghz power amplifier less transmitter in CMOS operating beyond fmax. Of course, QAM is preferable for its high data rate. But QAM will be destroyed by using frequency multiplier.

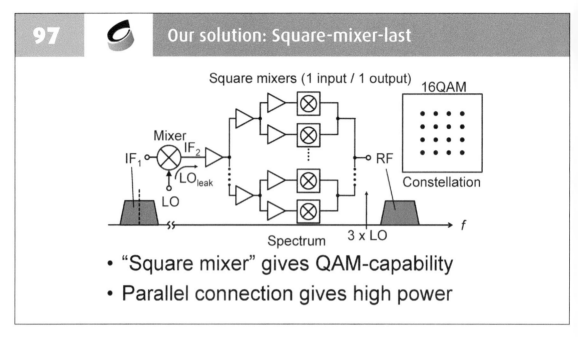

97 **Our solution: Square-mixer-last**

Square mixers (1 input / 1 output)

16QAM

Mixer

IF_1

IF_2

LO_{leak}

LO

RF

Constellation

Spectrum   3 x LO

f

- "Square mixer" gives QAM-capability
- Parallel connection gives high power

How can we reconcile these conflicting requirements, our solution is to use square mixer last architecture. There is one input and one output in square mixer. The input is modulated on IF signal by LO signal.

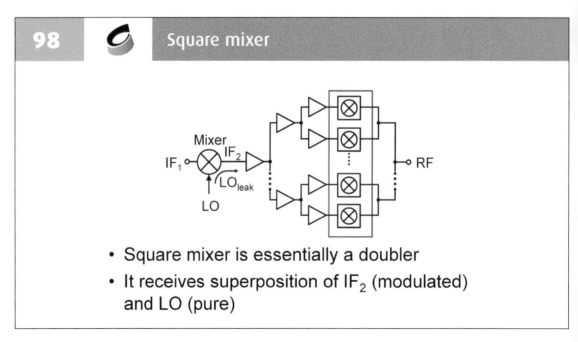

98 **Square mixer**

Mixer

IF_1

IF_2

LO_{leak}

LO

RF

- Square mixer is essentially a doubler
- It receives superposition of IF_2 (modulated) and LO (pure)

This is a summary of the square mixer. Square mixer is especially a doubler, but the operation is not doubled. It receives superposition of IF modulated IF signal and pure LO signal.

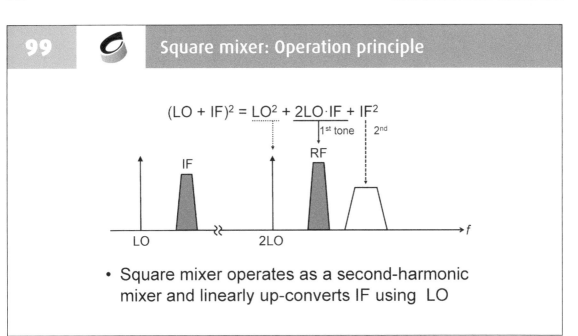

99 Square mixer: Operation principle

$$(LO + IF)^2 = LO^2 + 2LO \cdot IF + IF^2$$

- Square mixer operates as a second-harmonic mixer and linearly up-converts IF using LO

This slide shows the operation principle of square mixer. Circuit operation of the square mixer is a doubler. Both LO signal and IF signal is given to the doubler. This are three components, LO square, two times LO IF, and IF square. We don't want the first and the third components, therefore somehow we have to cancel these two components.

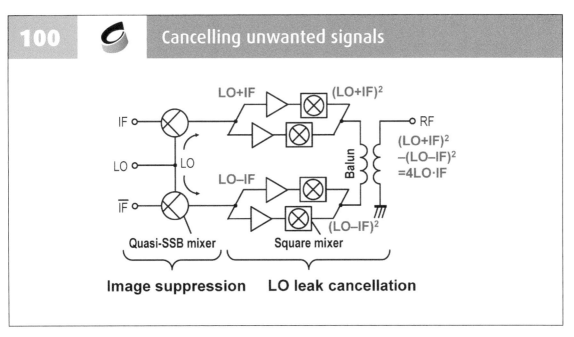

100 Cancelling unwanted signals

To cancel these components, we apply to two path architecture here. The output of the two path components are given to balun. The balun generate differential. Subtract from LO plus IF square by LO minus IF square then we obtain four times LO IF. LO square and IF square will be canceled by applying these two path architecture.

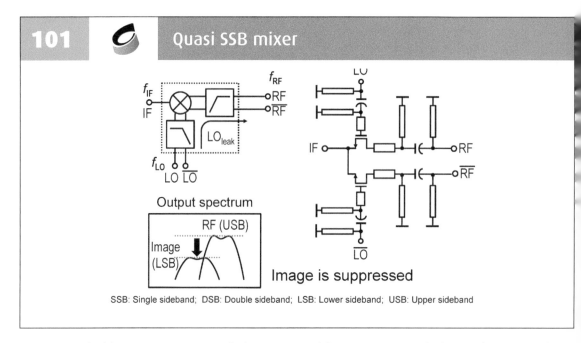

101 Quasi SSB mixer

Output spectrum

Image is suppressed

SSB: Single sideband; DSB: Double sideband; LSB: Lower sideband; USB: Upper sideband

SB is supplied by Quasi SSB mixer, which is composed by two mixers with the quadrature signal. But quadrature signal is not easy to apply in the high frequency so we apply different approaches to suppress LSB. Our technique is to make a filter in the mixer in the arrow signal. Low pass filter is applied, if the frequency of the LO is lower, amplitude of LO is high. A high pass filter is applied at the output compensate with robust filter. Then only the LSB signal is suppressed and RF USB signal is increased.

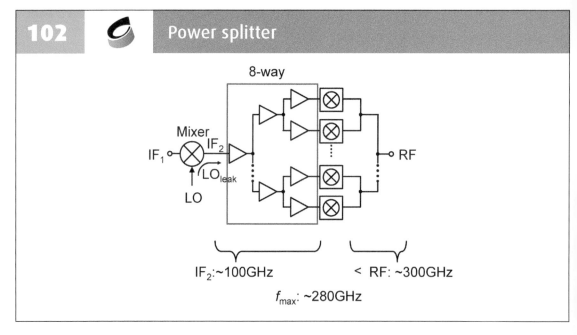

102 Power splitter

8-way

IF_2:~100GHz

< RF: ~300GHz

f_{max}: ~280GHz

Since we apply multiplier power architecture to increase the output power, we have to make a power splitter.

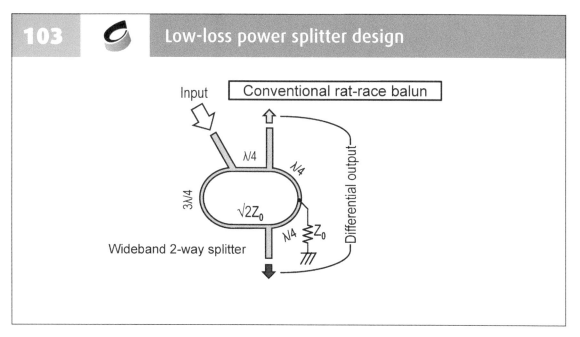

103 Low-loss power splitter design

We use a rat-race balun in the low-loss power splitter. The rat-race balun is usually used for single signal to differential signal.

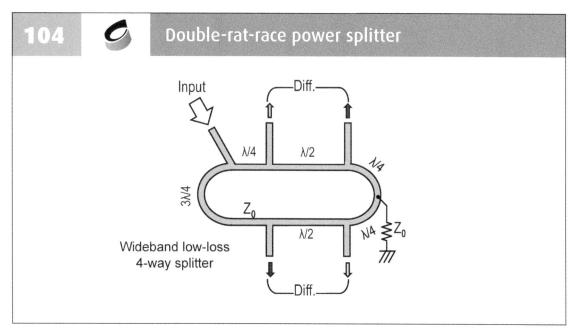

104 Double-rat-race power splitter

If we expand this rat race balun as shown in this slide, then we have two sets of differential signals. One wavelength is inserted in the balun so the phase is preserved, even though the total length is changed. Then we can obtain two sets of differential signals. By using this balun input signal is separate to four signals.

105 Schematic of 300GHz TX

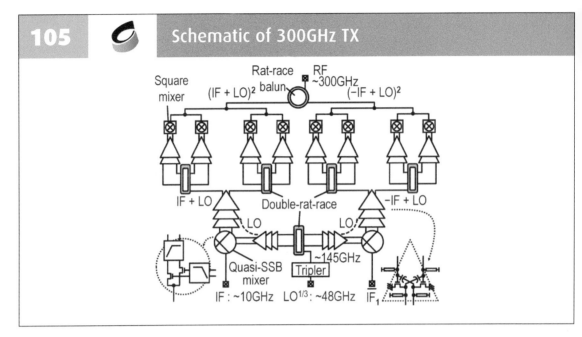

This is a final architecture of the 300ghz transmitter [xx]. Firstly, we, externally apply around 50ghz signal to the LO, the double rat race, the tripler. Then the tripled to around 150ghz. The signals are amplified by the amplifier with double rat race balun and finally those signals are given to square mixer. Finally, by applying this signal and this signal to our rat race balun, only wanted signal is generated.

106 Chip micrograph

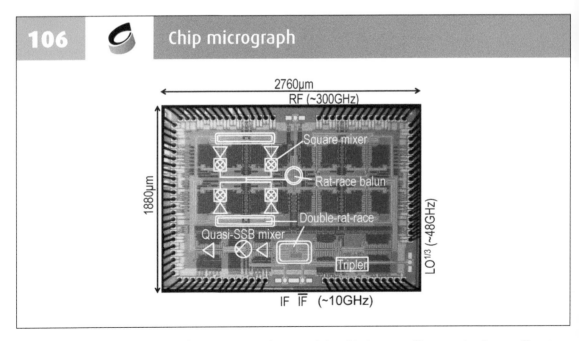

This is chip micro graph of 300ghz transmitter. The size of the chip is two millimeters by three millimeters. This is fabricated with 40nm CMOS processing.

107 Measurement setup

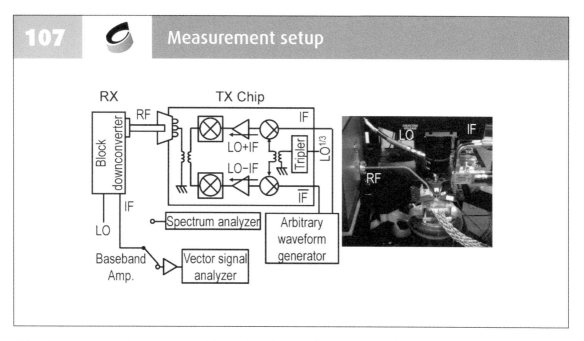

The chip is measured by a wave guides probe. The output signal of the wavelength probe is connected to the block downconverter. This 300ghz frequency is down converted to IF frequency. That IF signal is applied to vector signal analyzer and modulation is generated by arbitrary waveform generator.

108 Measured linearity

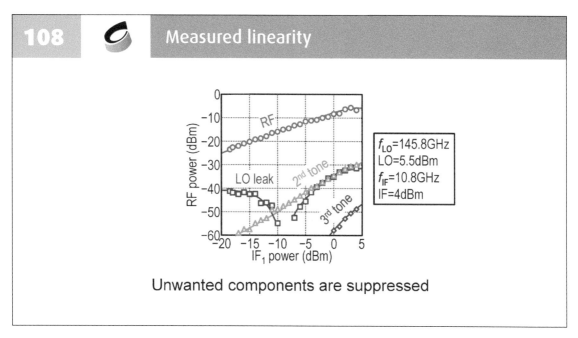

Unwanted components are suppressed

This is a measurement result, this is the output power as the functions of the input power, if input power is increasing, nearly increasing to output power like this. And LO leak and second tone is unwanted signal as I told you LO square and IF square, those signals are well suppressed like this.

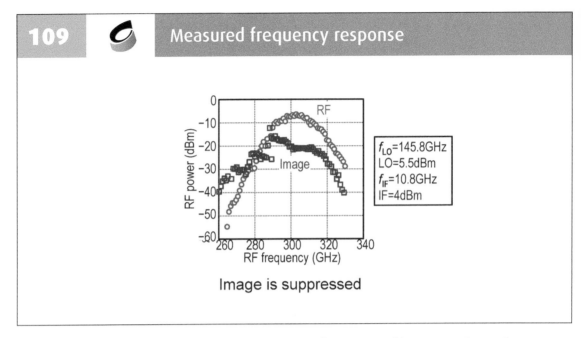

109 Measured frequency response

Image is suppressed

A nd this is the RF signal this is image signal, image signal is suppressed by SSB, Quasi SSB mixer.

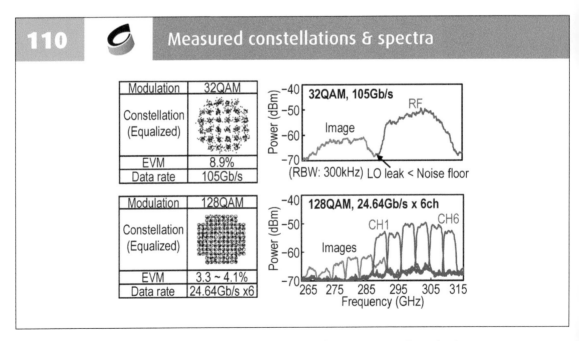

110 Measured constellations & spectra

This is a measure constellations and spectra, upper case is the 32QAM with 21Gbps with 100Gbps data rate. You can clearly see constellation even if the data rate is 105Gbps. The lower case is 128QAM with 24.64Gbps data rate and six channels.

111 Performance comparison

	[1]	[2]	[3]	[4]	[5]	This work	
Technology	250nm InP	35nm GaAs	35nm GaAs	0.13μm SiGe	40nm CMOS	40nm CMOS	
Freq. (GHz)	300	240	300	240	300	**302**	**289–311**
Modulation	QPSK	8PSK	QPSK	64QAM	16QAM	**32QAM**	**128QAM**
Pout (dBm)	–	-3.5	-4	7	-14.5	**-5.5**	
Pdc (W)	–	–	–	0.54	1.4	**1.4**	
Data rate (Gb/s)	50	96	64	1.02	28	**105**	**24.64 x 6**

[1] Song et al., TMTT, 2014.   [4] Sarmah et al., TMTT, 2016.
[2] Boes et al., IRMMW-THz, 2014.   [5] Takano et al., Electron. Lett., 2016.
[3] Kallfass et al., IEICE Trans., 2015.

This is performance comparison. This is our work we apply 100Gbps data rate or 25Gbps times or six channels. The highest data rate could be achieved using our transmitter.

112 Conclusion

- 105Gb/s transmission data rate at 300GHz (> f_{max}) using CMOS
  - Square-mixer-last architecture with double-rat-races gave –5dBm output power
  - Double-balanced square mixer realized LO cancellation
  - Quasi SSB mixer suppressed unwanted image

10. 300GHZ CMOS RECEIVER

113 **300GHZ CMOS RX challenge**

- *Wide bandwidth* TRXs capable of *QAM modulation* are required for a 300-GHz ultrahigh-speed wireless communications.

- THz front end realized by *Si CMOS technology* is favorable to embed baseband logic for an intensive signal processing.

This slide summarized 300ghz CMOS receiver challenge, wide bandwidth with transceivers capable of QAM modulation are required for 300ghz ultrahigh speed wireless communications.

114 **Typical RX architecture**

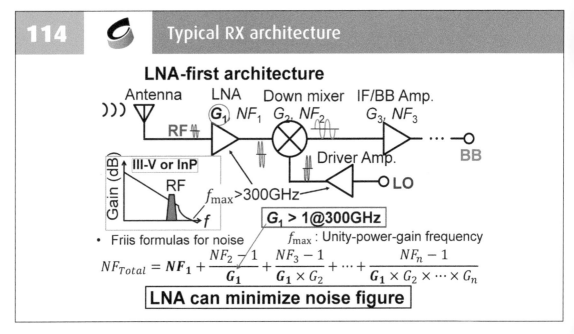

This is a typical receiver architecture. Usually IF signal is amplified by LNA. Large signal in is down converted by the down convert mixer. The driver amplifier can be also utilized and fundamental mixer can be used for the normal architect IF architecture and IF or inter frequency signal or base band signal is amplified by the base band amplifier and output signal is given here. This architecture can be applied in the III-V typically in the inp case. If the fmax is higher than 300ghz, this amplifier can be applied, them in a total NF can be minimized.

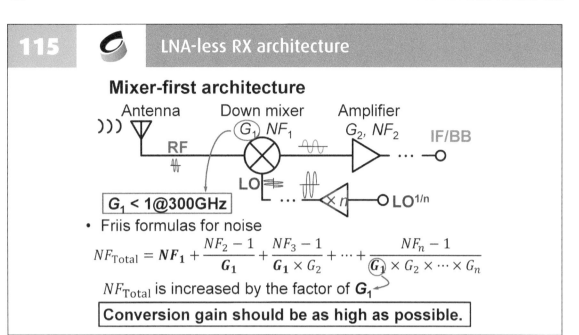

115 — LNA-less RX architecture

Mixer-first architecture

$G_1 < 1@300GHz$

- Friis formulas for noise

$$NF_{Total} = NF_1 + \frac{NF_2 - 1}{G_1} + \frac{NF_3 - 1}{G_1 \times G_2} + \cdots + \frac{NF_n - 1}{G_1 \times G_2 \times \cdots \times G_n}$$

NF_{Total} is increased by the factor of G_1

Conversion gain should be as high as possible.

If we can't use LNA, we have to apply mixer first architecture. Small signal is applied to down conversion mixer. The output is given to amplifier. But the gain of the down conversion mixer is not large. Therefore, total NF is degraded.

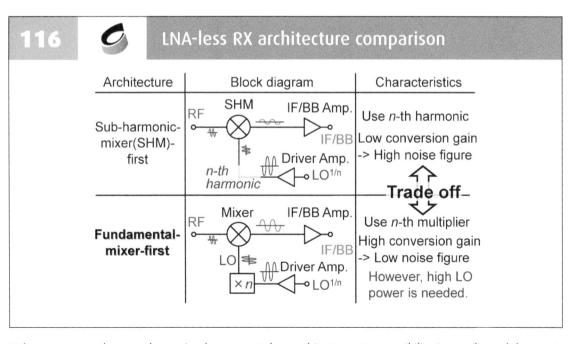

116 — LNA-less RX architecture comparison

Architecture	Block diagram	Characteristics
Sub-harmonic-mixer(SHM)-first	SHM, IF/BB Amp., IF/BB, Driver Amp., n-th harmonic, LO$^{1/n}$	Use n-th harmonic, Low conversion gain -> High noise figure
		Trade off
Fundamental-mixer-first	Mixer, IF/BB Amp., IF/BB, Driver Amp., LO, ×n, LO$^{1/n}$	Use n-th multiplier, High conversion gain -> Low noise figure, However, high LO power is needed.

There are several approaches to implement LNA less architecture. One possibility is to utilize sub harmonic mixer. Another case as shown in the lower figure.

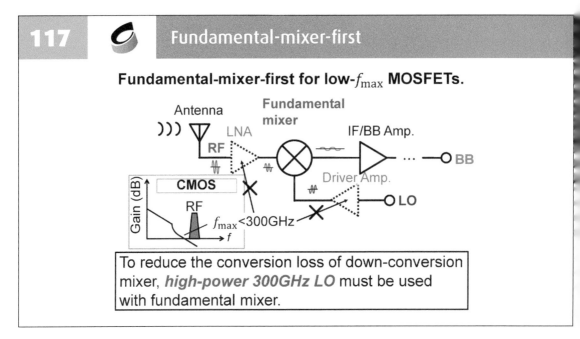

117 Fundamental-mixer-first

Fundamental-mixer-first for low-f_{max} MOSFETs.

To reduce the conversion loss of down-conversion mixer, *high-power 300GHz LO* must be used with fundamental mixer.

This is a summary of the issue: the receiver we can't apply LNA, nor apply drive amplifier. To reduce the conversion loss of the down conversion mixer, high output power 300GHz LO must be used. But how we apply higher output power without using a drive amplifier?

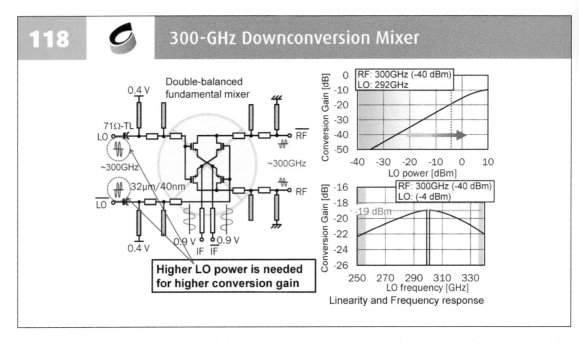

118 300-GHz Downconversion Mixer

This slide shows a simulation results of the conversion gain as a function of LO power and a schematic of the mixer. It compares the conversion gain of the mixer with the LO power.

119 LO multiplier chain

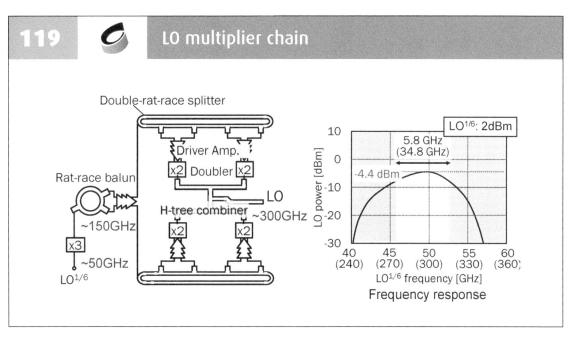

The solution is parallel architecture like transmitter. This is LO multiplier chain to obtain high output power LO. By connecting a parallel architecture, sufficiently high output power at the LO is available. The maximum output power of the LO is -4.4dbm.

120 RX block diagram

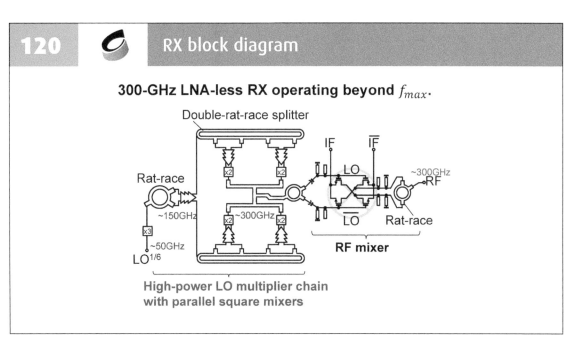

This is the final receiver block diagram, including the 300GHz LNA, the LO driver amplifier, the high power LO multiplier chain, and the mixer.

121 · Chip micrograph

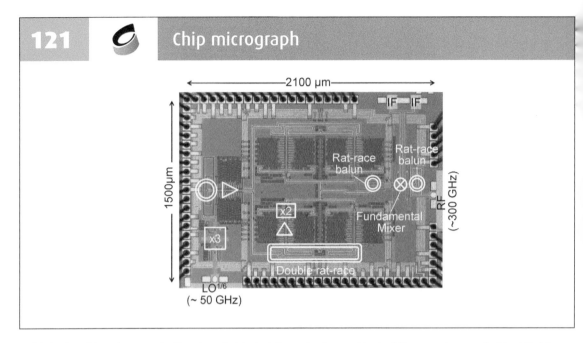

This is the chip micro graph. The size of ship is 1.5mm by 2mm. Most of the area is occupied by LO driver.

122 · Measurement setup

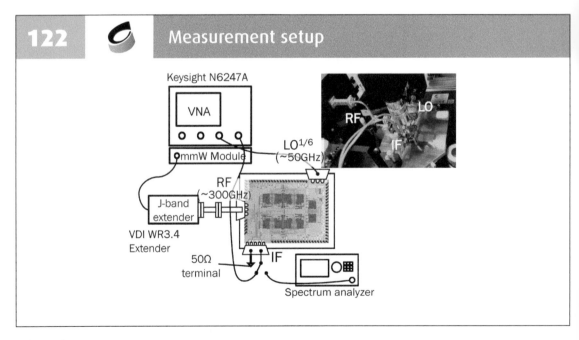

This is the measurement setup, using VNA.

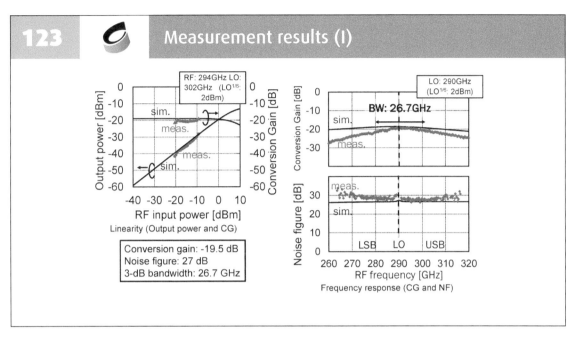

This is the measurement results. Output power is linearly proportional to the input power, and gain. The conversion gain at 294GHz is around -19dB. Noise figure is 27dB.

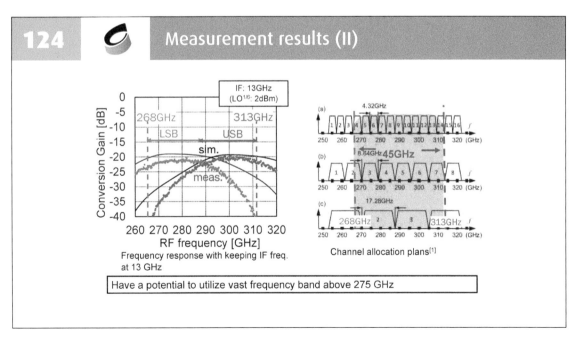

This receiver can receive the signal from 268GHz to 330GHz.

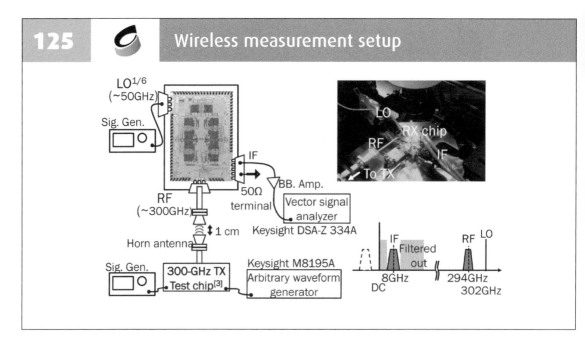

This is the wireless measurement setup. The modulation signal is given by our transmitter with a distance of one centimeter, since no measurement equipment exists for the modulate signal for the 300GHz. This modulate signal is applied to our signal and down conversion signal is analyzed by vector signal analyzer.

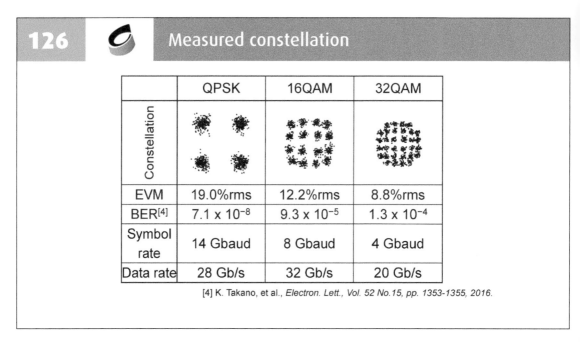

This is a result of the measured constellation, in the case of QPSK maximum data rate is reaching to 28Gbps. In the 16QAM case maximum data rate reaches 32Gbps. And 32QAM maximum data rate which is 20Gbps.

127 Performance comparison

Ref	[5]	[6]	[7]	[8]	[9]	This work
Process	35nm GaAs mHEMT	130nm SiGe BiCMOS	130nm SiGe HBT	90nm CMOS	65nm CMOS	40nm CMOS
RF (GHz)	240	240	320	200	240	290
3dB BW (GHz)	24	18	13	3	-	26.5
Conv. gain (dB)	3	11	−14	6.6	25	−19
NF (dB)	10	16	36	29.9	15	27
Modulation	8PSK	16QAM / 64QAM	-	BPSK / QPSK	BPSK /QPSK	16QAM /32QAM
Data rate (Gb/s)	96	0.7 / 1.0	-	4 / 2	10 / 16	32 / 20
DC (W)	-	0.87	0.216	0.063	0.26	0.65
Chip size (mm²)	-	1.57	0.92	0.375	2	3.15
Integration	-	ANT., LNA, IQ Mixer, LO chain	SHM, LO chain	Mixer, IF amp.	ANT.,Mixer IF Amp. LO chain	Mixer LO chain

[5] I. Kallfass, et al., *J. Infrared Millim. Terahz. Waves.*, 2015.
[6] N. Sarmah, et al., *MTT-S*, 2016.
[7] E. Öjefors., et al., *MTT-S*, 2012.
[8] M. Tytgat, et al., *Eur. Solid-State Circuits Conf.*, 2011.
[9] S. V. Thyagarajan, et al., *JSSC*, 2015.

This is a comparison of other works, this is a first 300GHz CMOS receiver.

128 Conclusions

- Present a 300-GHz LNA-less RX in 40-nm CMOS, operating above f_{max}
- Conversion gain of −19.5 dB, noise figure of 27 dB, and 3-dB bandwidth of 26.5 GHz
- Have a potential to utilize vast frequency band above 275 GHz
- Demonstrate high data rate of 28, 36, and 20 Gb/s with QPSK, 16QAM, and 32QAM

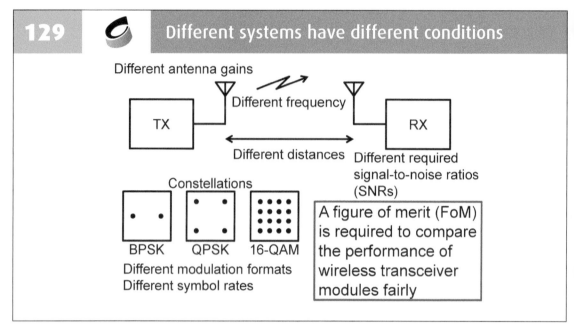

129 — Different systems have different conditions

Different antenna gains

Different frequency

TX   RX

Different distances

Different required signal-to-noise ratios (SNRs)

Constellations

BPSK   QPSK   16-QAM

Different modulation formats
Different symbol rates

A figure of merit (FoM) is required to compare the performance of wireless transceiver modules fairly

This is a summary of the measurement with different conditions. Different antenna was used with different frequency and different required SNR, different modulation scheme. How to compare the wireless transceiver fairly with these different conditions?

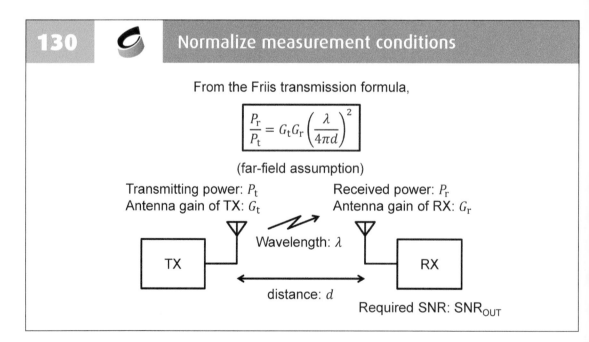

130 — Normalize measurement conditions

From the Friis transmission formula,

$$\frac{P_r}{P_t} = G_t G_r \left(\frac{\lambda}{4\pi d}\right)^2$$

(far-field assumption)

Transmitting power: P_t
Antenna gain of TX: G_t

Received power: P_r
Antenna gain of RX: G_r

Wavelength: λ

TX   RX

distance: d

Required SNR: SNR_{OUT}

The antenna gain is changed in the different communication measurement. Both the transmitting and receiving power is different. First of all, let's normalize antenna gain.

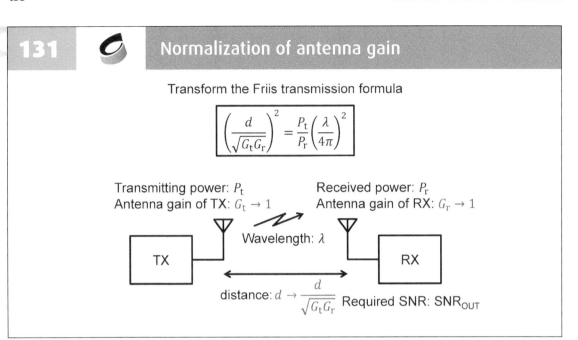

131   **Normalization of antenna gain**

Transform the Friis transmission formula

$$\left(\frac{d}{\sqrt{G_tG_r}}\right)^2 = \frac{P_t}{P_r}\left(\frac{\lambda}{4\pi}\right)^2$$

Transmitting power: P_t
Antenna gain of TX: $G_t \to 1$

Received power: P_r
Antenna gain of RX: $G_r \to 1$

Wavelength: λ

TX

RX

distance: $d \to \dfrac{d}{\sqrt{G_tG_r}}$ Required SNR: SNR_{OUT}

To normalize antenna gain, just divide antenna gain from right hand side and left hand side. Then in this formula, no parameters of gain existing more.

The antenna gain is normalized from right hand side and distance is changed by antenna gain using a d over square root of Gt Gr.

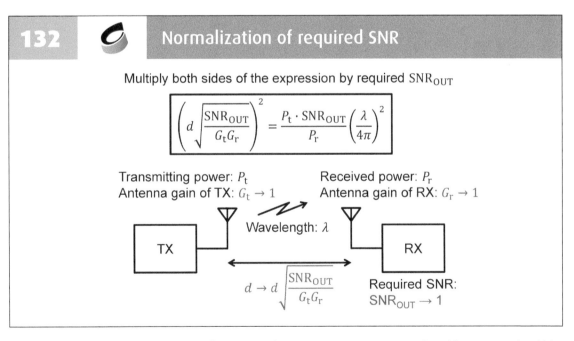

132   **Normalization of required SNR**

Multiply both sides of the expression by required SNR_{OUT}

$$\left(d\sqrt{\frac{\text{SNR}_{\text{OUT}}}{G_tG_r}}\right)^2 = \frac{P_t \cdot \text{SNR}_{\text{OUT}}}{P_r}\left(\frac{\lambda}{4\pi}\right)^2$$

Transmitting power: P_t
Antenna gain of TX: $G_t \to 1$

Received power: P_r
Antenna gain of RX: $G_r \to 1$

Wavelength: λ

TX

RX

$d \to d\sqrt{\dfrac{\text{SNR}_{\text{OUT}}}{G_tG_r}}$ Required SNR:
$\text{SNR}_{\text{OUT}} \to 1$

If we want to normalize SNR to one that means the transmission power is reduced by SNR. Pr should be divided by SNR.

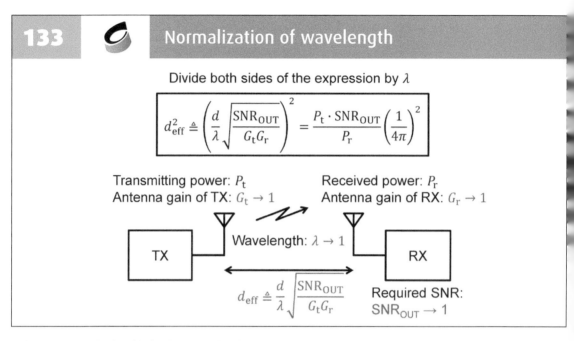

The wave length should also be normalized as shown in this slide. Then all the parameters are normalized.

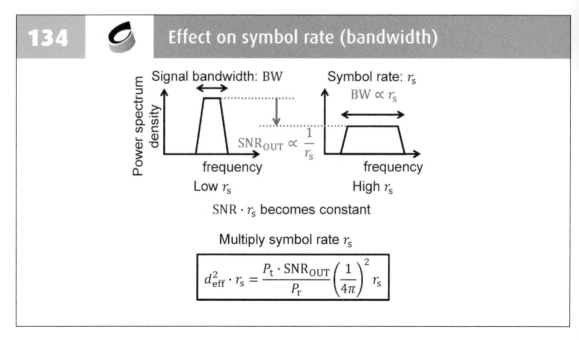

And also, we have to know the relationship between the symbol rate and communication distance. If we apply a high symbol rate, spectrum density of the transmitting power is reduced. SNR is proportional to one over symbol rate.

135 Derive FoM

$$P_r = \text{SNR}_{\text{OUT}} \cdot \text{BW} \cdot N_p \cdot \text{NF}$$
$$= \text{SNR}_{\text{OUT}} \cdot r_s \cdot N_p \cdot \text{NF} \quad (\text{assumed that BW} = r_s)$$

N_p: Thermal noise floor

NF: Noise figure of a receiver

$$d_{\text{eff}}^2 \cdot r_s = \frac{P_t}{N_p \cdot \text{NF}} \left(\frac{1}{4\pi} \right)^2$$

$$\text{FoM} \triangleq (4\pi d_{\text{eff}})^2 N_p \cdot r_s = \frac{P_t}{\text{NF}}$$

Various conditions are normalized

Even though we can't measure transmitting power at a noise figure in the real communication measurement, we can calculate effective distance and noise floor. Figure of merit can be calculated as well. It is related to transmitting power.

136 FoMfor wireless transceiver modules

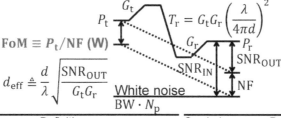

$$\text{FoM} \equiv P_t / \text{NF (W)}$$

$$T_r = G_t G_r \left(\frac{\lambda}{4\pi d} \right)^2$$

$$d_{\text{eff}} \triangleq \frac{d}{\lambda} \sqrt{\frac{\text{SNR}_{\text{OUT}}}{G_t G_r}} \quad \frac{\text{White noise}}{\text{BW} \cdot N_p}$$

Symbols	Definitions	Symbols	Definitions
P_t	Power fed into a transmitting antenna	SNR_{IN}	SNR at receiver input
P_r	Power available from a receiving antenna	SNR_{OUT}	SNR at receiver output
G_t	Transmitting antenna gain	BW	Signal bandwidth
G_r	Receiving antenna gain	N_p	Thermal noise floor
T_r	Transmission loss	λ	Signal wavelength
NF	Noise figure of a receiver	d	Transmission distance
		d_{eff}	Effective distance

This is a signal block diagram from transmitter to receiver.

137 · Measurement setup

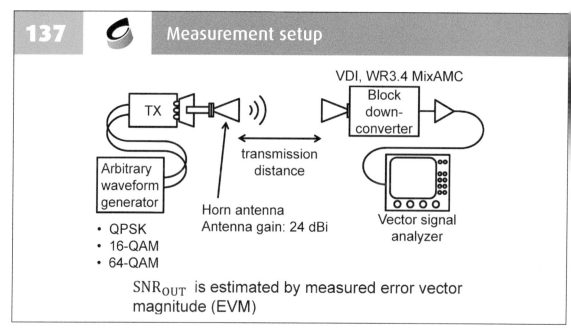

VDI, WR3.4 MixAMC

TX

Block down-converter

transmission distance

Arbitrary waveform generator
- QPSK
- 16-QAM
- 64-QAM

Horn antenna
Antenna gain: 24 dBi

Vector signal analyzer

SNR_{OUT} is estimated by measured error vector magnitude (EVM)

Then we measure using our transmitter. The AWG is applied. The communication distance is changing. The receiver is breaking down conversion which is not CMOS. Three types of modulation, QPSK, 16QAM and 64QAM, are applied.

138 · Performance of CMOS THz transceiver

	QPSK	16QAM
Distance	1m	5cm
Constellation		
Data rate	6Gb/s	56Gb/s
Center freq	302GHz	302GHz
EVM	32.7%	13.4%

This is the result. These are numbers can be normalized.

139   Performance comparison with FoM

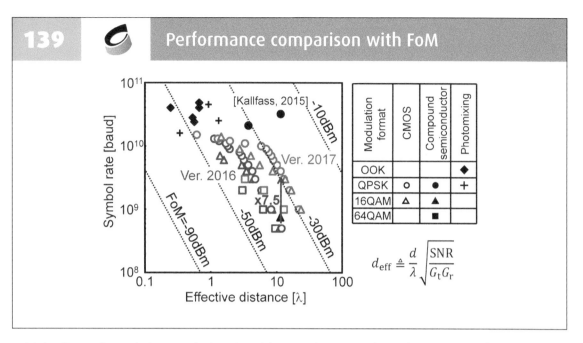

This is a figure of a symbol rate as the function of the effective distance. Effective distance is normalized by a wave length, SNR, antenna gains. Red markers show the result using a CMOS transmitter. The black and cross markers show compound semiconductor and photomixing case. The communication distances are almost the same for the three cases.

140   Conclusions

- FoM for comparison of terahertz wireless transceiver modules is introduced
- Wireless performances with 300 GHz CMOS transmitters are compared
  - Previous one: 16-QAM, 5cm, 28 Gbit/s
  - Improved one: 16-QAM, 5cm, 56 Gbit/s
- Our wireless transceiver modules has comparable FoMs with the other modules

12. THZ LINK TOWARDS SPACE

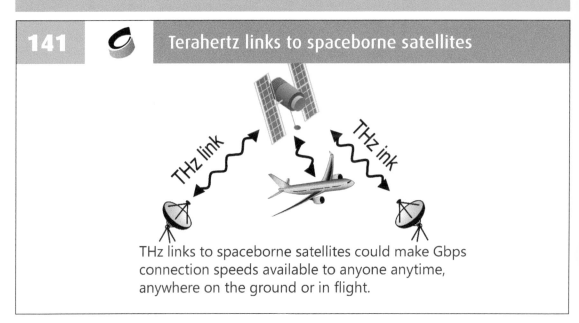

141 Terahertz links to spaceborne satellites

THz links to spaceborne satellites could make Gbps connection speeds available to anyone anytime, anywhere on the ground or in flight.

This slide shows the terahertz links to spaceborne satellites. Currently this kind of in-flight wifi is supported by satellite communication. The link to the satellite communication is not so wide. Thus, the speed of the wifi in flight is not so fast. If we can link from ground to satellite by terahertz or satellite to flight by terahertz, bandwidth will be increased a lot as well as the speed. Terahertz links to space satellites could make gigabit per second connection speed and it was available to anyone any time anywhere on the ground or in flight if we realize this kind of terahertz link.

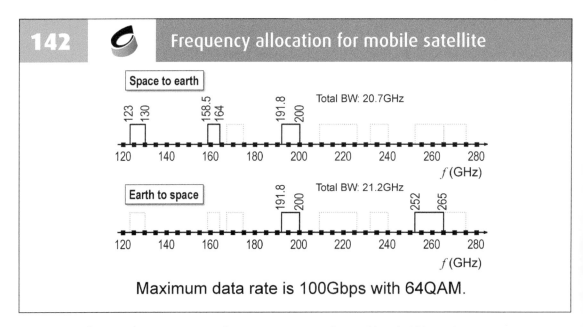

142 Frequency allocation for mobile satellite

Maximum data rate is 100Gbps with 64QAM.

Frequency allocation from space to earth is 100 to 300GHz. The total bandwidth can be 20GHz. A maximum data rate of 100Gbps with 64QAM can be achieved.

143 ## More applications with 100Gbps

Transceiver can send the following data.

- ### 750GB in 1 minute
  - − 90 dual layer Blu-ray discs
- ### 45TB in 1 hour
  - − Year's data of full HD video
- ### 1 PB (peta byte) in 1 day
  - − Two day's data of Facebook

If we can realize 100Gbps transceiver, it can send 750GB in one minute. Where can we use this kind of high speed transceiver?

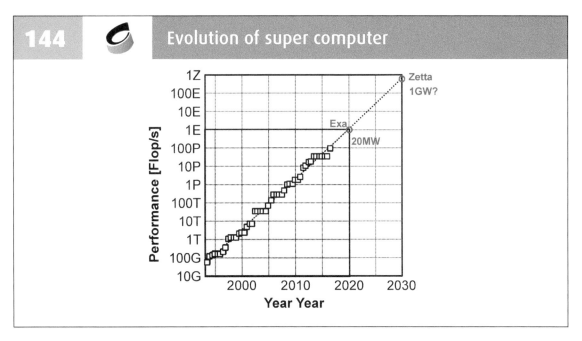

144 ## Evolution of super computer

One possibility is super computer. This is a performance trend of the super computer. Currently, exa-scale computer is studied, targeting year 2020. The exa-scale computer may consume that 20MW.

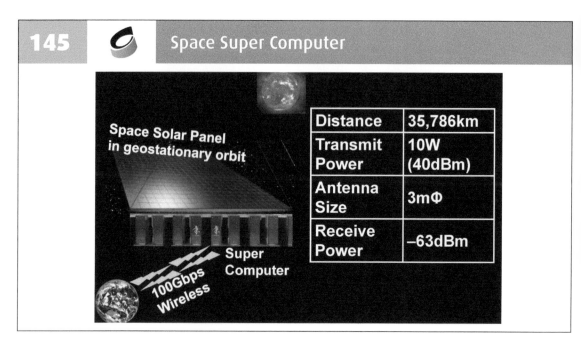

However maybe we can build space solar panel to generate power. This is required power to realize 100Gbps wireless communication.

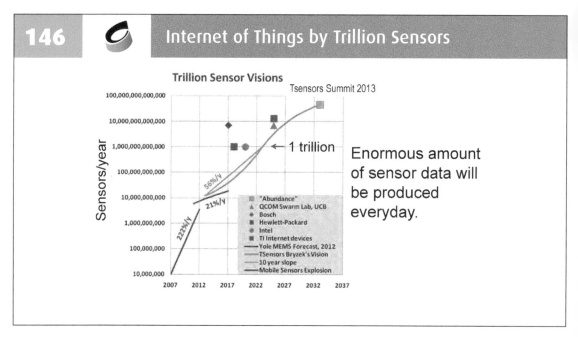

Another possibility is storage. In year 2020 or 2025, number of sensors reaches one trillion. Enormous amount of data is generated by trillion sensors and that data is produced every day. Most of the data is thrown away, but some data is stored. The data should be increased exponential.

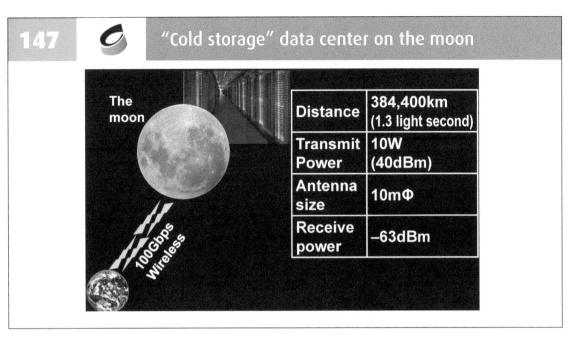

If we don't require high access time to all data, storage can be played on the moon, which is named a cold state data center on the moon. That distance from ground to moon is 384000km, 1.3 light second. The receiving power can be sufficient power to 100Gbps wireless condition with 10W transmitting power and 10m antenna size.

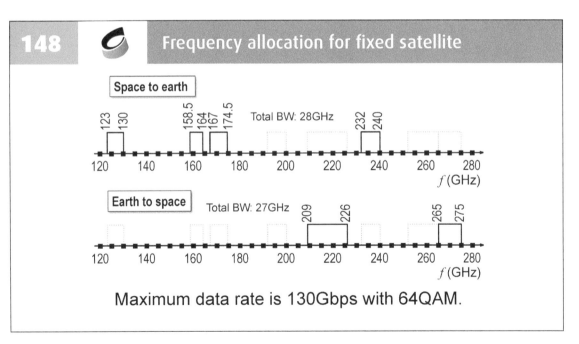

This is a frequency allocation for fixed satellite. The space to earth is the down link, while ground to space is the up link. Total bandwidth is 27GHz. Maximum data rate can reach to 130Gbps with 64QAM. So, high data rate link may be possible using terahertz communication.

149 Summary

- 100Gbps era appears in the 2020's
- Terahertz is capable of long-distance and mobile communication
  - Promising frequencies lie from 100 to 300GHz
  - Propagation loss decreases as frequency increases with the constant antenna size
  - Flight and space communication will be 100Gbps

100Gbps era will appears in the 2020 and terahertz is capable of long distance and mobile communication. Promising frequencies is lie from 100 to 300GHz. And propagation loss decreases as frequency increases with the constant antenna size. Do not register antenna size. And flight and space communication will be 100Gbps if we can realize high output power transmitter.

13. CONCLUDING REMARKS

150 The beginning of the end?

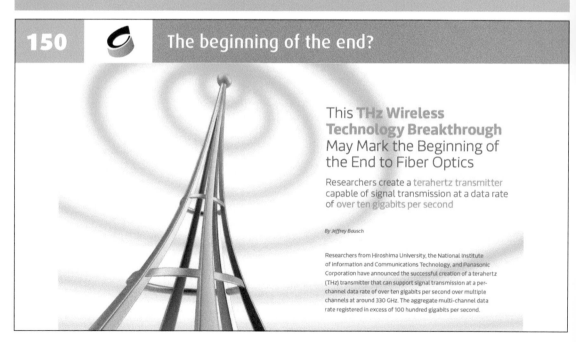

This **THz Wireless Technology Breakthrough** May Mark the Beginning of the End to Fiber Optics

Researchers create a terahertz transmitter capable of signal transmission at a data rate of over ten gigabits per second

By Jeffrey Bausch

Researchers from Hiroshima University, the National Institute of Information and Communications Technology, and Panasonic Corporation have announced the successful creation of a terahertz (THz) transmitter that can support signal transmission at a per-channel data rate of over ten gigabits per second over multiple channels at around 330 GHz. The aggregate multi-channel data rate registered in excess of 100 hundred gigabits per second.

This is a report published in 2016. It suggests that terahertz wireless technology may mark the beginning of the end of fiber optics. But, I don't think so.

151 — Who is the leader in communication research?

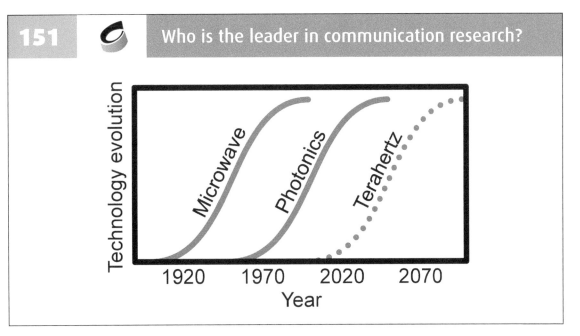

This is a slide of the leader in communication research, in year 1920, magnetron was invented. Research on microwave was increasing then. In 1967 fiber optics communication is invented, then the research on photonics is increased. After the research on photonics is increasing year over year, microwave still exist because the application is different. The difference of time year is 50 years. We are approaching the 2020, we should have another method for communication research. That would be terahertz.

152 — Gap between Radio Wave and Light

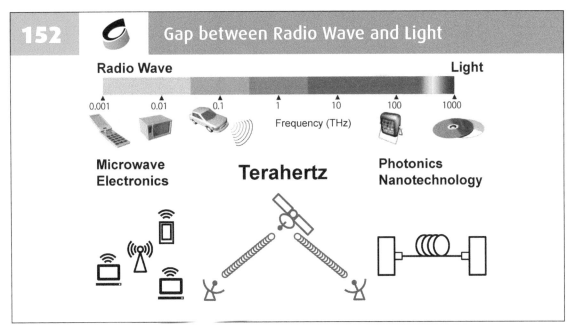

My wish is terahertz is the frequency between microwave and photonics, and terahertz application should fill the application gap between microwave and photonics.

153 References

1. M. Fujishima, M. Motoyoshi, K. Katayama, K. Takano, N. Ono, R. Fujimoto, "98 mW10 Gbpswireless transceiver chipset with D-band CMOS circuits," IEEE Journal of Solid-State Circuits, vol. 48, no. 10, pp. 2273-2284, 2013.

2. K. Katayama , K. Takano, S. Amakawa, S. Hara, A. Kasamatsu, K. Mizuno, K. Takahashi, T. Yoshida, M. Fujishima, "A 300GHz 40nm CMOS Transmitter with 32-QAM 17.5Gb/s/chCapability over 6 Channels," Digest of Technical Papers IEEE International Solid-State Circuits Conference (ISSCC) 2016, pp. 342-343 (1-3 Feb. 2016).

3. K. Katayama, K. Takano, S. Amakawa, S. Hara, A. Kasamatsu, K. Mizuno, K. Takahashi, T. Yoshida and M. Fujishima, "A 300 GHz CMOS Transmitter With 32-QAM 17.5 Gb/s/chCapability Over Six Channels," IEEE Journal of Solid-State Circuits, vol. 51, no,. 12, pp. 3037-3048, 2016.

4. K. Katayama, K. Takano, S. Amakawa, S. Hara, T. Yoshida and M. Fujishima, "CMOS 300-GHz 64-QAM transmitter," 2016 IEEE MTT-S International Microwave Symposium (IMS), pp. 1-4, 2016.

5. K. Takano, K. Katayama, S. Amakawa, T. Yoshida and M. Fujishima, "A 300-GHz 64-QAM CMOS transmitter with 21-Gb/s maximum per-channel data rate," 2016 11th European Microwave Integrated Circuits Conference (EuMIC), pp. 193-196, 2016

6. K. Takano, S. Amakawa, K. Katayama, S. Hara, R. Dong, A. Kasamatsu, I. Hosako, K. Mizuno, K. Takahashi, T. Yoshida, M. Fujishima, "A 105Gb/s 300GHz CMOS transmitter," 2017 IEEE International Solid-State Circuits Conference (ISSCC), pp. 308-309, 2017

7. S. Hara, K. Katayama, K. Takano, R. Dong, I. Watanabe, N. Sekine, A. Kasamatsu, T. Yoshida, S. Amakawaand M. Fujishima, "A 32Gbit/s 16QAM CMOS receiver in 300GHz band," 2017 IEEE MTT-S International Microwave Symposium (IMS) to be presented.

8. K. Takano, K. Katayama, S. Amakawa, T. Yoshida, M. Fujishima, "Wireless digital data transmission from a 300-GHz CMOS transmitter," Electronics Letters, 2 pp (3 June 2016)

9. K. Takano, K. Katayama, S. Amakawa, T. Yoshida, and M. Fujishima, "56-Gbit/s 16-QAM wireless linkwith 300-GHz-band CMOS transmitter," 2017 IEEE MTT-S International Microwave Symposium (IMS) to be presented.

10. M Fujishima, "Channel allocation of 300GHz band for fiber-optic-speed wireless communication," URSI Asia-Pacific Radio Science Conference (URSI AP-RASC), pp. 330-333 (22 Aug. 2016)

11. M. Fujishimaand S. Amakawa, "Integrated-circuit approaches to THz communications: challenges, advances, and future prospects," IEICE Trans. Electron., vol. 100, no. 2, pp. 516-523, 2017.

CHAPTER 06

Layout Design in Advanced Technologies

Alan Hastings

Texas Instrument

In order to create successful products, designers must not only consider circuits, but also layouts. Analog circuits require proper choice of components, careful attention to matching, and robust protection against minority carrier injection. Circuit design and layout are synergistic and thus require collaboration between circuit designers and layout staff. However, many circuit designers are not familiar with layout techniques such as common-centroid devices and minority carrier guard rings. This short course provides an introduction to these and other aspects of layout. Topics discussed include silicon processing; analog CMOS and BiCMOS process flows; layout of resistors, capacitors, and MOS transistors; device matching; and minority carrier guard rings.

1. FABRICATION

 1   **INTRODUCTION**

◆ *Layout* **creates the geometric patterns needed to fabricate an integrated circuit.**

   ◆ We can't understand layout unless we first understand fabrication.

   ◆ Wafer fabs are some of the most sophisticated manufacturing facilities ever created.

   ◆ Luckily, we only need a broad overview of the technologies involved.

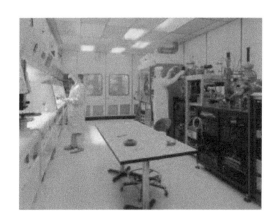

This chapter starts with a brief overview on how do you make an IC.

 2   **CRYSTAL GROWING**

◆ **The** *Czochralski process* **transforms purified polycrystalline silicon into monocrystalline silicon.**

   ◆ A silica crucible charged with polysilicon (*poly*) is heated until the silicon melts.

   ◆ A *seed crystal* lowered into the melt provides a surface upon which silicon atoms crystalize.

   ◆ Rotating the silicon crystal and slowly withdrawing it from the crucible creates a cylindrical silicon crystal, or *ingot*.

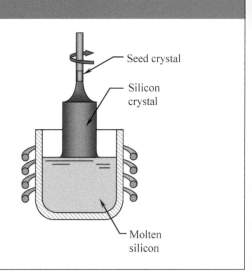

Seed crystal

Silicon crystal

Molten silicon

The first thing you need is some silicon. We need a single crystal. There's a process called the Czochralski process which is many years old that was not originally design for silicon, but it works for silicon.

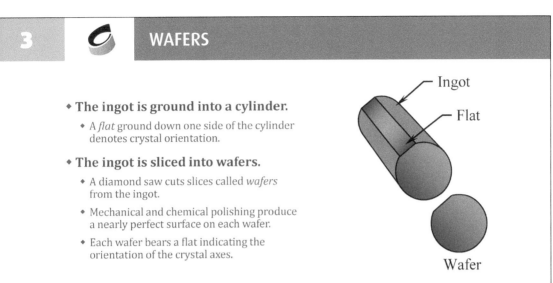

3 ⬤ **WAFERS**

- **The ingot is ground into a cylinder.**
  - A *flat* ground down one side of the cylinder denotes crystal orientation.
- **The ingot is sliced into wafers.**
  - A diamond saw cuts slices called *wafers* from the ingot.
  - Mechanical and chemical polishing produce a nearly perfect surface on each wafer.
  - Each wafer bears a flat indicating the orientation of the crystal axes.

Ingot

Flat

Wafer

The next step is to find the orientation of the crystal. You ground down the side of the ingot to make a flat and then that identify as crystal orientations. Wafers are sliced off from ingot.

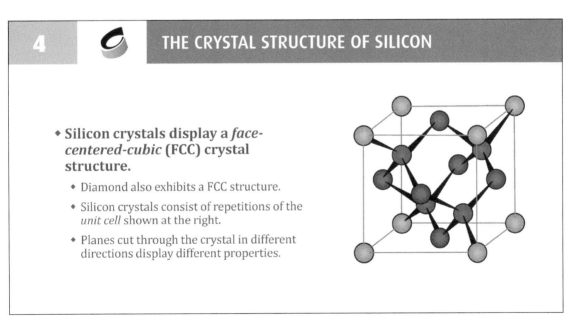

4 ⬤ **THE CRYSTAL STRUCTURE OF SILICON**

- **Silicon crystals display a *face-centered-cubic* (FCC) crystal structure.**
  - Diamond also exhibits a FCC structure.
  - Silicon crystals consist of repetitions of the *unit cell* shown at the right.
  - Planes cut through the crystal in different directions display different properties.

Silicon displays a face centered cubic crystal structure. This is the example of the atoms of a silicon lattice.

5   MILLER INDICES

* **The properties of a silicon surface depend upon its orientation with respect to the unit cell.**
  * Each plane surface intersecting the cubic unit cell can be described by a trio of numbers called *Miller indices.*
  * CMOS and BiCMOS processes usually employ wafers whose surfaces are (100) planes.

(100) face   Z   (111) face

X   Y

(100) face

(100) face

There is a set of nomenclatures called miller indices. Each plane surface intersecting the cubic unit cell can be described by miller indices. CMOS and BiCMOS processes usually employ wafers with surfaces of 100 planes. Standard bipolar employs wafers with 111 planes.

6   PHOTOLITHOGRAPHY

* ***Photolithography* allows selective deposition or removal of materials from the wafer surface.**
  * A thin film of *photoresist* (*resist*) is applied to the wafer.
  * Light shining through openings in a *photomask* (*mask*) exposes portions of the photoresist.
  * Flooding the wafer surface with *developer* creates a pattern of openings in the photoresist.

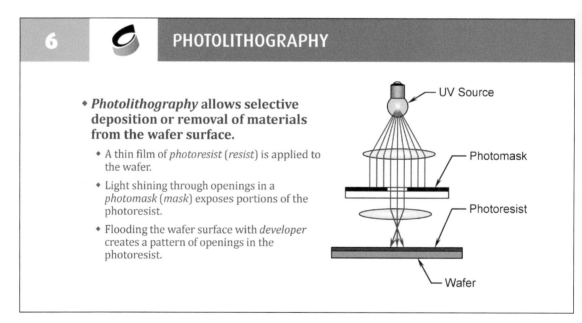

UV Source

Photomask

Photoresist

Wafer

Photolithography allows selective deposition or removal of materials from the wafer surface. The UV light source is a room sized equipment. Photo mask is applied in the middle between the wafer and the UV source. The photo mask has some areas that are transparent.

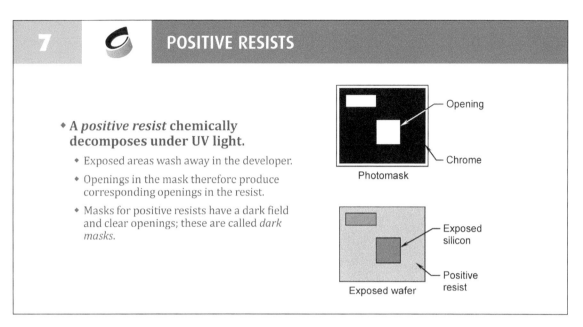

A positive resist chemically decomposes under the UV light. There are two type of resists. The positive resist decomposes when it is hit by a UV light.

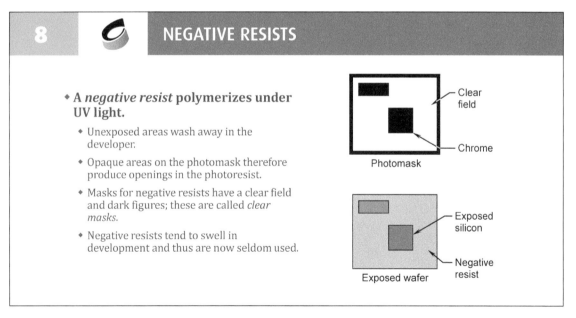

The Negative resist tend to swell in development and are seldom used.

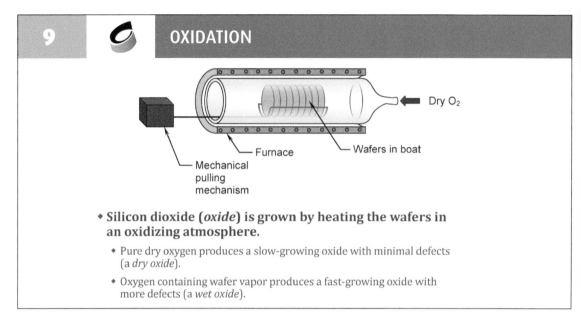

9   OXIDATION

Dry O₂ ← ... Wafers in boat — Furnace — Mechanical pulling mechanism

• **Silicon dioxide (*oxide*) is grown by heating the wafers in an oxidizing atmosphere.**

  • Pure dry oxygen produces a slow-growing oxide with minimal defects (a *dry oxide*).

  • Oxygen containing wafer vapor produces a fast-growing oxide with more defects (a *wet oxide*).

Only a couple of angstroms thick this so-called native oxide. To get it, you'd heat the tube with a heating man, it's all mechanical. Dry oxide is a very high quality low defect oxide good for gates in MOSFETS but it doesn't very thick. Squirt a little steam into the tube, Steam grows oxide very quickly and so this is called a wet oxide .

10   PROPERTIES OF OXIDE

Times Required to Grow 0.1µm (1000Å) of Oxide on (111) Silicon.

Ambient	800°C	900°C	1000°C	1100°C	1200°C
Dry O₂	30 hr	6 hr	1.7 hr	40 min	15 min
Wet O₂	1.7 hr	20 min	6 min		

• **Properties of silicon dioxide (oxide):**
  • Readily formed in films as thin as 10 Å (1 nm).
  • Adheres tenaciously to silicon.
  • Resists attack by most chemicals.
  • Readily dissolves by *hydrofluoric acid* (HF).
  • Excellent dielectric for capacitors and MOS transistors.
  • Extremely low defect density when grown on (100) silicon surfaces.

Wet oxides take less time, the table shows grow 0.1 microns of oxide on a 111 wafer. The growth rates are different on different cuts, oxide is an excellent dielectric for capacitors and MOSFETS and when grown on 100 silicon it's very low density.

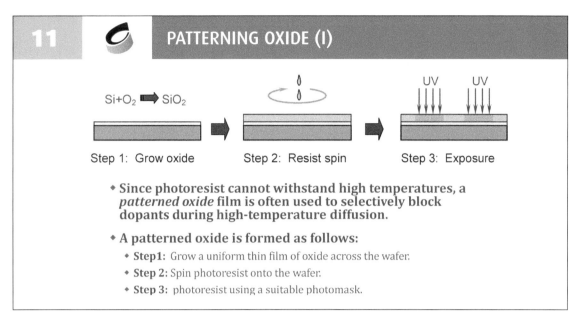

11 PATTERNING OXIDE (I)

$Si+O_2 \Rightarrow SiO_2$

Step 1: Grow oxide   Step 2: Resist spin   Step 3: Exposure

- **Since photoresist cannot withstand high temperatures, a *patterned oxide* film is often used to selectively block dopants during high-temperature diffusion.**
- **A patterned oxide is formed as follows:**
  - **Step1:** Grow a uniform thin film of oxide across the wafer.
  - **Step 2:** Spin photoresist onto the wafer.
  - **Step 3:** photoresist using a suitable photomask.

After grow a thin oxide film, we put some photoresist on it. Then spin the wafer around and drip it on and it sprays out across the wafer, this is called spinning. The next step, expose it with a suitable photo mask.

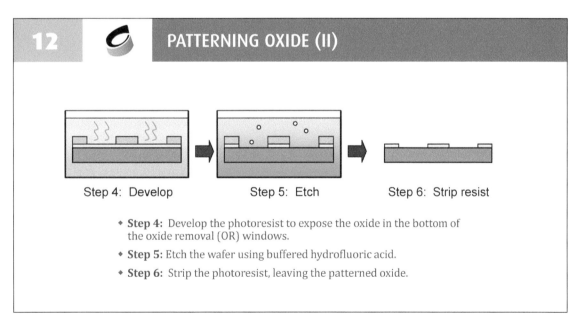

12 PATTERNING OXIDE (II)

Step 4: Develop   Step 5: Etch   Step 6: Strip resist

- **Step 4:** Develop the photoresist to expose the oxide in the bottom of the oxide removal (OR) windows.
- **Step 5:** Etch the wafer using buffered hydrofluoric acid.
- **Step 6:** Strip the photoresist, leaving the patterned oxide.

Then put it in a developer and wash away some portions of the resist and put it in the acid. Then removes the oxide in the areas not covered by the resist and strip away the resist. There are chemicals that will do that but today the strip is almost always done with a plasma, which is called dry ash.

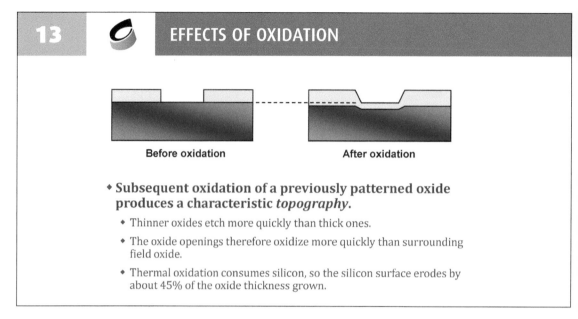

| 13 | | **EFFECTS OF OXIDATION** |

Before oxidation **After oxidation**

- ♦ **Subsequent oxidation of a previously patterned oxide produces a characteristic *topography*.**
  - ♦ Thinner oxides etch more quickly than thick ones.
  - ♦ The oxide openings therefore oxidize more quickly than surrounding field oxide.
  - ♦ Thermal oxidation consumes silicon, so the silicon surface erodes by about 45% of the oxide thickness grown.

If we put it back in the furnace, the area exposed will oxidize very quickly because the silicon surface is exposed. So actually we get a surface discontinuity after do an oxidation on a patterned wafer, this is a problem because modern lithography has a very narrow depth of field.

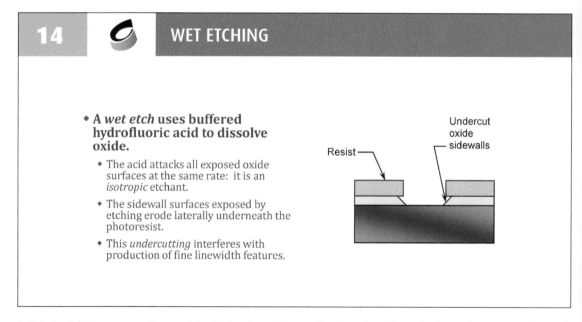

| 14 | | **WET ETCHING** |

- ♦ **A *wet etch* uses buffered hydrofluoric acid to dissolve oxide.**
  - ♦ The acid attacks all exposed oxide surfaces at the same rate: it is an *isotropic* etchant.
  - ♦ The sidewall surfaces exposed by etching erode laterally underneath the photoresist.
  - ♦ This *undercutting* interferes with production of fine linewidth features.

Resist

Undercut oxide sidewalls

Wet etching use a form of hydrofluoric acid usually buffered with ammonia and it's a tax the outside in all directions oxide is not crystalline, so etch is in all directions of the same ray which is unfortunate because it starts hatching near the surface and as it etches down it also inches out underneath the mask. Lateral etching makes the structure bigger and size is everything in modern processes. How can we avoid it?

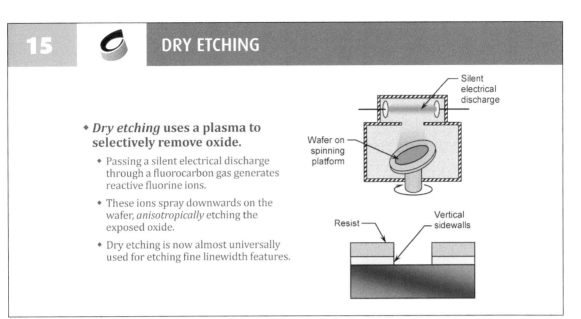

15   DRY ETCHING

• *Dry etching* uses a plasma to selectively remove oxide.

  • Passing a silent electrical discharge through a fluorocarbon gas generates reactive fluorine ions.

  • These ions spray downwards on the wafer, *anisotropically* etching the exposed oxide.

  • Dry etching is now almost universally used for etching fine linewidth features.

Dry etching is the primary etching process normally used. We form a plasma in a fluorocarbon, the molecules apart and releases flurrying gas, an extraordinarily reactive substance throws it down onto the wafer. It shot down onto the wafer vertically and so they etch preferentially in a vertical direction and we get an anisotropic different rates in different direction.

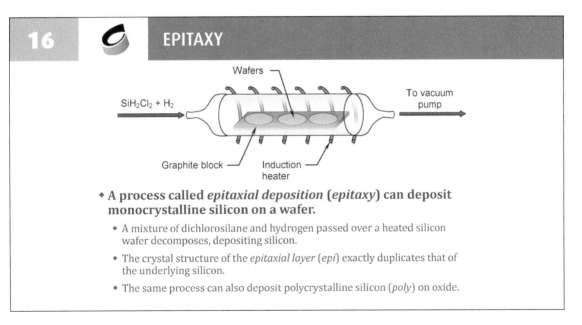

16   EPITAXY

• A process called *epitaxial deposition* (*epitaxy*) can deposit monocrystalline silicon on a wafer.

  • A mixture of dichlorosilane and hydrogen passed over a heated silicon wafer decomposes, depositing silicon.

  • The crystal structure of the *epitaxial layer* (*epi*) exactly duplicates that of the underlying silicon.

  • The same process can also deposit polycrystalline silicon (*poly*) on oxide.

We can deposit silicon on the surface of the crystal, this is called epitaxial deposition or epitaxy. We can deposit silicon either mono crystalline on mono crystalline silicon or polycrystalline on oxide.

17 · DIFFUSION (I)

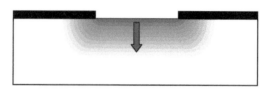

♦ **Dopant atoms become mobile and can *diffuse* through silicon at sufficiently high temperatures.**

- ♦ Diffusion of dopants from an external source through oxide openings allows selective doping of specific regions.
- ♦ Diffusing dopants move downwards and spread outwards from the oxide opening.

Dope silicon uses the planar process. The idea is open a window in an oxide film. Put on some kind of dopant source as phosphorus. Heat it up and the phosphorous will slowly diffuse into the silicon very slowly, about millimeters a year. This is selective because we have oxide blocking the dopants.

18 · DIFFUSION (II)

Representative Junction Depths, in Microns
(10^{20} cm^{-3} source, 10^{16} cm^{-3} background, 15min deposition, 1hr drive)

Dopant	950°C	1000°C	1100°C	1200°C
Boron	0.9μm	1.5μm	3.6μm	7.3μm
Phosphorus		0.5	1.6	4.6
Antimony			0.8	2.1
Arsenic			0.7	2.0

♦ **Dopants diffuse at different rates.**

- ♦ *Boron* and *phosphorus* diffuse relatively quickly.
- ♦ *Arsenic* and *antimony* diffuse much more slowly.
- ♦ All dopants diffuse more rapidly at higher temperatures.

Boron is a p-type dopant, phosphorus antimony and arsenic are all n-type dopants. And it turns out that antimony and arsenic are very slow diffusing, which is handy.

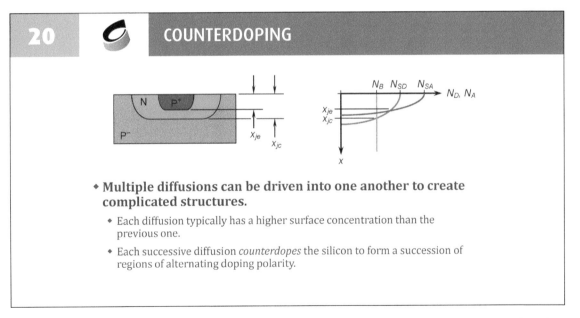

19   PLANAR DIFFUSIONS

- **A diffusion driven in from the surface of the wafer is called a *planar diffusion*.**
  - Planar diffusions are usually patterned by oxide windows.
  - Dopants diffuse laterally (*outdiffusion*) underneath the edges of the oxide window to about 80% of the junction depth.

This is a cross-section of the wafer. In this case we started with a P minus lightly doped p-type wafer and drew drove in n-type dopants. The curve shows uniformly doped with an NB number of dopant atoms per cubic centimeter.

20   COUNTERDOPING

- **Multiple diffusions can be driven into one another to create complicated structures.**
  - Each diffusion typically has a higher surface concentration than the previous one.
  - Each successive diffusion *counterdopes* the silicon to form a succession of regions of alternating doping polarity.

Anytime you change the polarity of the silicon from n to p or p to n, by means of doping, it's called counter doping. Now we started with lightly doped p-type silicon, dope and make it n-type silicon, moderately doped. Put in some more P and made a very handily vote, P region PNP. That's the bipolar transistor we just created.

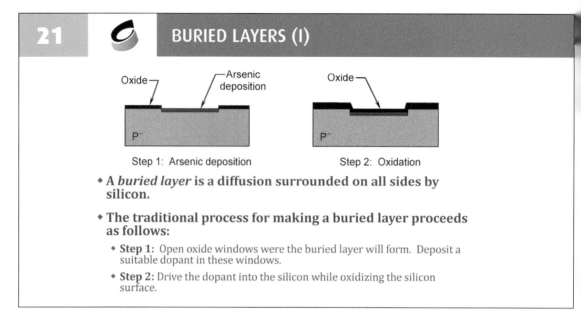

21   **BURIED LAYERS (I)**

Step 1: Arsenic deposition   Step 2: Oxidation

- ◆ A *buried layer* is a diffusion surrounded on all sides by silicon.

- ◆ The traditional process for making a buried layer proceeds as follows:
  - ◆ **Step 1:** Open oxide windows were the buried layer will form. Deposit a suitable dopant in these windows.
  - ◆ **Step 2:** Drive the dopant into the silicon while oxidizing the silicon surface.

Buried layer is a dopant layer that is trapped beneath the surface of the silicon. There's two ways to make it. You can make it by an implantation process or you can do it by epitaxial process that can put it 50 microns down. We start with a silicon wafer, oxidize it, open a window, deposit arsenic , oxidize it up a little bit, strip away that oxide.

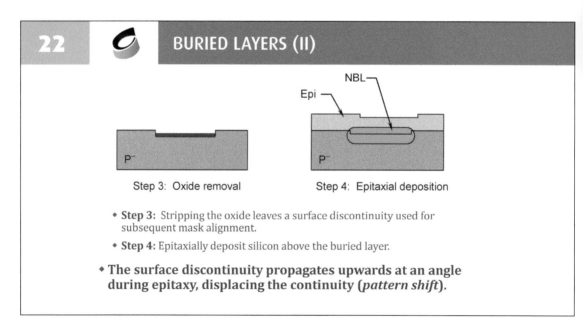

22   **BURIED LAYERS (II)**

Step 3: Oxide removal   Step 4: Epitaxial deposition

- ◆ **Step 3:** Stripping the oxide leaves a surface discontinuity used for subsequent mask alignment.
- ◆ **Step 4:** Epitaxially deposit silicon above the buried layer.

- ◆ The surface discontinuity propagates upwards at an angle during epitaxy, displacing the continuity (*pattern shift*).

It arose during the oxidation that leaves the surface discontinuity, that propagates up during the epitaxy leaving a shadow which is called NBL. This can be a problem because surface discontinuities can affect device matching and so if you're in a process with a buried layer, and if it lies across a critical device, it may create a mismatch.

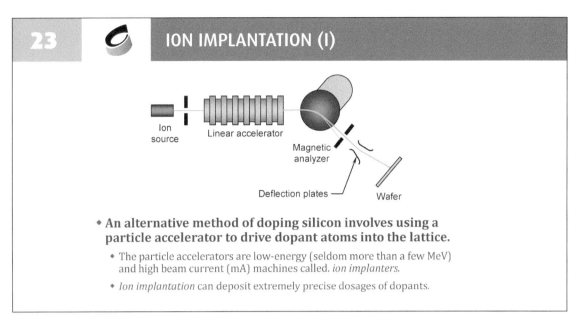

23 ⬤ **ION IMPLANTATION (I)**

- **An alternative method of doping silicon involves using a particle accelerator to drive dopant atoms into the lattice.**
  - The particle accelerators are low-energy (seldom more than a few MeV) and high beam current (mA) machines called. *ion implanters.*
  - *Ion implantation* can deposit extremely precise dosages of dopants.

The diffusion processes described above allow a lot of dopant to be put in silicon, but they're not very accurate. There's another process that is very accurate, it is hard to put a lot of dopant but you'll know exactly how much you put in and how far down it wins. The process uses essentially a particle accelerator.

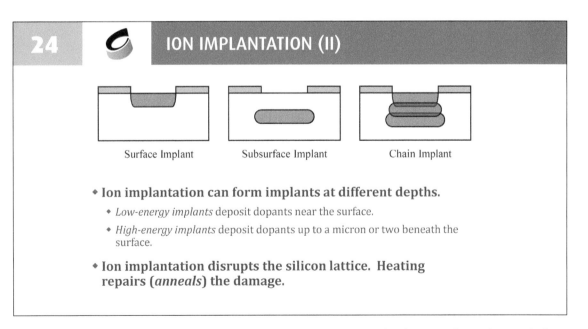

24 ⬤ **ION IMPLANTATION (II)**

- **Ion implantation can form implants at different depths.**
  - *Low-energy implants* deposit dopants near the surface.
  - *High-energy implants* deposit dopants up to a micron or two beneath the surface.

- **Ion implantation disrupts the silicon lattice. Heating repairs (*anneals*) the damage.**

If want a deep layer, you can stack several of these one on top of another to form a chain. Unfortunately ion implantation rips the silicon. Luckily, if you heat it up, it repairs the damage, the undamaged silicon around the implant acts as a seed and the silicon re-crystallizes unbelievable perfection.

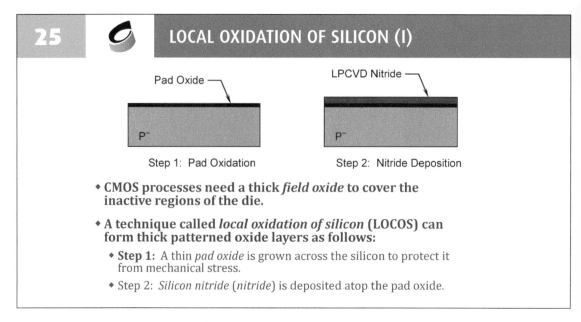

25 — LOCAL OXIDATION OF SILICON (I)

Pad Oxide

Step 1: Pad Oxidation

LPCVD Nitride

Step 2: Nitride Deposition

- ◆ CMOS processes need a thick *field oxide* to cover the inactive regions of the die.
- ◆ A technique called *local oxidation of silicon* (LOCOS) can form thick patterned oxide layers as follows:
  - ◆ **Step 1:** A thin *pad oxide* is grown across the silicon to protect it from mechanical stress.
  - ◆ **Step 2:** *Silicon nitride* (*nitride*) is deposited atop the pad oxide.

Local oxidation of silicon (LOCOS) is the way to form very thick oxides, it is an interesting process. You grow a very thin oxide and then you grow something else low pressure chemical vapor deposited (LPCVD) nitride silicon nitride.

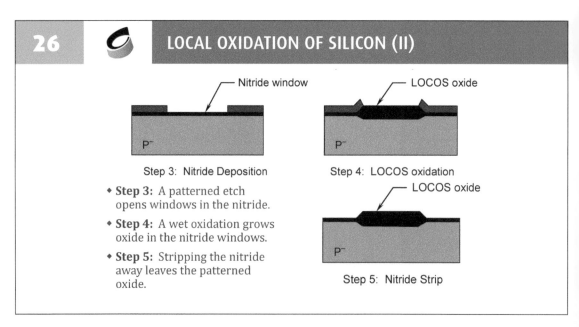

26 — LOCAL OXIDATION OF SILICON (II)

Nitride window

Step 3: Nitride Deposition

LOCOS oxide

Step 4: LOCOS oxidation

LOCOS oxide

Step 5: Nitride Strip

- ◆ **Step 3:** A patterned etch opens windows in the nitride.
- ◆ **Step 4:** A wet oxidation grows oxide in the nitride windows.
- ◆ **Step 5:** Stripping the nitride away leaves the patterned oxide.

So then we open a nitride window down to the oxide surface, there are etching that etch nitride. We then just heat it up and spray some steam in. Only in the opening, because the oxide or the oxygen has a hard time getting through nitride. And so here is covered by nitride don't grow more oxide strip away the nitride when you now have a locally oxidized piece of silicon.

27 🎲 SHALLOW TRENCH ISOLATION

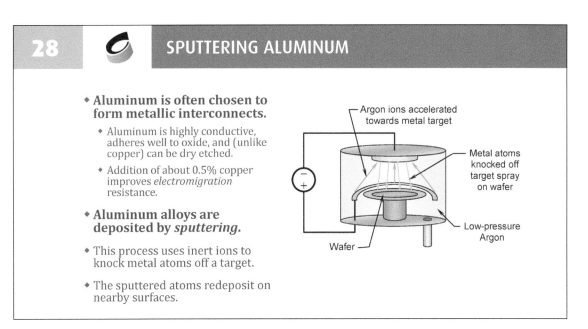

LOCOS   STI

- **LOCOS cannot support deep-submicron processes because:**
  - Most forms of LOCOS generate significant topography.
  - The transition region between LOCOS and thin oxide (*bird's beak*) increases minimum spacings.

- ***Shallow trench isolation* (STI) eliminates these problems:**
  - The anisotropically etched trench has nearly vertical sidewalls.
  - The trench is filled with oxide and then polished back to the surface.

The problem is a transition region between the thin and the thick. This transition looks like a cross-section of a bird with a beak pointed out, so it is called bird's beak. Instead, people have turned to a process where you etch down into the silicon and isotropically. We can make very steep sidewalls, then deposit or grow oxide, that is shallow trench isolation.

28 🎲 SPUTTERING ALUMINUM

- **Aluminum is often chosen to form metallic interconnects.**
  - Aluminum is highly conductive, adheres well to oxide, and (unlike copper) can be dry etched.
  - Addition of about 0.5% copper improves *electromigration* resistance.

- **Aluminum alloys are deposited by *sputtering*.**
  - This process uses inert ions to knock metal atoms off a target.
  - The sputtered atoms redeposit on nearby surfaces.

Argon ions accelerated towards metal target

Metal atoms knocked off target spray on wafer

Low-pressure Argon

Wafer

How do we metalize the devices? We spray aluminum on the surface of the wafer, patterning, remove the aluminum where you don't want. The usual process uses an alloy of aluminum which is the conductive metal, a little bit of copper for electro migration resistance.

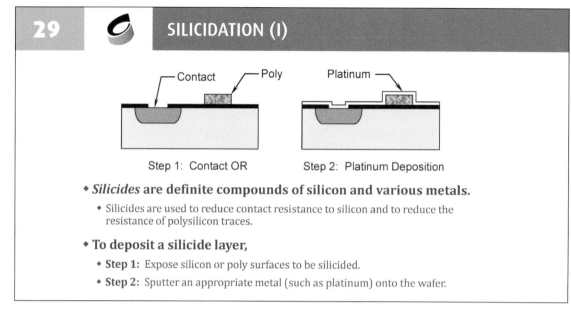

29 — SILICIDATION (I)

Step 1: Contact OR   Step 2: Platinum Deposition

- ♦ *Silicides* **are definite compounds of silicon and various metals.**
  - ♦ Silicides are used to reduce contact resistance to silicon and to reduce the resistance of polysilicon traces.

- ♦ **To deposit a silicide layer,**
  - ♦ **Step 1:** Expose silicon or poly surfaces to be silicided.
  - ♦ **Step 2:** Sputter an appropriate metal (such as platinum) onto the wafer.

Silicidation or Silicides is a process in which you take a metal and make it react with silicon to form a silicides which forms a good contact then you put your aluminum on top of that. You place the mental on there, you react it to form the silicide where it touches silicon. You use some sort of an etchant to remove the unreacted silicide leaving the silicide.

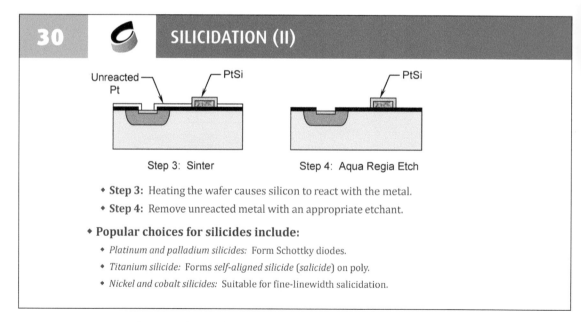

30 — SILICIDATION (II)

Step 3: Sinter   Step 4: Aqua Regia Etch

- ♦ **Step 3:** Heating the wafer causes silicon to react with the metal.
- ♦ **Step 4:** Remove unreacted metal with an appropriate etchant.

- ♦ **Popular choices for silicides include:**
  - ♦ *Platinum and palladium silicides:* Form Schottky diodes.
  - ♦ *Titanium silicide:* Forms *self-aligned silicide* (*salicide*) on poly.
  - ♦ *Nickel and cobalt silicides:* Suitable for fine-linewidth salicidation.

We want a very planar wafer surface, so we need chemical mechanical polishing. We use a alkaline liquid that softens the surface of the oxide and then the little pad comes along and scrubs off the softened oxide in the high spots. Modern wafers are covered with little dummy metal and dummy poly shapes which fill up the space to avoid dishing.

2. PROCESSES

The second unit is processes. What is a process? A process is a set of fabrication steps that make a photo of wafers.

31 STANDARD BIPOLAR

- **Standard bipolar was the first widely used analog integrated circuit process.**
  - Although new designs are seldom implemented in this process, many standard bipolar devices are still manufactured today.

- **Standard bipolar uses many features that are now common components of other processes:**
  - Epitaxy
  - Buried layers
  - Isolation tanks
  - Vertical bipolar transistors

Standard bipolar is a very old process, today most people have forgotten it, but it still exists because it was the beginning of analog integrated circuit design and it shows a lot of features that are very common.

32 NBL

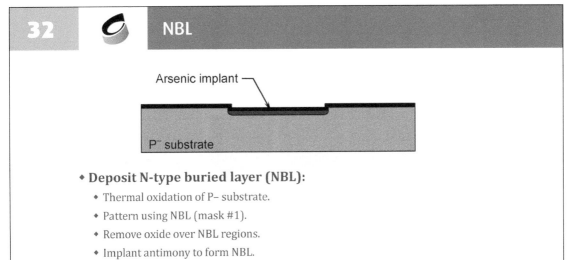

- **Deposit N-type buried layer (NBL):**
  - Thermal oxidation of P– substrate.
  - Pattern using NBL (mask #1).
  - Remove oxide over NBL regions.
  - Implant antimony to form NBL.
  - Anneal NBL implant while growing thermal oxide.

Standard bipolar starts with a P– substrate, which is 111 oriented and thermally oxidized. Pattern was applied using the first mask that in Bury layer mask, then basically you stripped out the oxide implanted antimony.

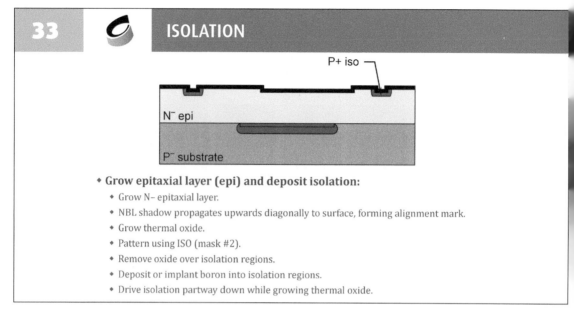

33 ISOLATION

* **Grow epitaxial layer (epi) and deposit isolation:**
  * Grow N– epitaxial layer.
  * NBL shadow propagates upwards diagonally to surface, forming alignment mark.
  * Grow thermal oxide.
  * Pattern using ISO (mask #2).
  * Remove oxide over isolation regions.
  * Deposit or implant boron into isolation regions.
  * Drive isolation partway down while growing thermal oxide.

EPI is whatever you want. When we were doing the anneal of the NBL that means there's a surface discontinuity and that propagates up at an angle to create an NBL shadow which we align the next basket to mask alignment is crucial. Then we grow some thermal oxide patterning, using the isolation mask remove oxide over the areas, where we're going to have isolation depositor implant boron.

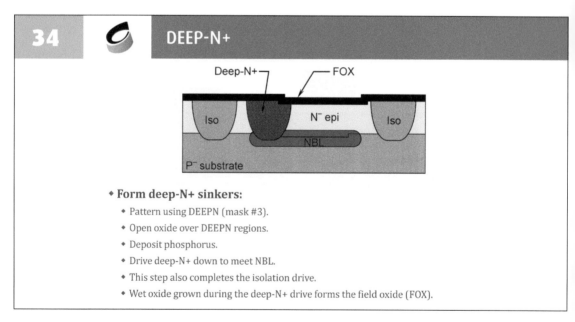

34 DEEP-N+

* **Form deep-N+ sinkers:**
  * Pattern using DEEPN (mask #3).
  * Open oxide over DEEPN regions.
  * Deposit phosphorus.
  * Drive deep-N+ down to meet NBL.
  * This step also completes the isolation drive.
  * Wet oxide grown during the deep-N+ drive forms the field oxide (FOX).

Using the mask open oxide deposit lots and lots of phosphorus that is almost certainly going to be spun on glass of some sort dry. We're driving down the ISO. the NBL moves a little, but it's a slow moving stuff layer. We probably need a thickness of 10 to 15 microns. It's pretty thick and we have to drive down through that depth.

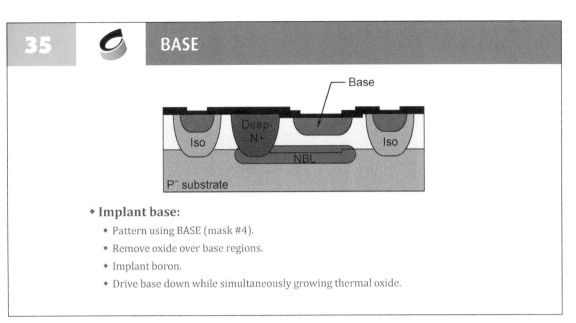

35   **BASE**

* **Implant base:**
  * Pattern using BASE (mask #4).
  * Remove oxide over base regions.
  * Implant boron.
  * Drive base down while simultaneously growing thermal oxide.

The next step is to pattern the wafer use the base mask which creates the base region of the NPN transistor. We plant some boron, control is crucial to maintaining accurate bipolar beta. It's actually very thin and flat.

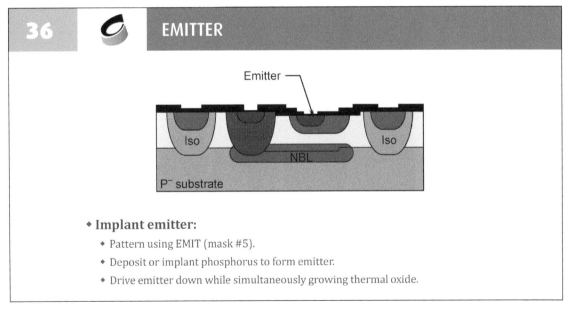

36   **EMITTER**

* **Implant emitter:**
  * Pattern using EMIT (mask #5).
  * Deposit or implant phosphorus to form emitter.
  * Drive emitter down while simultaneously growing thermal oxide.

Then we open a window, using the emitter mask in the middle of the base . This gives us a heavily doped region. There's our NPN transistor forming and then see the NBL and the deep-end which form a low resistance connection to the vertical transistor that's the basic structure.

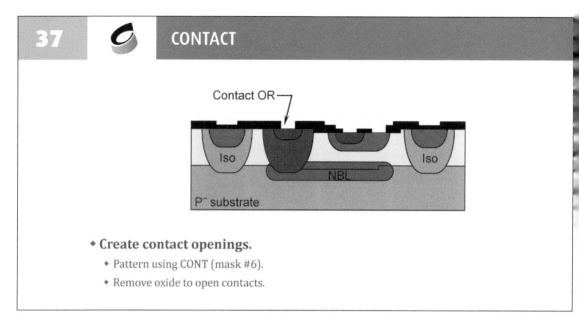

37 **CONTACT**

⬥ **Create contact openings.**
  ⬥ Pattern using CONT (mask #6).
  ⬥ Remove oxide to open contacts.

This process is built to build and finally we open up some contacts where we want to put contacts down.

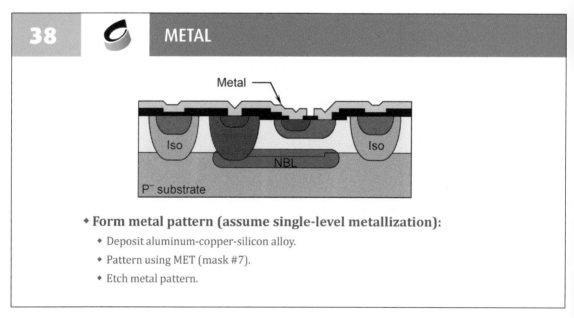

38 **METAL**

⬥ **Form metal pattern (assume single-level metallization):**
  ⬥ Deposit aluminum-copper-silicon alloy.
  ⬥ Pattern using MET (mask #7).
  ⬥ Etch metal pattern.

We put in some metal. The emitter should be thick enough, so that the aluminum doesn't eat through it. We add some silicon to the aluminum alloy to prevent the eating, because there's a certain solubility in aluminum of silicon, this is helpful.

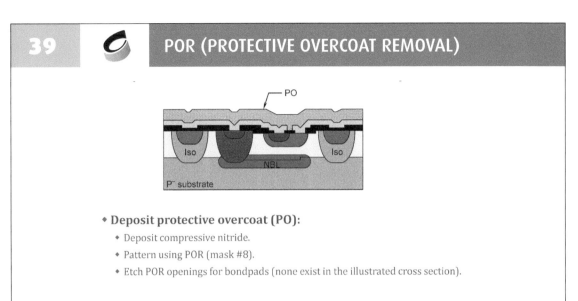

39 — POR (PROTECTIVE OVERCOAT REMOVAL)

* **Deposit protective overcoat (PO):**
    * Deposit compressive nitride.
    * Pattern using POR (mask #8).
    * Etch POR openings for bondpads (none exist in the illustrated cross section).

Now we need a protective overcoat. As wafers are very fragile and the metal was so delicate, once you put a good overcoat on top and slide them into the envelope, no harm done. The overcoat is tough and hard glassy Rock like stuff not squishy toothpaste like metal pure aluminium is very soft.

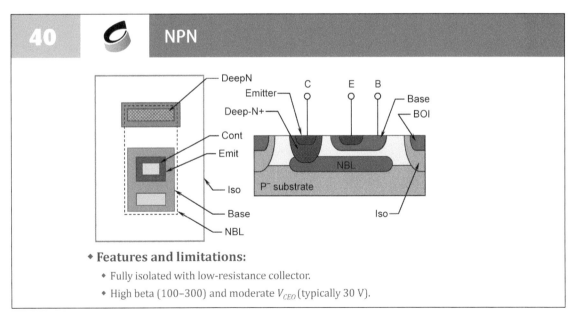

40 — NPN

* **Features and limitations:**
    * Fully isolated with low-resistance collector.
    * High beta (100–300) and moderate V_{CEO} (typically 30 V).

NPN is the premier component from standard bipolar. It is a pretty good device. It's fully isolated low resistance collector pretty high beta 100 to 300. Moderate veto desk collector to emitter breakdown with base terminal open about 30 volts.

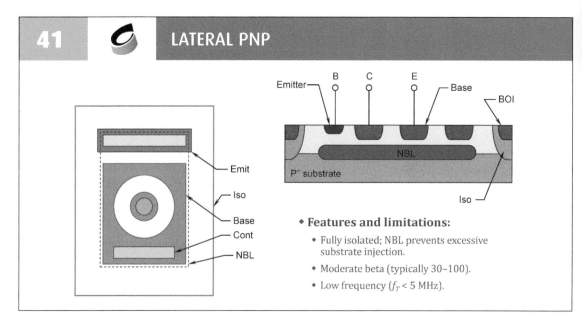

Features and limitations:

- Fully isolated; NBL prevents excessive substrate injection.
- Moderate beta (typically 30–100).
- Low frequency (f_T < 5 MHz).

Lateral PNP: We use base diffusion to create an emitter region, the collector is also made out of base diffusion. The emitter is a circle or square placed in a hole. The carriers move from emitter to collector radically outwards in all directions. So it's called a lateral transistor. The beta goes from 1 or less to something like thirty to a hundred, and the frequency response is poor, because the base is the entire region.

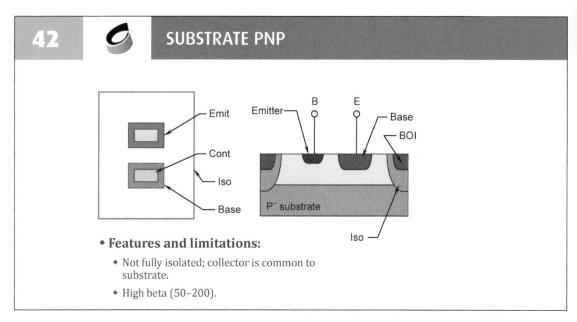

Features and limitations:

- Not fully isolated; collector is common to substrate.
- High beta (50–200).

We take a square of emitter that forms contact take a square base that forms an emitter. That's the vertical device with the collector common to the substrate. In other words, grounded collector that's a substrate PNP, it's usually got a pretty good beta and somewhat hired the reason is that it's a much smaller base.

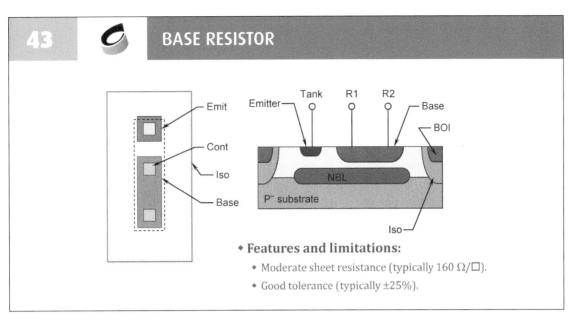

43 BASE RESISTOR

* **Features and limitations:**
  * Moderate sheet resistance (typically 160 $\Omega/\square$).
  * Good tolerance (typically ±25%).

In standard bipolar, we don't have deposited resistors. We take a strip of base, put two contacts, sheet resistance about 160 ohms. So you can build 50K Ohms or on your dies. However you have to make sure that you keep the PN Junction between the body of your resistor and the enclosing tank reverse bias.

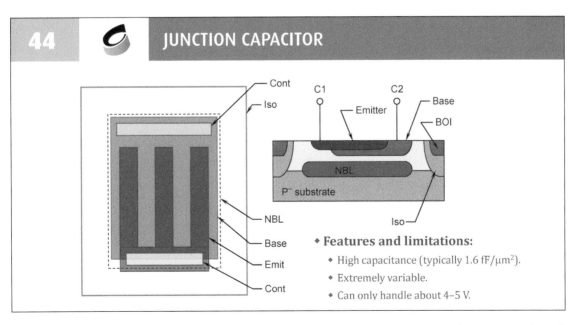

44 JUNCTION CAPACITOR

* **Features and limitations:**
  * High capacitance (typically 1.6 fF/μm^2).
  * Extremely variable.
  * Can only handle about 4–5 V.

Junction capacitor is so easy to do. Every PN Junction is a capacitor because the depletion region itself is a dielectric and so while it's a voltage dependent. The base emitter Junction is kept reverse biased. The forward bias that you get a diode in conduction. That's probably not what you want.

45   POLY-GATE CMOS

- **The self-aligned polysilicon-gate CMOS process is the basis for most modern analog CMOS and BiCMOS processes.**

- **Many features of this process are still widely used:**
  - P+ substrates
  - Self-aligned polysilicon gates
  - LOCOS field oxidation
  - Channel stop and threshold adjust implants
  - Silicidation and aluminum metallization

The self-aligned polysilicon gate CMOS process is the basis for most CMOS process. The very early CMOS processes used aluminum gate and they were not very good, control was typically between 2V and 4V. Now we use poly silicon gate, it's really low current. Another feature is low cost field oxidation which is today often replay with STI.

46   NWELL

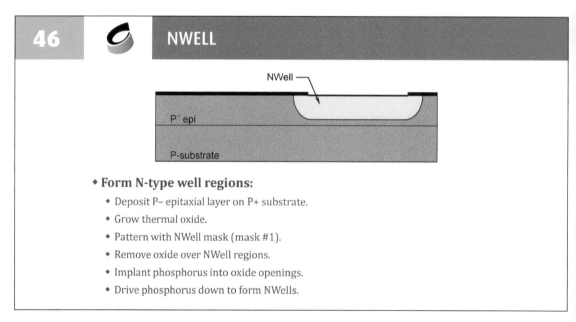

- **Form N-type well regions:**
  - Deposit P– epitaxial layer on P+ substrate.
  - Grow thermal oxide.
  - Pattern with NWell mask (mask #1).
  - Remove oxide over NWell regions.
  - Implant phosphorus into oxide openings.
  - Drive phosphorus down to form NWells.

We start with a P+ substrate and deposit P– EPI, the EPI quality is higher than the substrate quality, no oxygen precipitates. Then grow a thermal oxide patterning with the first mask for the n-type. This is a deep lightly doped region, then you remove the oxide over the in well regions and plant a little phosphorus drive.

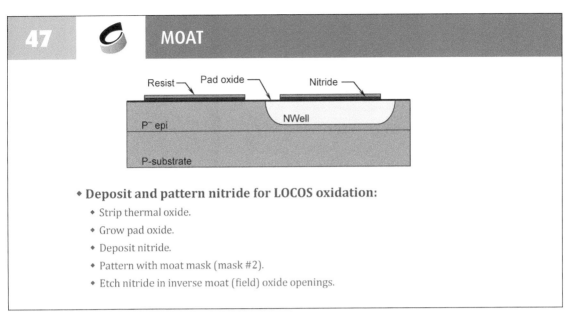

47   MOAT

- **Deposit and pattern nitride for LOCOS oxidation:**
  - Strip thermal oxide.
  - Grow pad oxide.
  - Deposit nitride.
  - Pattern with moat mask (mask #2).
  - Etch nitride in inverse moat (field) oxide openings.

The next step, we need some field oxide, but we strip away the thermal oxide we had, before through a thin pad oxide, the purpose of the pad oxide is to really reduce the problem of stress. So we deposit nitride pattern with the mote mask and etch the nitride, using the so called inverse moat mask.

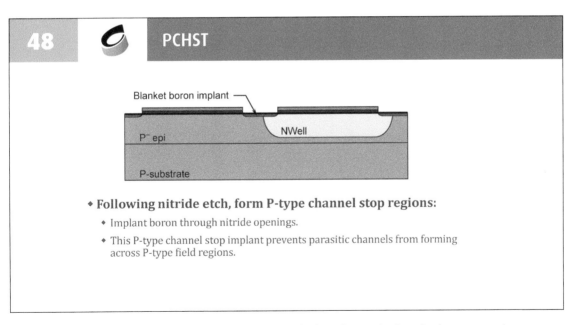

48   PCHST

- **Following nitride etch, form P-type channel stop regions:**
  - Implant boron through nitride openings.
  - This P-type channel stop implant prevents parasitic channels from forming across P-type field regions.

Then we shoot in some p-type implant through the openings. Why would we do that? This is a CMOS, we can invert with a gate electrode but what stops a metal wire from inverting the silicon? we could thicken the oxide, but that's not enough. We want to be able to work at 20V or 30V. So shoot in extra implant stuff in there to dope the surface up.

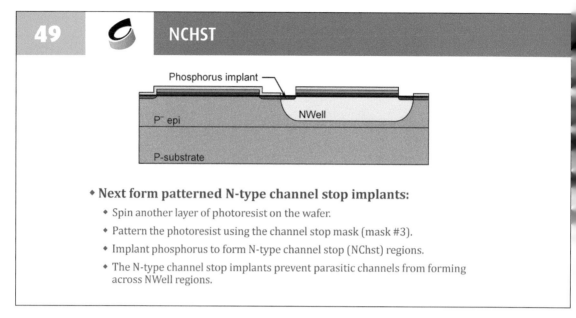

49   NCHST

* **Next form patterned N-type channel stop implants:**
    * Spin another layer of photoresist on the wafer.
    * Pattern the photoresist using the channel stop mask (mask #3).
    * Implant phosphorus to form N-type channel stop (NChst) regions.
    * The N-type channel stop implants prevent parasitic channels from forming across NWell regions.

Using a mask and we add extra P to the regions, we don't want transistors under the field. So we're going to add extra N, this raises the VT of the final transistors up to 40V or something and we don't have to worry, only where we want transistors do they form these are so-called channel stub implants.

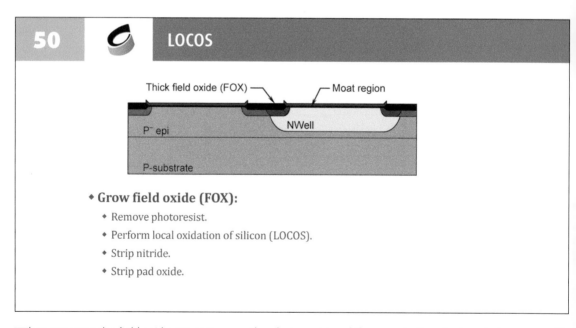

50   LOCOS

* **Grow field oxide (FOX):**
    * Remove photoresist.
    * Perform local oxidation of silicon (LOCOS).
    * Strip nitride.
    * Strip pad oxide.

Then we grow the field oxide. We strip away the photo resist and then we do the nice long hot process with a little steam, then we strip away the nitride strip off the pad oxide.

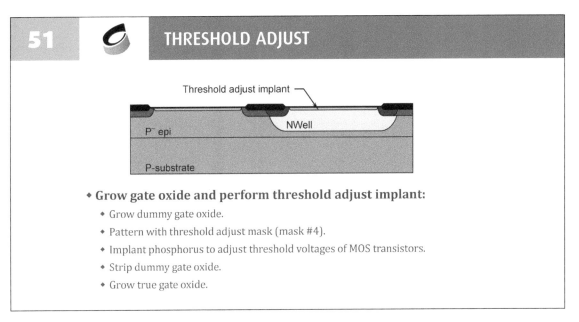

51   THRESHOLD ADJUST

Threshold adjust implant

- **Grow gate oxide and perform threshold adjust implant:**
  - Grow dummy gate oxide.
  - Pattern with threshold adjust mask (mask #4).
  - Implant phosphorus to adjust threshold voltages of MOS transistors.
  - Strip dummy gate oxide.
  - Grow true gate oxide.

The next thing is grow some gate oxide, but when people try doing this, they found that the low-cost process which uses steam water screwed up the gate oxide. The solution is, the nitride has reluctantly oxidized away. So we grow an oxide and throw it away, grow another oxide. It'll be better.

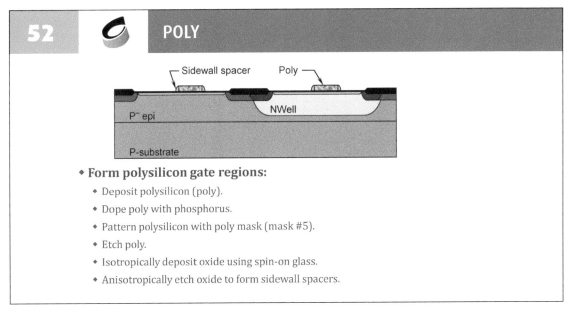

52   POLY

Sidewall spacer   Poly

- **Form polysilicon gate regions:**
  - Deposit polysilicon (poly).
  - Dope poly with phosphorus.
  - Pattern polysilicon with poly mask (mask #5).
  - Etch poly.
  - Isotropically deposit oxide using spin-on glass.
  - Anisotropically etch oxide to form sidewall spacers.

Then we shoot in something called a threshold adjust. This is a implant that Tunes the VT to the value. We want the same VT adjust for both NMOS and PMOS. We go ahead and strip off dummy gate oxide, grow true gate oxide, so the MOSFET transistors forming. The next step is deposit oxide with a spin on glass. It forms kind of an equal thickness of oxide everywhere.

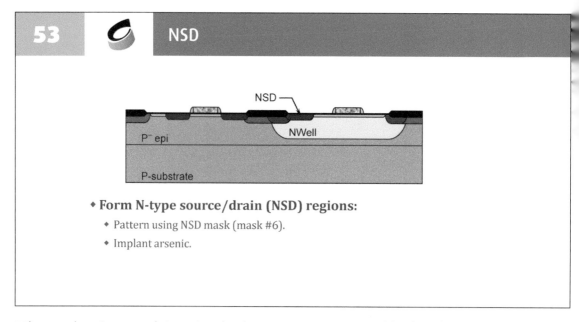

53   NSD

* **Form N-type source/drain (NSD) regions:**
  * Pattern using NSD mask (mask #6).
  * Implant arsenic.

Then we form in source drain regions by shooting an implant into the die and that looks that aligns with the poly. Because the NSD can't through the poly or the sidewall spacers. and the source and drain regions are mask by the poly itself, so it's called self aligns. This forms source trains perfectly aligned to the gate with minimum overlap capacitance.

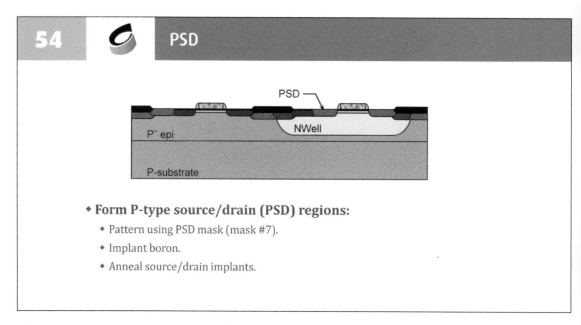

54   PSD

* **Form P-type source/drain (PSD) regions:**
  * Pattern using PSD mask (mask #7).
  * Implant boron.
  * Anneal source/drain implants.

Then we go into p-type source strength and that gives us source and drain regions for a PMOS transistor as well as a back gate contact for my n-type transistor. Just as an n-type source drain region forms a back gate contact for PMOS transistor, and now have all the diffusions in place.

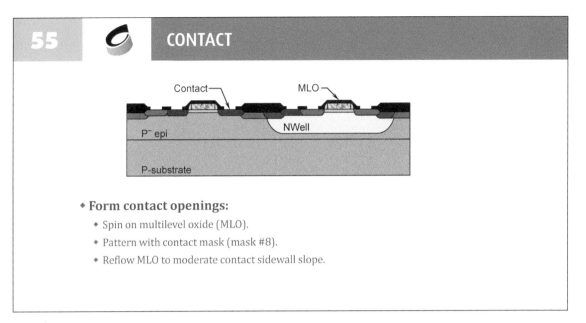

55   **CONTACT**

* **Form contact openings:**
  * Spin on multilevel oxide (MLO).
  * Pattern with contact mask (mask #8).
  * Reflow MLO to moderate contact sidewall slope.

Then we form some contacts, as described before.

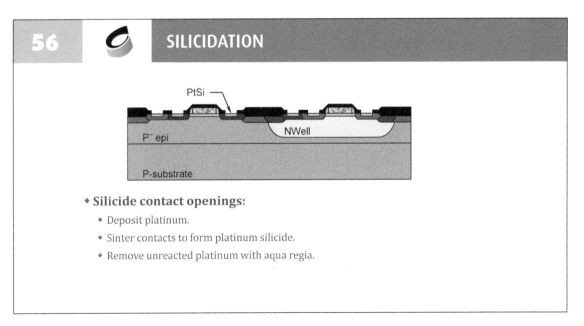

56   **SILICIDATION**

* **Silicide contact openings:**
  * Deposit platinum.
  * Sinter contacts to form platinum silicide.
  * Remove unreacted platinum with aqua regia.

The early processes of silicidation is platinum silicate, then they become titanium silicide, which can reduce the resistance of poly gate.

57   METAL

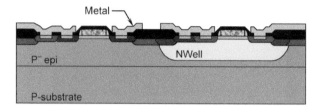

* **Form metal system:**
  * Deposit refractory barrier metal (RBM).
  * Deposit aluminum doped with 0.5% copper.
  * Pattern using metal mask (mask #9).
  * Etch metal.

To form metal system, we use a layer of so-called refractory barrier metal to prevent the aluminum from interacting with my silicide that might be titanium nitride. Now put aluminum dope with a bit of copper on top edge for the metal mask slap on protests of nitride overcoat open protective.

58   PROTECTIVE OVERCOAT

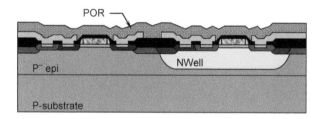

* **Deposit protective overcoat:**
  * Deposit compressive nitride.
  * Pattern using protective overcoat removal (POR) mask (mask #10).
  * Etch openings in overcoat for bondpads (none shown in cross section).

Illustrate the step of deposit protective overcoat (PO), as described before. Then open protective overcoat removable opening for bond pads.

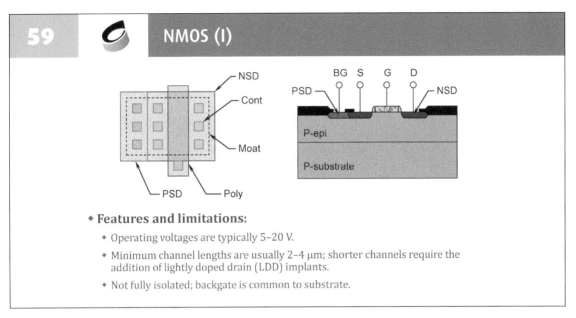

Features and limitations:

- Operating voltages are typically 5–20 V.
- Minimum channel lengths are usually 2–4 μm; shorter channels require the addition of lightly doped drain (LDD) implants.
- Not fully isolated; backgate is common to substrate.

Now are shown here and we complete in 10 masks, in standard bipolar it is 7 masks. There's a pattern here as technology marches on the mask count grows. So if we looked at a modern process very advanced process for BiCMOS, it might have 35 mask in it or 40 mask in it.

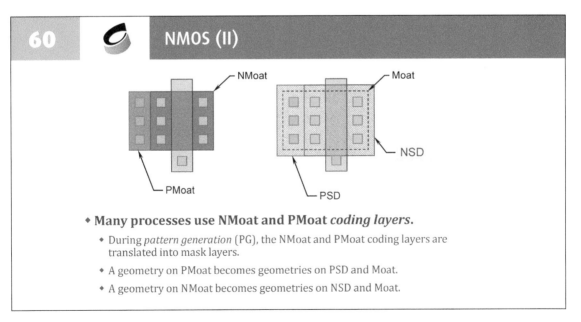

Many processes use NMoat and PMoat *coding layers*.

- During *pattern generation* (PG), the NMoat and PMoat coding layers are translated into mask layers.
- A geometry on PMoat becomes geometries on PSD and Moat.
- A geometry on NMoat becomes geometries on NSD and Moat.

This is a NMOS transistor, we create a Moat region, which is the dotted line, we cross the Moat with a piece of poly. When you cross Moat with poly, you are forming a MOS transistor, the active gate region of the MOS transistor is the intersection of the poly and the Moat. CMOS processes typically had lots of little square contacts rather that large arbitrary shaped contact in standard bipolar.

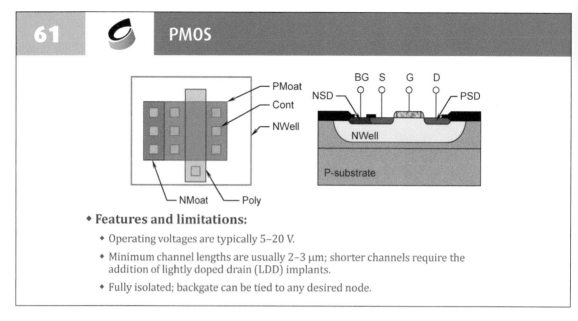

61 PMOS

* **Features and limitations:**
  * Operating voltages are typically 5–20 V.
  * Minimum channel lengths are usually 2–3 μm; shorter channels require the addition of lightly doped drain (LDD) implants.
  * Fully isolated; backgate can be tied to any desired node.

A PMOS is a kind of inverse of the NMOS. The process Voltage is 5V to 20V, P channel makes usually a trifle shorter than the NMOS, because holes are sluggish and electrons carried problems aren't quite so bad. We could put a different voltage on the back gate than the source back gate modulation.

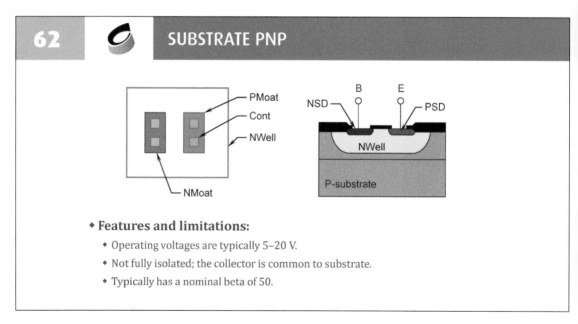

62 SUBSTRATE PNP

* **Features and limitations:**
  * Operating voltages are typically 5–20 V.
  * Not fully isolated; the collector is common to substrate.
  * Typically has a nominal beta of 50.

We can also create a substrate PNP, a PSD region in NWELL sitting in a P substrate. It's not a good PNP, it is limited in connect and the beta is about 5 in modern process, but you can use it to build bandgap to generate a stable voltage, usually within between 1/2 percent and 2 percent accurate.

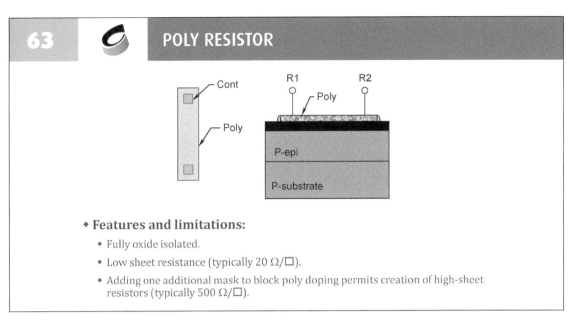

63   POLY RESISTOR

♦ Features and limitations:

- ♦ Fully oxide isolated.
- ♦ Low sheet resistance (typically 20 $\Omega/\square$).
- ♦ Adding one additional mask to block poly doping permits creation of high-sheet resistors (typically 500 $\Omega/\square$).

There are poly resistors, they are wonderful, no junction and don't have to worry about keeping reverse biased. The poly resistance used for gate is not silicided, it probably has a resistance of about 20 ohms a square, adding an additional mask allows us to change the doping in the poly body and can get 500 ohms a square, this is cool by having very narrow poly strips.

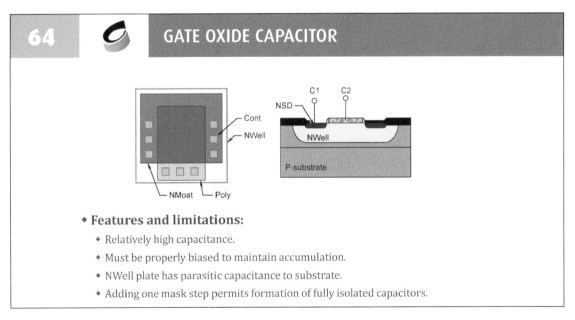

64   GATE OXIDE CAPACITOR

♦ Features and limitations:

- ♦ Relatively high capacitance.
- ♦ Must be properly biased to maintain accumulation.
- ♦ NWell plate has parasitic capacitance to substrate.
- ♦ Adding one mask step permits formation of fully isolated capacitors.

We can see in well to dielectric very thin oxide to gate poly. So this is a gate oxide capacitor as long as I keep it biased properly. So that I accumulate and do not deplete this region under the gate electrode, this will form a nice capacitor.

3. PASSIVES

This unit is called passives. The technical definition of a passive device is a device that does not exhibit power gain. So technically resistors capacitors inductors and for most purposes diodes are all passive devices.

---

65 **PASSIVE DEVICES**

* **Passive devices include resistors, capacitors, and inductors.**
  * Milliohms to megohms of resistance are easily integrated.
  * Capacitors of up to a few hundred picofarads can be integrated.
  * Inductors don't lend themselves to easy integration and won't be discussed (although some RF designs do make use of very small integrated inductors).

* **Diodes are technically passive devices, but we will consider them later when we discuss bipolar transistors.**

We will only talk about resistors and capacitors. The reason is, it is possible to create an integrated inductor but you can't make a very big one and it's not a very good one, and they do involve spirals of wires but not the silicon.

---

66 **RESISTIVITY (I)**

* **Consider a simple rectangular slab having width *W*, length *L*, and thickness *t*. Its resistance *R* equals**

$$R = \rho \frac{L}{Wt}$$

* **The constant of proportionality ρ is called the *resistivity* and has units of $\Omega \cdot$cm.**

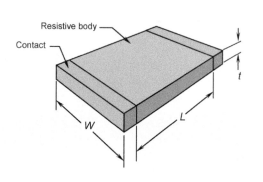

There's a fundamental property of matter called resistivity. Semiconductors have high, but not very high resistivity and those resistivities can be varied by changing the doping The resistivity times the length divided by the width divided by the thickness, this gives a value of resistance.

67   RESISTIVITY (II)

Material	Resistivity, $\Omega \cdot cm$ (25°C)
Copper, bulk	$1.7 \cdot 10^{-6}$
Gold, bulk	$2.4 \cdot 10^{-6}$
Aluminum, thin film	$2.7 \cdot 10^{-6}$
Aluminum (2% silicon)	$3.8 \cdot 10^{-6}$
Platinum silicide	$3.0 \cdot 10^{-5}$
Silicon, N-type ($N_d = 10^{18} cm^{-3}$)	0.25
Silicon, N-type ($N_d = 10^{15} cm^{-3}$)	48
Silicon, intrinsic	$2.5 \cdot 10^{5}$
Silicon dioxide	$\sim 10^{14}$

- ◆ *Conductors*, such as metal, have low resistivities.

- ◆ *Semiconductors*, such as silicon, have moderate resistitivies that depend strongly upon doping levels.

- ◆ *Insulators*, such as oxide, have extremely high resistivities.

All types of matter have some quantity of resistivity. Insulators have high resistivity, conducting metals have low resistivity.. We can see thin film resistivities are frequently a little higher than the bulk resistivities, because of surface, interactions aluminum with a little silicon added to it is more resistive.

68   SHEET RESISTANCE

- ◆ **Integrated resistors usually consist of diffusions or depositions of constant thickness.**
  - ◆ We can combine resistivity and thickness into a single term called *sheet resistance*, R_S, where $R_S = \rho/t$.
  - ◆ The equation for resistance now becomes:

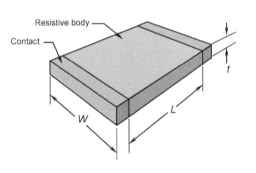

$$R = R_s \left(\frac{L}{W} \right)$$

We know that the thickness of the resistor is the same everywhere and the layout person doesn't set it, the layout person sets the length and width. So we just pull that into this proportional constant, and produce a constant that we call the sheet resistance which is the resistivity divided by the thickness.

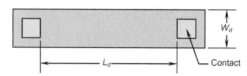

69   **WIDTH AND LENGTH BIASES**

- **The *drawn width W_d* and *drawn length L_d* are those entered into the database.**
  - The true width and length differ from the drawn width and length because of outdiffusion, overetching, process size adjustments, etc.
  - Therefore we define a *width bias W_b* and a *length bias L_b* such that

$$R = R_S \left(\frac{W_d + W_b}{L_d + L_b} \right)$$

The drawn dimensions are what the layout designer puts into a database. A lot of processing is in the middle and so what comes out is not quite what we sent in. They shift because of effects like out diffusion over etching changes in photo mask processing light interaction with the sidewalls. We can define Wb and Lb and say whatever your drawing width is draw add the width bias.

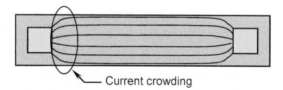

70   **CURRENT CROWDING**

Current crowding

- **Current crowds inwards towards the contacts, increasing resistance by an amount**

$$\Delta R = \frac{R_s}{\pi} \left[\frac{1}{k} \ln \left(\frac{k+1}{k-1} \right) + \ln \left(\frac{k^2 - 1}{k^2} \right) \right]$$

where $k = W_d / (W_d - W_c)$, in which W_c is the width of the contact.

We'd like to believe the current flows one end of the resistor to the other end uniformly. but what about the contact doesn't span the entire width of the resistor? The contact increases the resistance at the ends of the resistor. There are various ways to account for, this is a very famous way, we add a correction to whatever value you have and they call it like a head correction.

71 CONTACT RESISTANCE

- **Current also does not flow uniformly into the contact in the vertical direction.**
  - This is usually modeled by assuming each contact adds a *contact resistance* R_c to the resistor.
  - For a homogenous resistor whose contact material has resistivity ρ_c,

$$R_c = \frac{\sqrt{R_s \rho_c}}{W_C} \coth\left(L_C \sqrt{R_s / \rho_c}\right)$$

There's also vertical current crowding. The contact on most resistors is at the surface. So the lines of current flow have to bend up and exit through the surface and if the material is homogeneous.

72 TEMPCO

$$R(T) = R(T_0)[1 + 10^{-6} TC(T - T_0)]$$

- **Resistivity varies with temperature.**
  - The actual relationship is nonlinear, but we can linearize it.
  - The constant of proportionality *TC*, called the *coefficient of resistivity*, has units of ppm/°C.
  - *TC* is often called the *tempco*.

Material	TCR, ppm/°C
Copper, bulk	+4000
Gold, bulk	+3700
Aluminum, bulk	+3800
160 Ω/□ base diffusion	+1500
7 Ω/□ emitter diffusion	+600
5 kΩ/□ base pinch diffusion	+2500
2 kΩ/□ HSR implant (P-type)	+3000
500 Ω/□ polysilicon (4 kÅ N-type)	−1000
25 Ω/□ polysilicon (4 kÅ N-type)	+1000
10 kΩ/□ NWell	+6000

Temperature coefficient: the resistance of our resistors shift temperature. If you do thin-film aluminum with about two percent silicon and a half percent copper, the number is about 3300 per million, and that's a very special number that is roughly, the number of parts per million that a voltage proportional to absolute temperature.

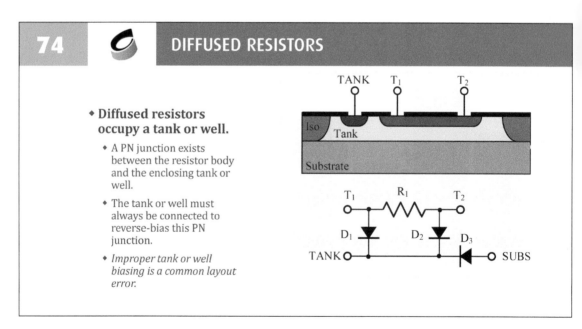

73 — DEPOSITED RESISTORS

* **Deposited resistors (such as poly resistors) are fully insulated from the silicon.**
  * The only important parasitics are therefore capacitances.

We can also make deposited resistors, they don't have any junctions. But they still have parasitic capacitance between the deposited layer and the substrate and also between the deposited layer in any metal beads you put on top of them.

74 — DIFFUSED RESISTORS

* **Diffused resistors occupy a tank or well.**
  * A PN junction exists between the resistor body and the enclosing tank or well.
  * The tank or well must always be connected to reverse-bias this PN junction.
  * *Improper tank or well biasing is a common layout error.*

Diffused resistor is a piece of base, sitting in a tank with a tank contact, instead of just being capacitors. We have diodes PN diodes between the body and the tank and also between the tank and the substrate. Why would one use a diffused resistor? Maybe because you like the TC. Maybe because it's embedded in mono crystalline silicon which is a pretty good thermal conductor.

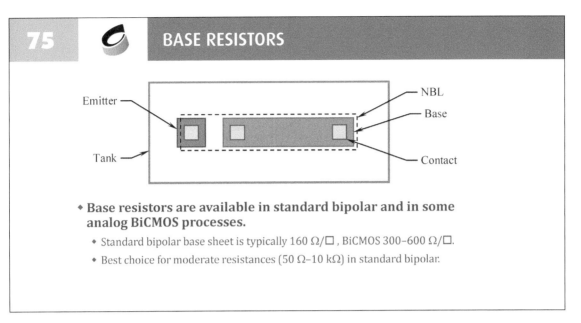

75   **BASE RESISTORS**

* **Base resistors are available in standard bipolar and in some analog BiCMOS processes.**
  * Standard bipolar base sheet is typically 160 $\Omega/\square$, BiCMOS 300–600 $\Omega/\square$.
  * Best choice for moderate resistances (50 Ω–10 kΩ) in standard bipolar.

This is a layout of a base resistor typically. We strip a base, a tank contact, two   contacts on either end of the base strip, and usually we will throw some NBL in underneath, that reduces the tank resistance.

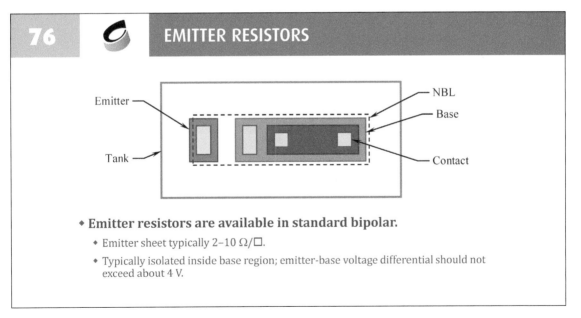

76   **EMITTER RESISTORS**

* **Emitter resistors are available in standard bipolar.**
  * Emitter sheet typically 2–10 $\Omega/\square$.
  * Typically isolated inside base region; emitter-base voltage differential should not exceed about 4 V.

We can also make an emitter resistor, placing it inside base, making a reverse bias. Junction here putting the base inside a tank, using a tank contact, and make sure they're all reverse biased and don't forget that the base emitter junction breaks down. It's something like maybe 7V.

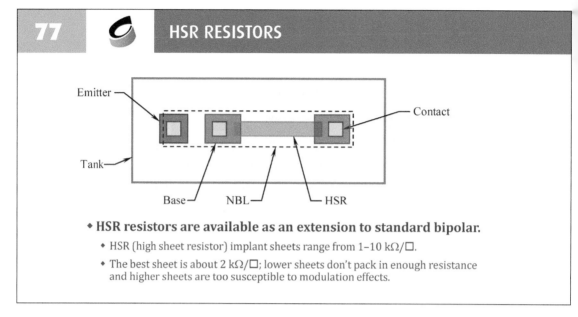

77 HSR RESISTORS

* **HSR resistors are available as an extension to standard bipolar.**
    * HSR (high sheet resistor) implant sheets range from 1–10 kΩ/$\square$.
    * The best sheet is about 2 kΩ/$\square$; lower sheets don't pack in enough resistance and higher sheets are too susceptible to modulation effects.

There's also a thing called a HS resistor. Assume you bought standard bipolar with 8 masks and you don't like the only resistors, they will sell you another feature if you'll pay for it, buy a mask and they give you another implant. This is an example of a process option.

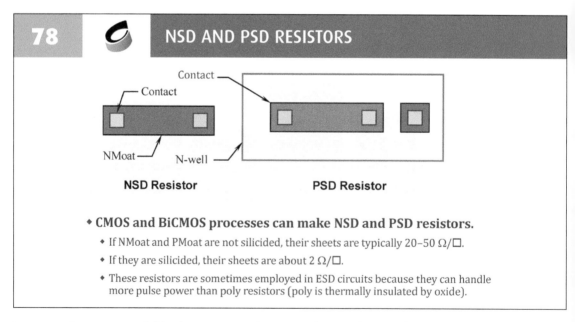

78 NSD AND PSD RESISTORS

NSD Resistor   **PSD Resistor**

* **CMOS and BiCMOS processes can make NSD and PSD resistors.**
    * If NMoat and PMoat are not silicided, their sheets are typically 20–50 Ω/$\square$.
    * If they are silicided, their sheets are about 2 Ω/$\square$.
    * These resistors are sometimes employed in ESD circuits because they can handle more pulse power than poly resistors (poly is thermally insulated by oxide).

NSD and PSD are available doping materials in CMOS process and can create diffused resistors. If not silicided, the sheet resistance is 20-50 Ohm. However very well heat sunk by this silicon die and so for ESD devices, these are very frequently used, because they can take more pulse power than a isolated device, sitting up in the oxide like poly.

79 HSR POLY RESISTORS

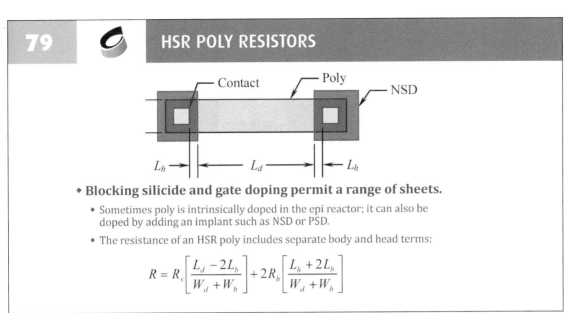

* **Blocking silicide and gate doping permit a range of sheets.**
  * Sometimes poly is intrinsically doped in the epi reactor; it can also be doped by adding an implant such as NSD or PSD.
  * The resistance of an HSR poly includes separate body and head terms:

$$R = R_s \left[\frac{L_d - 2L_b}{W_d + W_b} \right] + 2R_h \left[\frac{L_h + 2L_b}{W_d + W_b} \right]$$

There are also HSR poly resistors and this is where you take and buy another mask process extension. You can block the gate into it whatever the gate poly is doped with. There's only a little bit of dopant that comes from the EPI reactor that they call that in situ doping and so this is now a highly resistive layer, maybe 500 ohms a square.

80 POLY RESISTORS

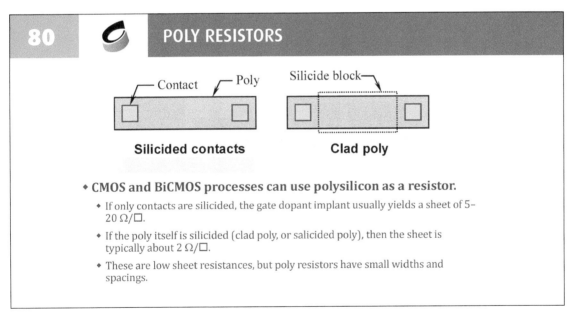

Silicided contacts   **Clad poly**

* **CMOS and BiCMOS processes can use polysilicon as a resistor.**
  * If only contacts are silicided, the gate dopant implant usually yields a sheet of 5–20 $\Omega/\square$.
  * If the poly itself is silicided (clad poly, or salicided poly), then the sheet is typically about 2 $\Omega/\square$.
  * These are low sheet resistances, but poly resistors have small widths and spacings.

This is an example of a self-aligned structure, it would be the same if that we had two contacts of the structure and set the length of the resistor. You could buy a mask called the silicide block that will block the silicide from a selected part of the resistor.

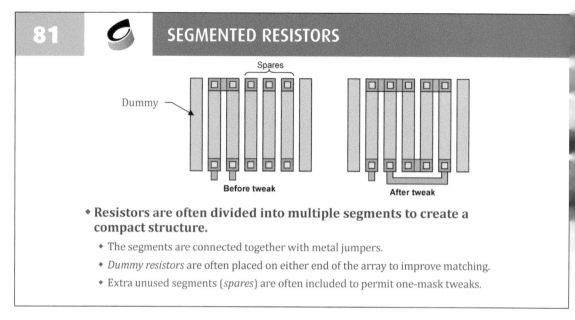

81   SEGMENTED RESISTORS

Before tweak

After tweak

♦ **Resistors are often divided into multiple segments to create a compact structure.**

  ♦ The segments are connected together with metal jumpers.

  ♦ *Dummy resistors* are often placed on either end of the array to improve matching.

  ♦ Extra unused segments (*spares*) are often included to permit one-mask tweaks.

If there is a large resistor, we can cut it into pieces called sections or fingers or segments and parallel them up all in a row. And then, we jump them together with metal jumpers, we'll talk about it later, the correct way is to jump them back and forth and back and forth.

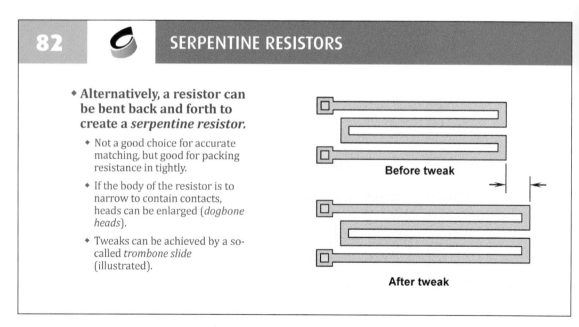

82   SERPENTINE RESISTORS

♦ **Alternatively, a resistor can be bent back and forth to create a *serpentine resistor*.**

  ♦ Not a good choice for accurate matching, but good for packing resistance in tightly.

  ♦ If the body of the resistor is to narrow to contain contacts, heads can be enlarged (*dogbone heads*).

  ♦ Tweaks can be achieved by a so-called *trombone slide* (illustrated).

Before tweak

After tweak

If you don't want a bunch of segments. You can just bend it back and forth. We call these serpentine resistance. There's a trick that if you make your wafer that you set aside right before poly pattern rather than metal pattern. You can go back and cut a new poly mask. You take the cool turns of the resistor and pull them out like a trombone gets pulled out. and it gets bigger.

83 PARALLEL-PLATE CAPACITOR

* **A parallel-plate capacitor consists of an insulating layer called a *dielectric* sandwiched between two conductive electrodes.**

  * The capacitance C equals approximately

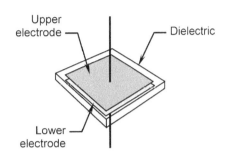

$$C \cong 0.0885 \frac{A\varepsilon_r}{t}$$

where A is the area of either electrode in μm^2, t is the thickness of the dielectric in Angstroms (Å), and ε_r is a dimensionless constant called the *relative permittivity* or *dielectric constant*.

In most IC processes, we can't expect to have an large capacitor but we can expect to have small but high quality capacitors like 1pF. As we need area for them to work. Almost all of our capacitors are really parallel plate capacitors now. ?r is a dimensionless constant called the relative permittivity. The relative permittivity of most materials is greater than one.

84 DIELECTRICS

Material		Relative permittivity	Dielectric strength (MV/cm)
Silicon		11.8	30
Silicon dioxide (SiO_2)	Dry oxide	3.9	11
	Plasma	4.9	3–6
	TEOS	4.0	10
Silicon nitride (Si_3N_4)	LPCVD	6–7	10
	Plasma	6–9	5

* **The maximum voltage V_{max} a capacitor can withstand equals**

$$V_{max} = 0.01 t E_{crit}$$

where t is dielectric thickness in Angstroms and E_{crit} is the dielectric strength in MV/cm. For reliable operation, voltages should not exceed about a third of V_{max}.

Silicon is the highest of the substances we normally use at around 12 and so that would be the dielectric in a junction. Dry oxide is a good quality oxide around 3.9. Nitride is varies because the composition of the nitrite varies quite a lot between 6 and 9. So dry oxides is very stable and predictable.

85 · UNEQUAL PLATE AREAS

- **The effective area of a parallel plate capacitor equals the area of intersection of its plates.**
  - The illustrated capacitor has an area of 300 μm².
  - The actual capacitance is slightly larger due to so-called *fringing fields*.
  - Fringing fields don't contribute much to capacitors with dimensions of more than a couple of microns.

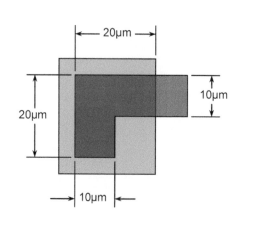

Anyway in a parallel capacitor with a thin dielectric, it's the overlap of the two plates that calculates the area. In this case, some pieces stick out over other pieces. They don't really contribute much to the capacitance, for large capacitors, you can ignore the fringing effects.

86 · JUNCTION CAPACITANCE (I)

- **A reverse-biased PN junction can be used as a capacitance.**
  - The area of such capacitors is hard to compute because of the contributions of sidewalls.
  - The capacitance decreases with increasing reverse bias because of the increasing width of the depletion region.
  - These capacitors are only used where accuracy is not required; for example, for compensation capacitors.

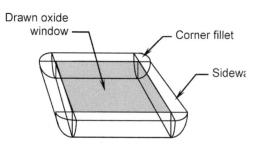

Junction capacitance is actually a three dimensional surface, and the junction is not flat. So if we draw an oxide window and diffuse through, we can calculate with various formulas correction factors. If you make the window large and all dimensions then essentially, you can ignore the edge effects.

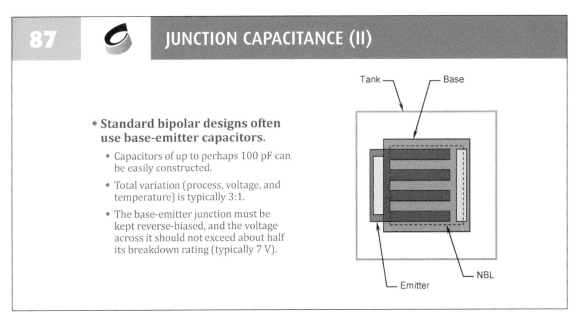

87 JUNCTION CAPACITANCE (II)

* **Standard bipolar designs often use base-emitter capacitors.**
  * Capacitors of up to perhaps 100 pF can be easily constructed.
  * Total variation (process, voltage, and temperature) is typically 3:1.
  * The base-emitter junction must be kept reverse-biased, and the voltage across it should not exceed about half its breakdown rating (typically 7 V).

When we make a base-emitter capacitor, we are not ignoring the perimeter, but counting on them. We realize these sidewalls occur and if I can make the fingers narrow and closely spaced. I will get more capacitance off the fingers than off of a plate.

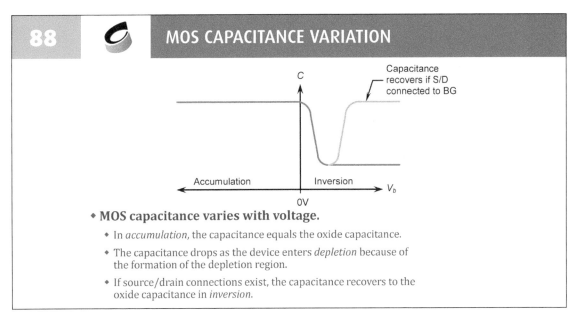

88 MOS CAPACITANCE VARIATION

* **MOS capacitance varies with voltage.**
  * In *accumulation*, the capacitance equals the oxide capacitance.
  * The capacitance drops as the device enters *depletion* because of the formation of the depletion region.
  * If source/drain connections exist, the capacitance recovers to the oxide capacitance in *inversion*.

There are capacitors where we use essentially a MOSFET, the problem is their voltage dependent. If the back gate is p-type and I place a negative on the poly, that attracts more holes up to the surface it makes it look like it's more heavily though that's accumulation capacitance remains constant. If I change the gate voltage to repel the carrier's away from the surface, the capacitance decreases.

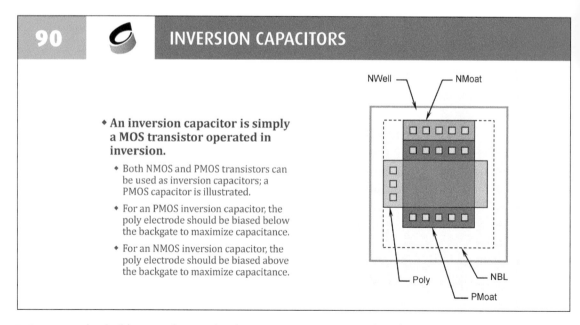

89 ACCUMULATION CAPACITORS

• **Most MOS capacitors are operated in accumulation.**

- • A typical accumulation capacitor uses NWell as its diffused electrode (called its *backgate*).
- • NMoat contacts provide electrical connection to the backgate.
- • This structure looks like a MOS transistor, but it isn't one.
- • To maintain maximum capacitance, the poly electrode should always be biased above the backgate.

NWell — NMoat
Poly — NBL

Accumulation capacitor looks like a MOSFET, but it is not a MOSFET, because the well isn't any well, there are contacts but not source or drain at all, they are contacts for accumulation. It is in the accumulation and this is a fine cap.

90 INVERSION CAPACITORS

• **An inversion capacitor is simply a MOS transistor operated in inversion.**

- • Both NMOS and PMOS transistors can be used as inversion capacitors; a PMOS capacitor is illustrated.
- • For an PMOS inversion capacitor, the poly electrode should be biased below the backgate to maximize capacitance.
- • For an NMOS inversion capacitor, the poly electrode should be biased above the backgate to maximize capacitance.

NWell — NMoat
Poly — NBL
PMoat

We can also build a cap that works always an inversion, it is inversion capacitor. We just took a MOSFET up, keep it beyond the Vt and operate in inversion. So there's always a channel and we have a capacitor.

91   DEPOSITED CAPACITORS

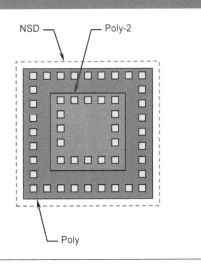

- **Many analog CMOS and BiCMOS processes offer deposited capacitors.**
    - These capacitors are fully oxide isolated and have no parasitic PN junctions.
    - The dielectric material can be optimized for maximum capacitance at a desired operating voltage.
    - Dielectrics with higher dielectric constants (like nitride) provide more capacitance per unit area.

This is a deposited capacitor, it uses two layers of poly one and poly two, a lot of more modern examples use metal or nitride layers to form the capacitor but it doesn't really matter. Poly is awesome, if you want to build some sort of 18 bit data converter and you need absolutely guaranteed stable capacitance over voltage, they deplete a little over voltage.

92   LATERAL FLUX CAPACITORS

- **Metal-metal capacitors can be constructed using the interlevel oxide.**
    - Typically one interdigitates fingers to maximize the lateral as well as the vertical capacitance between conductors.
    - All available metal layers are used to maximize capacitance.
    - Even with these measures, these so-called *lateral flux capacitors* are quite large, and capacitances are thus limited to a few tens of picofarads.

There are also so-called lateral flux caps, this is a cross section through. The capacitor they are layers of metal assay four layers of metal. C1 is one end of the capacitor and C2 is the other, everybody who's red hooks to C2 and everybody who's blue hooks to C1. It's no longer a parallel plate capacitor, It's same thing as fingering. We can get better capacitance with this than with just plain place.

4. MATCHING

Now we get into some of the more interesting aspects of layout. Here's some resistors and capacitors. What do we care about when we put them in circuits? We care about matching most of the time. Everybody uses the same process and the same components. Everybody put together something that passes the DRC and the LVS. Here's where the good layout designer beats the bad one!

PART I

93 INTRODUCTION

* **Analog integrated circuits depend upon device matching.**
  * The tolerances of most integrated components aren't very good; for example, integrated resistors typically vary by ±20%.
  * The matching between adjacent components can be much better; for example, integrated resistors can easily match within ±0.1%
  * Many analog circuits that benefit from matching have been developed specifically for integrated applications.

* **In order to achieve accurate matching, we must understand something about the mechanisms that cause mismatch.**

We got resistors, they are not very good, they vary by +-20% and that's a 6 Sigma number. Hold on, many circuits don't care what the absolute value of is, they care about the tracking of one resistor to another.

94 MATCHING VERSUS TOLERANCE

* **Consider resistor divider R$_1$–R$_2$.**
  * If $R_1 = R_2$, then $V_{out} = \frac{1}{2} V_{in}$.
  * The values of R_1 and R_2 don't affect this relationship.
  * Instead, the ratio $R_2/(R_1+R_2)$ matters.

* **Components whose values track are said to _match_.**
  * Suppose R$_1$ and R$_2$ are supposed to have the same value.
  * If we measure $R_2 = 1.01 R_1$, then the two resistors mismatch by 1%.

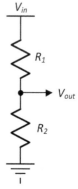

An example of matching two resistors in the resistor divider. We have got an input voltage and make an output voltage, if both resistors are 20 percent large, the output voltage is still the same, it doesn't matter. But if one of the resistors varies from the other, they don't track each other, it matters.

95   RANDOM MISMATCH

* **Resistance, capacitance, and other electrical parameters exhibit random variation.**
  * **This variation can be quantified by a *mean, m,* and a *standard deviation, s.***
  * **For a sample of *N* measurements,**

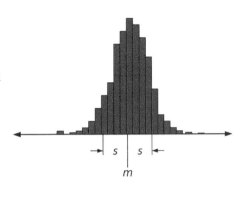

$$m = \sum_{i=1}^{N} x_i \qquad s = \sqrt{\sum_{i=1}^{N} \frac{(x_i - m)^2}{N - 1}}$$

So let's talk about how to measure matching. Everyone knows that quantities are distributed kind of Gaussian. We can apply statistics one thing, the average is the mean. There's another thing called the standard deviation which is a measure of how broad the distribution is how much things scattered.

 96   PELGROM'S LAW

* **Most electrical parameters exhibit random variations due to areal fluctuations.**
  * If this is the case, then the standard deviation of an electrical parameter, say resistance R, equals

    $$s_R = \frac{c_R}{\sqrt{A_R}}$$

    where c_R is a constant and A_R is the active area of the component, in this case the active area of a resistor.
  * **The difference ΔR between two resistors R_1 and R_2 exhibits a standard deviation $s_{\Delta R}$,**

    $$s_{\Delta R} = \sqrt{s_{R1}^2 + s_{R2}^2}$$

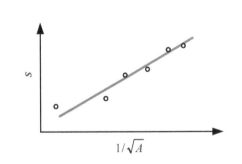

Each resistors varies a percentage of their total resistance, deviation is inversely proportional to the square root of the area. But what about match? If you have a resistor divider, you can't just add their variances, there are random quantities you have to add. The plot of mismatch, making a bunch of measurements data and we can draw a line through it and we got Pelgrom's law.

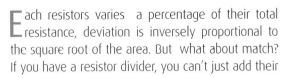

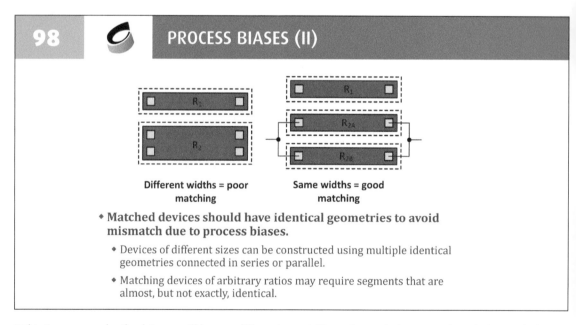

97   PROCESS BIASES (I)

Process Bias = −0.1μm

- **Actual dimensions seldom match drawn dimensions.**
  - Causes include overetching, outdiffusion, straggle, etc.
  - The difference between actual and drawn dimensions is called *process bias*.
  - Process biases cause systematic mismatches.

Process biases: you drew a certain size of shape and you didn't get what you drew. Actual dimensions never match the drawn dimensions. There is a width reduction factor between drawn width and silicon width. Everything is affected by process bias resistors, capacitors, MOSFETs and bipolars.

98   PROCESS BIASES (II)

Different widths = poor matching

Same widths = good matching

- **Matched devices should have identical geometries to avoid mismatch due to process biases.**
  - Devices of different sizes can be constructed using multiple identical geometries connected in series or parallel.
  - Matching devices of arbitrary ratios may require segments that are almost, but not exactly, identical.

This is an example, the fat one will have a different reduction in percent points than the thin one. But if I taking divided into two resistors of the same width and match it to another, it is much better. Same principle applies to every kind of component.

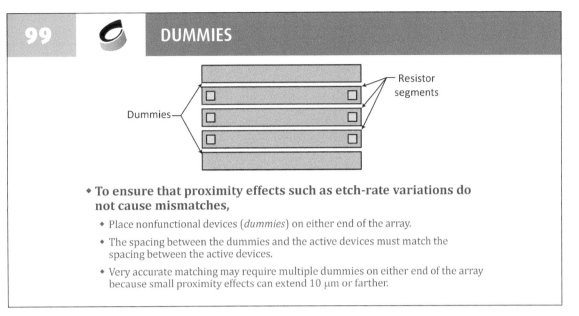

99   **DUMMIES**

* **To ensure that proximity effects such as etch-rate variations do not cause mismatches,**
    * Place nonfunctional devices (*dummies*) on either end of the array.
    * The spacing between the dummies and the active devices must match the spacing between the active devices.
    * Very accurate matching may require multiple dummies on either end of the array because small proximity effects can extend 10 µm or farther.

Dummy is very important. If I'm going to make any kind of resistor. The outside resistors are called dummies, we don't even putting contacts in them, but make sure that the spacings are exactly the same. The outside segments inch more than the inside. So that's only for extreme matching when you're looking for, pointed 1% matching that becomes very important.

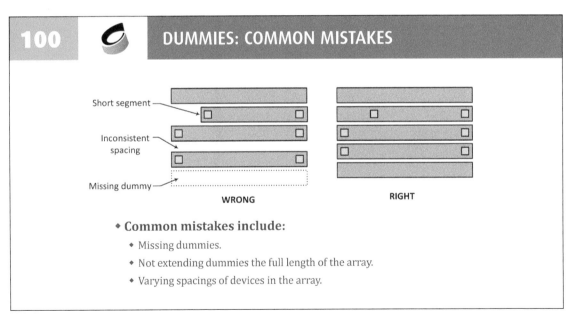

100   **DUMMIES: COMMON MISTAKES**

* **Common mistakes include:**
    * Missing dummies.
    * Not extending dummies the full length of the array.
    * Varying spacings of devices in the array.

The left one is wrong and the right one is right. They put a dummy on one side and none on the other that's wrong thing. Spacing must be consistent. Placing short segment is wrong, you can extend the segment out beyond its contact.

101 DUMMY CAPACITORS

- Dummies

- **Deposited capacitors can also benefit from dummies.**
  - Capacitors are usually drawn as squares or rectangles of low aspect ratio.
  - Width and length are therefore of similar importance.
  - This means dummies must be placed on all four sides of the array.
- **Dummies don't have to be full size.**
  - Their function is mostly to maintain spacings between adjacent geometries.
  - For very accurate matching, the dummies should be at least 10 μm wide.

Capacitors also can benefit from dummies because they also have over etching. This is a poly cap array of four capacitors, you can put dummies around it. We don't know what circuit is around my cap array in the layout, so we always use dummies.

102 HYDROGENATION

Metal getters hydrogen

Hydrogen

Hydrogen doesn't reach poly under metal

- **Hydrogen affects the values of certain components.**
  - Hydrogen ties off dangling bonds at the oxide interface, reducing surface state charge and shifting MOS threshold voltages.
  - Hydrogen also ties off dangling bonds at grain boundaries in poly resistors, slightly shifting the value of high-sheet poly resistors.
- **The compressive nitride overcoat releases hydrogen during final anneal, but metal patterns can block this hydrogen from reaching the surface.**

Hydrogenation, the first people discovered it were people that made lateral PNP transistors, they found those bipolar laterals had a gain of 3 before there are 300, that is hydrogenation affects the silicon surface very strongly. It matters hugely in MOSFETs, and also matters in poly resistors particularly in boron doped high sheet poly resistors.

103 — METALLIZATION-INDUCED MISMATCHES

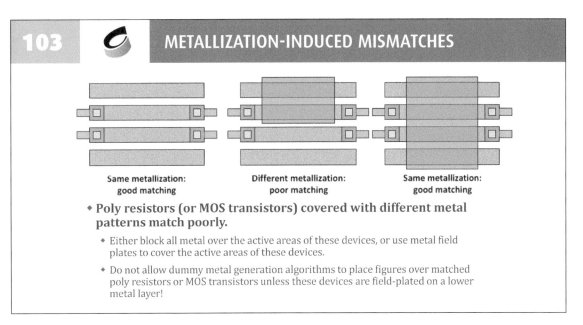

| Same metallization: good matching | Different metallization: poor matching | Same metallization: good matching |

- **Poly resistors (or MOS transistors) covered with different metal patterns match poorly.**
  - Either block all metal over the active areas of these devices, or use metal field plates to cover the active areas of these devices.
  - Do not allow dummy metal generation algorithms to place figures over matched poly resistors or MOS transistors unless these devices are field-plated on a lower metal layer!

If you don't code dummy block and you're using chemical mechanical polish as planarization as most new processes are, little metal spare figures may spangled all over your resistors. So, coating a dummy block geometries around the resistors to block dummy metal generation or putting a metal one plate over the resistors.

PART II

104 — GRADIENTS

- **Many processing and environmental parameters vary gradually across the surface of the die.**
  - Examples include oxide thickness, temperature, and mechanical stress.
  - These gradual variations can be quantified by what mathematicians call a *gradient.*
- **Gradient-induced mismatches are a serious concern for analog integrated circuits.**
  - All circuits suffer from processing gradients, but these are usually small.
  - Devices containing power devices can suffer *large* thermal gradients.
  - Plastic-packaged dice can suffer *large* mechanical stress gradients.

The thickness variation of oxide is not big today, because our processes are so good. But temperature and mechanical stress these are not gone. They are very serious problems. There's a mathematical thing called gradient that you can compute, and it is a measure of how it changes over space.

105 VISUALIZING A GRADIENT

+ **This drawing attempts to visualize a gradually varying parameter.**
  + The dark areas represent a higher value of the parameter, and the light areas a lower value.
  + A gradual and continuous change is observed as one moves across the die.

+ **Mathematically, this parameter can be described as a *scalar field*, P. Its gradient equals**

$$\nabla P = \frac{\partial P}{\partial x}\vec{i} + \frac{\partial P}{\partial y}\vec{j} + \frac{\partial P}{\partial x}\vec{k}$$

  where $\vec{i}, \vec{j},$ and $\vec{k}$ are unit vectors directed along the x, y, and z axes.

A gradient could be imagined as a color distribution. Stress is high down and low up. There's no easy way to draw that. So I'm going to simplify my mechanical stress analysis here grossly and   I'm talking about the average value of the stress and not the orientation of the stress. So really all that we care about is x and y.

106 THE IMPACT OF A GRADIENT

+ **Suppose we wish to quantify the difference in parameter P between two locations A and B.**
  + We can erect a vector $\vec{d}$ from A to B and compute the quantity

$$\Delta P = \nabla P \cdot \vec{d}$$

  + Alternatively, we can measure the distance d in the direction of the gradient and use the amplitude of the gradient to compute ΔP.

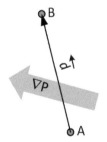

There is a gradient in the direction of the line between these two things. The magnitude of whatever that is along the line times the distance along the line, that's the impact of the gradient, we all understand dot products and all of that. If two things could be at the same spot well then the distance separating them would be zero, and gradient effect goes away.

107 CENTROIDS

* **Real components aren't mere point-devices.**
  * However, a real component will **behave** as if it were a point device placed in a specific location, called the *centroid* of the device.
  * If the gradient is essentially linear, then the centroid lies somewhere on every axis of symmetry bisecting the active area of the device.
  * Simply find two axes of symmetry and the centroid lies at their intersection.

Even we're talking about a world that's nonlinear, but on a small scale with device. Distance is very small we're on placing device away from major gradients, because I don't want to be near big gradients. I can usually assume that most of the impact is linear, then I can apply centroid.

108 CENTROIDS OF SEGMENTED DEVICES

Centroid of $R_1 + R_2$

* **Arrayed devices contain multiple identical sections.**
  * Such an arrayed device has a centroid.
  * This centroid can be found like any other: it lies on every axis of symmetry passing through the array.
  * However, the centroid of an arrayed device doesn't necessarily fall inside any of the segments.
  * This suggests that it is possible for two arrayed devices to have the same centroid....

We can make a centroid of a device that has two separate areas. Let's say a resistor formed of two sections. There is a section, and we add the other section as well sure it lies on the axis of symmetry of the overall shape. So if I took one and folded it across horizontal axis, one would fall right on top of two. We eliminate the stress gradients mismatches and that's called common centroid layout.

109   COMMON CENTROID ARRAYS

* **A common-centroid array consists of two or more devices that have been arrayed so that their centroids fall at the same point.**
  * Consider the array at the right.
  * R_1 and R_2 each have two segments, R_{1A}/R_{1B} and R_{2A}/R_{2B}.
  * By arranging these segments symmetrically about the center of the array, R_1 and R_2 can be made to have a common centroid.

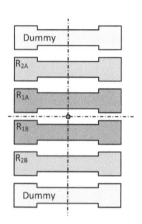

Resistors R1A and R1B are two segments of R1, R2A and R2B are two segments of R2. The centroid of R2A and R2B is at the green circle, R1A and R1B lie symmetrically about the axis. So the centroid of R1A and R1B lie at the green circle, the centroids of R1 and R2 align they coincide. They are common come and centroid latent. Dummies are placed on top and bottom.

110   BENEFITS OF COMMON-CENTROID ARRAYS

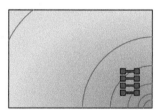

* **Gradients should have no effect upon common-centroid devices because they effectively occupy the same location.**
  * This is only approximately true because we assumed the gradient was linear in order to find the centroids.
  * The largest nonlinear component of the gradient is usually the quadratic.
  * Therefore the residual mismatch caused by nonlinearities tends to scale as the square of the distance across the array. ***Smaller arrays match better!***
  * Arrays placed in areas with low gradients will also match better.

It's the linear component of the gradients effect on the electric parameter which is that common centroid layout cancels the constant part, What about the rest any nonlinear parameter? Generally, the enemy is probably the quadratic. As the common centroid array is bigger the distances across which gets bigger, this quadratic residues impact gets bigger.

- *Coincidence:* The centroids of the matched devices should coincide at least approximately. Ideally, they should coincide exactly.

- *Symmetry:* The array should be symmetric around both the horizontal and vertical axes.

- *Dispersion:* When possible, a large array should be subdivided into as many smaller arrays that are possible that each satisfy the rules of coincidence and symmetry. If this can be achieved, then the larger array need not satisfy these rules; only the subarrays that comprise it need do so.

- *Compactness:* The array (or each of its subarray) should be as compact as possible.

The centroids of the match devices should coincide at least approximately. It's generally a good idea that your array be symmetric around its common centroid point in both x and y, avoid asymmetric looking arrays.

112 EXAMPLE: RESISTOR ARRAY

FAIR: Array not optimally compact

NOTE: Dummies aren't shown, but are necessary.

BETTER: Array compact, suboptimal dispersion

BEST: Array compact, optimal dispersion

- **The three above arrays all use common-centroid layouts, but they aren't equally good layouts.**
  - Compactness and dispersion are important considerations, especially in large arrays where the nonlinear residues of the gradients are large.

This is a reasonable array, ABBA dummies not shown, but expected. AABBBBAA is more compact, but it's not very dispersed, ABBAABBA bingo best. You get small sub arrays of the common centroids, good dispersion best layout.

113   2D COMMON-CENTROID ARRAYS

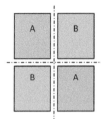
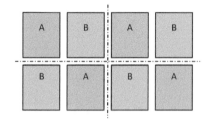

- **Compact devices, such as capacitors, can benefit from two-dimensional common-centroid layouts.**
  - The simplest such array, shown at the left, is often called a *cross-coupled pair.*
  - Multiple cross-coupled pairs can be combined into a larger array with good dispersion.

You can also do this across two axes. Resistors are hard to do that, because they are long and spindly, you know things like MOSFETs and capacitors. It's often useful to array them in two dimensions, because by folding the one dimensional array over, I can make it more compact.

114   STRESS GRADIENTS

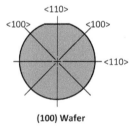
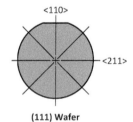

- **Mechanical stress can alter resistivity (piezoresistivity).**
  - N-type (100) silicon exhibits minimum piezoresitivity along <110> axes.
  - P-type (100) silicon exhibits minimum piezoresistivity along <100> axes.
  - (111) silicon, and polysilicon, do not exhibit an orientation dependence.

Here is mechanical stress, separated mechanical stress can affect resistance of transistors, affect on MOSFETs, transconductance and it will also affect the resistance of a resistor, which is called piezo resistivity. If you apply stress to silicon, you can cause mismatches, percent point kind of mismatches, it turns out that these mismatches vary depending on orientation.

115 CAUSES OF MECHANICAL STRESSS

Material	Coefficient of thermal expansion
Epoxy encapsulation	24 ppm/°C (typical)
Copper alloys	16–18
Alloy 42	4.5
Molybdenum	2.5
Silicon	2.5

- **Differences in the coefficient of thermal expansion of different materials stress a packaged die.**
  - Plastic encapsulation is cured at about 175 °C.
  - As the packaged device cools, the encapsulant shrinks more than the silicon die.
  - Therefore the die experiences compressive stresses that increase at low temperatures and decrease at high temperatures.

Where does mechanical stress come from? Here is how you make a plastic package die, you take the chips and mount them on a lead frame your wire line and put it in a molding machine, it squirts hot molten plastic in actually from beneath, hot about 175 degrees, and then plastic metal and silicon does everything shrink at the same rate as, which puts enormous stress onto the die.

116 STRESS GRADIENTS

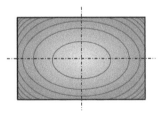

- **Stress gradients on the die are highest at the edges and especially in the corners.**
  - The actual stress gradients are more complex than this illustration suggests, but the rules are obvious:
  - Do not place matched devices along edges, or in corners.
  - For best matching, place matched devices near the center of the die such that an axis of symmetry of the array coincides with an axis of symmetry of the die.

The stress is so approaching the yield point of silicon, the levels of stress of hundreds of millions of Pascal. Now silicon has the same structure and centered cubic crystal as diamond. It is a very strong but brittle material brittle, so it is not surprising then that parameters of devices might change about a percent.

117 ⟲ THERMAL GRADIENTS

Power transistor

Okay (near edge, on axis of symmetry)

Best (away from edge and heat source, on axis of symmetry)

Okay (near power device, on axis of symmetry)

* **Ideally, matched arrays should lie on an axis of symmetry of major power devices, and as far from them as possible.**

  * Compromises must be made between locations optimal for stress and those optimal for temperature.

If we heat the die back up to about 175 degrees, almost all the strengths go away, if we cool it down, more stress. We know the plastic grips onto the die and creates gradients. Stay away from the edges if you have a stress sensitive component. The other thing is power device, which locally heats up the die. So the best location, make a compromise between heat gradients large near the power device and stress gradients large near edges of the die.

118 ⟲ THE SEEBECK EFFECT

V_c

$+$ $-$

Al   N⁻ Si

* **A voltage called the *contact potential* exists between any two dissimilar materials in contact with each other.**

  * Contact potentials vary with temperature.
  * The constant of proportionality is called the *Seebeck coefficient.*
  * The Seebeck coefficient of aluminum to silicon is typically 0.4 mV/°C.
  * This large Seebeck coefficient can cause mismatches in improperly connected resistors.

There is something called Seeback effect. This is also known as thermoelectric effect, when any two dissimilar materials are in contact, a voltage is generated between them, the Seeback coefficient of aluminum to Silicon is on the order of a half millivolt degree.

119 THERMOELECTRICS IN CONTACTS

Bad:
Thermoelectrics add

Good:
Thermoelectrics cancel

* **In a series-connected resistor array, the contact potentials add.**
  * Improper connection causes a large thermoelectric voltage to accumulate due to addition of the contributions of each segment.
  * Proper connection eliminates this problem.

The left one is wrong, because some of the contacts are cold and some of the contacts are hot, here I add a little difference and add some more and add some more and add some more, don't do that. Do the right one, where I exactly cancel it by running the other way for the next segment.

120 SUMMARY

* **Although this presentation focused upon resistors, most of the same principles apply to all matched components.**
  * Deposited capacitors aren't very susceptible to mechanical stress or temperature gradients.
  * MOS and bipolar transistors, on the other hand, are **very** sensitive to both of these types of gradients.
  * All matched components deserve careful consideration, but only the most critical such devices can occupy the best locations and consume sufficient area to minimize random mismatch.

* **No matter how carefully a circuit designer constructs their circuit, poor layout of its matched devices can ruin it!**

We focus on resistors but most of what I've said will apply to other devices. Deposited capacitors aren't very susceptible to mechanical stress or thermal gradients, diffuse capacitors very sensitive to thermal gradients not too sensitive to stress, bipolars very sensitive to both thermal and stress gradients. You have to make trade-offs, it's so important in crucially matched circuits that a IC may have to be entirely reworked merely, because there's a thermal stretch or a problem or a stress problem, it makes a part unsellable.

5. TRANSISTORS

This section introduce the feature and the tips for CMOS layout.

SECTION I

121 MOS TRANSISTORS

* **The self-aligned poly-gate CMOS transistor has become the mainstay of most analog CMOS and BiCMOS processes.**
  * Virtually all processes offer complementary NMOS and PMOS transistors.
  * The NMOS usually has a higher transconductance than the PMOS due to electrons having a higher mobility than holes.
  * These devices are extremely versatile. They can perform most analog functions as well as implement digital logic.

CMOS,Complementary metal oxide semiconductor, which means PMOS and NMOS have been designed to be similar, not equal but similar in their behavior. The NMOS usually has higher transconductance than the PMOS, because in silicon electrons are more mobile than holes.

122 THE MOS TRANSISTOR: A FOUR-PAGE REVIEW

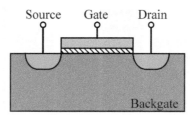

* *The metal-oxide-semiconductor* (MOS) transistor is a four-terminal active device.
  * The gate controls current flow from the source to the drain, both of which are embedded in the backgate.
  * An NMOS has N-type source/drain regions and uses electron conduction.
  * A PMOS has P-type source/drain regions and uses hole conduction.

CMOS is a four terminal active device, and it can operate in three mode. These pages briefly review the basic knowledge of CMOS in different mode. This is actually active devices, the MOSFETs were made for digital logic gates and they work really well for that, but they also can do a lot of analog, they're quite versatile.

123 — THE NMOS TRANSISTOR: *CUTOFF*

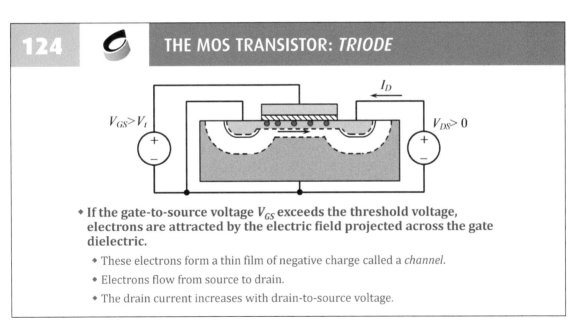

$V_{GS} < V_t$

V_{DS}

+ −

+ −

- **If the gate-to-source voltage V_{GS} of an NMOS is less than the *threshold voltage V_t*, no channel forms.**
  - Depending upon gate biasing, an accumulation layer or depletion region may form beneath the gate.
  - However, only leakage currents flow.

If I adjust Vgs less than the threshold voltage, there's no electrically conductive path from source to drain, the device has been cut off. As I increase the voltage closing in on the threshold voltage? a little tiny current begins to flow and that is called sub threshold leakage.

124 — THE MOS TRANSISTOR: *TRIODE*

I_D

$V_{GS} > V_t$

$V_{DS} > 0$

+ −

+ −

- **If the gate-to-source voltage V_{GS} exceeds the threshold voltage, electrons are attracted by the electric field projected across the gate dielectric.**
  - These electrons form a thin film of negative charge called a *channel*.
  - Electrons flow from source to drain.
  - The drain current increases with drain-to-source voltage.

So I crank the voltage a little higher on my Vgs, and I now attract up from the bulk electrons. They form a thin film and charge underneath the gate oxide and I place a voltage on the drain end. So the electrons slide underneath the surface of the dielectric moving from the source to the drain. I've attracted carriers and inverted the silicon and now this kind of looks like a little resistor.

125 THE MOS TRANSISTOR: *SATURATION*

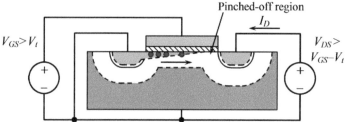

♦ **At higher drain-to-source voltages, the channel pinches off.**

  ♦ The depletion region widens towards the drain end of the channel, meaning less charge needs to reside in the channel itself.

  ♦ At higher drain-to-source voltages, the channel vanishes (*pinches off*) near the drain.

  ♦ Further voltage increases appear across the pinched-off region.

  ♦ The drain current therefore ceases to increase with drain-source voltage.

As I increase the drain voltage, the drain depletion region widens, as I have a bigger reverse bias. Now this is a little bit strange, because the current does not change, even though the drain end of the channel is thinner. As I increase the drain voltage, the current increases, it increases and then suddenly stops increasing. You've saturated the current, so that's saturation.

126 THE SELF-ALIGNED POLY-GATE NMOS

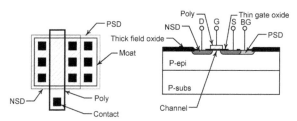

♦ **Most modern MOS processes use self-aligned polysilicon gates.**

  ♦ The poly serves as a masking layer to block the source/drain implants from the channel region.

  ♦ Sidewall spacers formed along the edges of the gate hold the implants slightly away from the actual gate.

  ♦ The resulting structure has minimal overlap capacitances.

Modern digital devices done with poly silicon gates, the problem with advanced materials is that you can't make a gate aluminum self-aligned. A strip of poly crossing a moat region creates an active region for the transistor. the portion over the poly is blocked by the poly, creating source and drain terminations.

127   DRAWING THE MOS TRANSISTOR

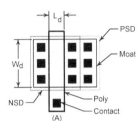 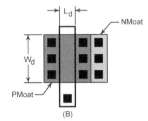

- **An NMOS can be drawn with *mask layers*.**
  - NSD and PSD are the N-type and P-type source/drain implants, respectively.
  - NSD and PSD should overlap the moat slightly to ensure that their edges are defined by the field oxide (this is another example of self-alignment).

- **Alternatively, an NMOS can be drawn with *coding layers*.**
  - NMoat generates NSD and moat; PMoat generates PSD and moat.

We can draw the MOS transistors using individual layers. These are so-called coding layers, because they allow you to draw or code the device with layers that are not ones that are actually mask layers.

128   TYPES OF PROCESSES

- **An N-well process places the PMOS in an N-well and the NMOS in the P-epi.**
  - This was a popular choice in the 1990s.
  - The use of a P-type epi (and by extension, a P-type substrate) allowed a negative ground.

- **A P-well process places the NMOS in a P-well and the PMOS in the N-epi.**
  - Requires a positive ground (unpopular).

- **A twin-well process places the NMOS in a P-well and the PMOS in an N-well.**
  - The most common choice for low-voltage processes.

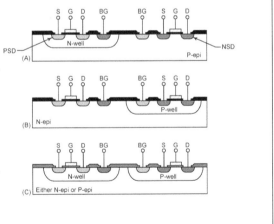

There are different choices for how you might put the two type of transistors NMOS and PMOS into a process, thus first and most obvious approach is to use a p-type EPI and grow a n-type well in it. That's very common. You could reverse everything and use a n-type EPI. The third choice is use separate wells for both the N and the PMOS, this requires an extra mask, but it allows you to tailor the doping.

129 THE ISOLATED NMOS

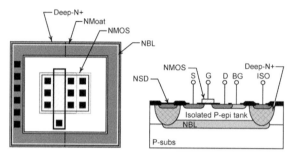

* **Normally, either NMOS or PMOS (usually NMOS) share a common backgate.**
    * BiCMOS processes that include a deep-N+ sinker and an N-type buried layer (NBL) can place an NMOS in an isolated *tank*.
    * A *tank* is a region cut off by means of diffusions; a *well* is just a deep diffusion.

There is isolated NMOS if you have a Bi-CMOS process, the key is to have an NBL layer which is not part of a standard CMOS process. You've seen it as standard bipolar but you could put one into a CMOS process. With the addition of two mask steps, we can fully isolate transistors in mouths.

SECTION II

130 BIPOLAR TRANSISTORS

* **Almost all processes can create bipolar transistors:**
    * Standard bipolar only fabricates bipolars.
    * Analog CMOS fabricates substrate PNP transistors.
    * Most BiCMOS processes can fabricate NPN and PNP transistors as well as MOS transistors.

* **Bipolar transistors still have their uses:**
    * They offer better voltage matching than MOS transistors.
    * For a given current, they generate more small-signal transconductance.
    * Ratioed collector currents generate voltages dependent upon absolute temperature.
    * Bipolars make superior ESD devices because they dissipate power in relatively large volumes of silicon.

Most Bi-CMOS processes can create NPN and PNP. The combination bipolar transistors still have their uses, they match better for voltage matching than MOSFETs. Most ESD devices as the grounded gate NMOS is actually an NPN transistor hiding inside the structure of the NMOS, the back gate is the base the source and the drain are the emitter and the collector.

131   THE BIPOLAR TRANSISTOR: A FOUR-PAGE REVIEW

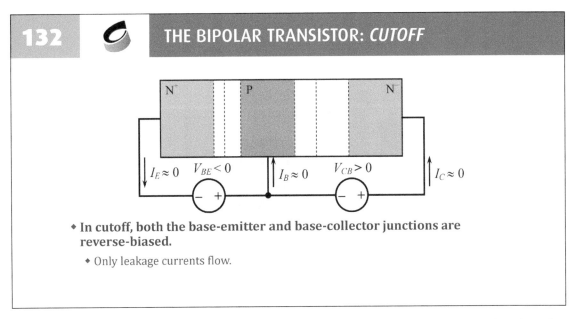

- **A *bipolar transistor* consists of a thin layer of silicon of one polarity called the *base* sandwiched between two layers of opposite polarity called the *emitter* and the *collector*.**
  - An NPN has a P-type base sandwiched between N-type emitter and collector.
  - A PNP has an N-type base sandwiched between P-type emitter and collector.

There are many biasing schemes, for our purposes of analysis we bias them with VBE and VCB. There are emitter base and collector currents. This is difference from MOS in which there is no constant gate current.

132   THE BIPOLAR TRANSISTOR: *CUTOFF*

- **In cutoff, both the base-emitter and base-collector junctions are reverse-biased.**
  - Only leakage currents flow.

The first mode is very similar to the MOSFET counterpart, if both the base emitter and base collector junction are reverse biased nothing but leakage flows, the device is cut off.

133   THE BIPOLAR TRANSISTOR: *FORWARD ACTIVE*

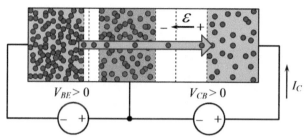

- **In the forward active region, the base-emitter junction becomes forward-biased while the base-collector junction remains reverse-biased.**
  - The emitter injects minority carriers into the base.
  - Those carriers that do not recombine diffuse to the collector.
  - The base-emitter bias therefore controls the collector current.

If forward bias the base emitter junction while reverse biasing the base collector junction the device enters the forward active region which corresponds to saturation in a MOSFET. The carriers diffuse from the emitter into the base where there are less numerous across the forward bias junction, most of them reach the collector base depletion region and drawn across into the collector.

134   THE BIPOLAR TRANSISTOR: *SATURATION*

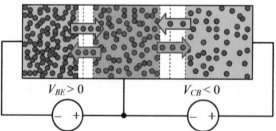

- **In saturation both the base-emitter and the base-collector junctions become forward-biased.**
  - Minority carriers injected into the collector take time to recombine, slowing turn-off.
  - In integrated transistors, minority carriers may traverse the collector to the substrate. This represents a major challenge for bipolar circuit designers.

The third mode is called saturation, in which the emitter base and collector base junction both forward biased, the direction of current flows through, the device whether it goes from emitter to collector or collector to emitter. Bipolar transistor saturation is kind of like linear mode in a MOSFET.

135 · THE PARASITIC PNP INSIDE THE VERTICAL NPN

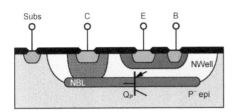

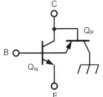

- **Saturating a vertical NPN activates the parasitic substrate PNP.**
  - This parasitic transistor diverts base drive to substrate.
  - If circuit designers don't consider this effect, it can cause circuits to malfunction.
  - To prevent it, avoid saturating the transistor, or add hole-blocking guard rings.

Now the problem is the presence of a parasitic PNP inside the vertical NPN. If I forward bias the base collector junction of vertical NPN which is the same as the base emitter junction of the parasitic PNP, so current from my base goes through the parasitic PNP and into the substrate, the parasitic PNP is inevitable but you can deal with it.

136 · THE STANDARD BIPOLAR VERTICAL NPN

- **The vertical NPN is the best transistor in standard bipolar.**
  - Vertical conduction typically gives a beta of 100-300.
  - Deep-N+ sinker can be added to minimize collector resistance.
  - Many alternative layouts exist; for example, one can swap the locations of base and emitter.

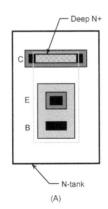

(A)

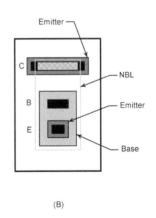

(B)

Standard bipolar vertical NPN look like this. there's an emitter of base in a collector but there's many alternative layouts. The emitter could go on the bottom of the base or on the top, this is only the beginning. I could add multiple emitters multiple base contacts all sorts of wild variations on this and a standard Bipolar designer would do so, of course.

137   THE STANDARD BIPOLAR SUBSTRATE PNP

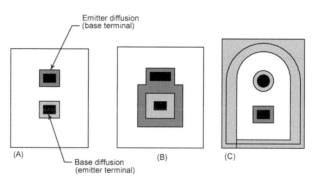

* **A substrate PNP uses the substrate as its collector.**
  * The simplest style uses plugs of base and emitter inside a tank.
  * Many other styles exist that purport to have better performance (often, they don't).

There are several forms of substrate PNP. The simplest is you put an emitter blob and a base blob and a tank. There are alternative forms, the middle one tries to improve vertical and supress lateral effects by ringing. The right one is a so-called tombstone transistor, it benefits from both lateral injection and vertical injection, these were popular once upon a time.

138   THE STANDARD BIPOLAR LATERAL PNP

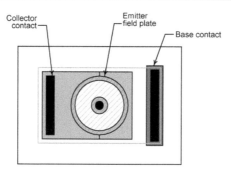

* **A lateral PNP trades off performance for an isolated collector.**
  * Lateral PNP transistors have large parasitic capacitances and are thus slow.
  * Always cover the exposed base with metal to prevent surface channel formation.

This is a lateral PNP, including the emitter field plate, when you build a lateral. There's an exposed base region surrounding. In the emitter inside the collector, please coat this with a layer of metal, the reason is this is a relatively lightly doped region and any electric fields over it can modulate the carrier flow through it.

139   THE BICMOS CDI NPN

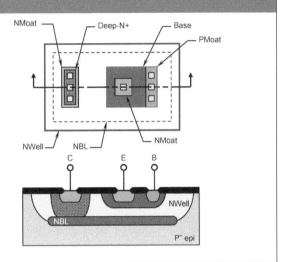

* **Many analog BiCMOS processes offer a relatively simple *collector-diffused-isolation* (CDI) NPN.**

  * This device uses either a dedicated base or a shallow P-well as its base.

  * A deep N-well forms the collector, which also isolates the device from substrate.

  * An N-buried layer must be added to the process; optionally deep-N+ sinker can also be added.

The Bi-CMOS NPN. This is the simplest way to do it. take a deep NWELL, put an NBL, use a base diffusion, the emitter could be NSD. If you look at the cross-section, it looks identical to the vertical standard bipolar.

140   TRANSISTOR MATCHING

* **Transistors obey the same principles of matching as resistors and capacitors.**

  * Both MOS and bipolar transistors are very sensitive to thermal gradients.

  * Bipolar transistors are very sensitive to stress gradients.

  * Use common-centroid layouts and consider device placement carefully.

  * MOS transistors usually benefit from careful use of dummy devices.

Transistor matching pretty much follow the same rules. We've already discussed the aware that MOS and Bipolar are very sensitive to thermal gradients. So you need to use common-centroid layouts and you need to carefully consider where to place the devices. Narrow dummy devices are often very useful for improving matching.

141 MULTIPLE FINGERS, MULTIPLE DEVICES

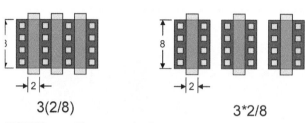

3(2/8)   3*2/8

* $N(W/L)$ specifies a single device containing N fingers, each having width W and length L.

* $N*W/L$ specifies N separate devices, each having width W and length L.
  * Matched devices are most easily constructed as separate devices.
  * If separate devices share sources or drains, then they can be merged together just as if they were multiple fingers of one device.

So multiple finger device, if you don't like having a long skinny MOSFET, you simply cut it into several fingers, each of these has a certain width and length. If they are separated from each other, they will take considerably more room. Which way is better? it depends on what your goal is.

142 MATCHING RULES: IDENTICAL SECTIONS

Good   Bad

* **Devices having different widths or lengths don't match well.**
  * The effective silicon dimensions W_e and L_e equal the drawn dimensions W_d and L_d offset by width and length reduction factors W_r and L_r,

$$W_e = W_d + W_r \qquad L_e = L_d + L_r$$

  * The only way to avoid mismatches due to uncertainties in width in length reduction factors is to use identical geometries.

Matching rules: You should probably whenever possible try to use identical sections in MOSFETs. So good is here's a 1x transistor and a 2x transistor. Bad is here's a 1x and here's a 2x, they're not the same.

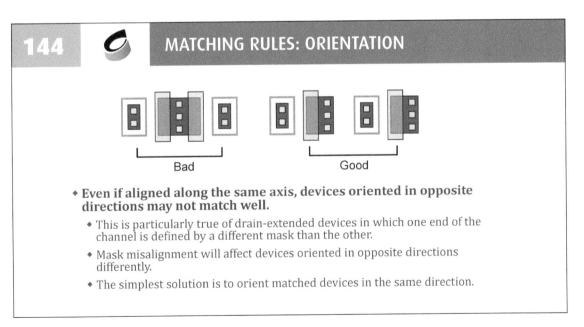

143 MATCHING RULES: ALIGNMENT

Good   Poor   Terrible!

- **Align devices in the same direction.**
  - Silicon becomes anisotropic when subjected to mechanical stress.
  - Wafers are sometimes cut slightly off-axis to minimize pattern shift.
  - Implants may be shot into the wafer at a tilt.
  - Devices aligned in different directions do not match well.

Alignment, You should try to orient the transistors in the same direction, aligning the transistors like the left one is good. Aligning them like others are bad. So looking maybe isotropic under no stress, but when it is stressed, it become not isotropic and the matching are not guaranteed. Let alone those at some other angle.

144 MATCHING RULES: ORIENTATION

Bad   Good

- **Even if aligned along the same axis, devices oriented in opposite directions may not match well.**
  - This is particularly true of drain-extended devices in which one end of the channel is defined by a different mask than the other.
  - Mask misalignment will affect devices oriented in opposite directions differently.
  - The simplest solution is to orient matched devices in the same direction.

Orientation can be a problem. These are devices of the type called drain extended transistors. So the left, for example, the wells were to shift a little to the left, it would affect both transistors. The right one is much better, each transistor oriented in the same direction.

145   MATCHING RULES: PROXIMITY

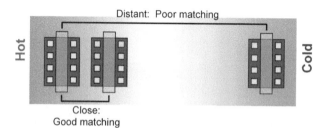

Distant: Poor matching

Hot   Cold

Close:
Good matching

● **Proximity minimizes the effects of stress and thermal gradients.**

  ◆ MOS device transconductance is sensitive to mechanical stress.

  ◆ MOS threshold voltage is sensitive to temperature.

  ◆ Bipolar transistors are even more sensitive to gradients because of their high small-signal transconductance.

Proximity, you don't want to put two devices far apart from on another because one will be hot and one will be cold. Placing them close together improves that common centroid layout is the best solution.

146   MATCHING RULES: COMPACTNESS

Bad

Good

● **Compact layouts experience less gradient-induced mismatches.**

  ◆ Proximity, and even common-centroid layout, can only do so much.

  ◆ The nonlinear components of the gradients will produce mismatches whose magnitude increases roughly as the square of distance.

  ◆ Compact layouts of matched devices will minimize the impact of these nonlinear residual variations.

Compactness: So the left one is a very long and skinny array, the quadratic residue is going to come in and beat us up, if we place the devices side by side, even though one end is hot the other is cold, both transistors see the same effect. So just reorienting the transistor can greatly improve the matching.

147 MATCHING RULES: METALLIZATION

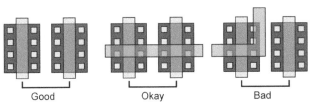

Good   Okay   Bad

- **Hydrogenation creates mismatches between devices with different amounts of metal coverage.**
  - Hydrogen reacts with dangling bonds at the silicon surface and thus reduces the surface state charge.
  - Metal blocks the diffusion of hydrogen down to the silicon surface.
  - Therefore MOS transistors with different metal coverage exhibit different threshold voltages.

Metallization: Keeping metal off of MOSFETs is a very good idea, because it means the amount of hydrogen blockage. Blocking one transistor without blocking the other is deadly, because the typical shift caused by this can be a VT shift of 10 mV. What about dummy metal? Best thing to do is to code dummy metal block over both devices, that way you assume you're sure that there will be no metal over either device. The next best thing if you can't do a dummy block is put a metal plate over the whole thing.

148 MATCHING RULES: DUMMIES

Dummy   A   B   B   A   Dummy

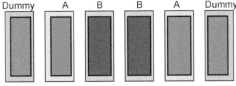

- **Etch rate variations affect devices on the ends of an array differently than those in the center.**
  - Arrays of interdigitated MOS transistors are very vulnerable to poly etch-rate variations.
  - Add dummy devices to either end of the array.
  - Maintain identical poly-to-poly spacings between active devices and from active devices to dummies.
  - Accurately matched short-channel devices may benefit from more than one dummy on either end of the array.

Dummy is a good thing. The reason is if you have a dummy shape, the initial shape is light shading and the final shape is dark shading. You can see the outside edges get eroded away more than the inside edges. So bottom line of this picture is a little hard to follow, because perhaps it's not that well drawn but bottom line is put dummies on either side of the main array.

149 THE FIVE RULES OF MOS COMMON-CENTROID LAYOUT

- **Coincidence:** The centroids of the matched MOS transistors should coincide at least approximately. Ideally, they should coincide exactly.

- **Symmetry:** The array should be symmetric around both the horizontal and vertical axes.

- **Dispersion:** When possible, a large array should be subdivided into as many smaller arrays that are possible that each satisfy the rules of coincidence and symmetry. If this can be achieved, then the larger array need not satisfy these rules; only the subarrays that comprise it need do so.

- **Compactness:** The array (or each of its subarray) should be as compact as possible.

- **Orientation:** Matched MOS transistors should possess the same orientation.

So the rules of MOS come and centrally layout are identical to the rules that I've already described with one addition. Orientation flips down at the bottom match MOS transistors should possess the same orientation ideally. All drains should go up, this rule isn't particularly important, but if you're doing drain extended devices, Orientation becomes absolutely crucial and assuring that they're properly matched in that case.

150 SAMPLE MATCHED LAYOUT: CURRENT MIRROR

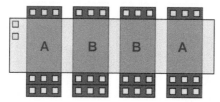

- **Current mirrors usually require only moderate matching.**
  - Dummies aren't used because they're hard to construct and they don't have too much impact on long-channel devices.
  - A common gate poly geometry saves space.
  - Merged backgate contacts have been added to enhance latchup immunity.
  - Extend poly beyond moat a bit more than required by rules.

Drawing a current mirror, put the transistors down ABBA, slap a gate segment across it and call it a day. If I want to think about it a little bit more, extend the poly a little beyond the outer transistor more than the rules required, because the corners on shapes round off so plant things out a little improves match.

151 SAMPLE MATCHED LAYOUT: CROSS-COUPLED PAIR

- **Cross-coupled pairs are often used for diff pairs.**
  - This is a small cross-coupled pair, but it shows the concept.
  - Notice so-called *half dummies* have been used. These employ narrow strips of poly and terminate the moat partway across the poly.
  - Half dummies will usually suffice for moderate matching, but wider poly strips may be needed for accurate matching.

This is a cross couple common centroid layout. I probably should have cut the dummies and had contacts to each dummy. I did not do that. So it's not perfect but it's close, it's a common centroid array with dummies attached.

152 COMMON-CENTROID NPN LAYOUT

- **Bandgap references require *N*:1 matched NPNs.**
  - To create a common-centroid layout, use an even value of *N*.
  - The emitter geometry determines the size of a vertical NPN, so multiple emitters can reside in a common base.
  - Slightly increase emitter-emitter spacing and base overlap of emitter to ensure that crowding does not alter emitter areas.

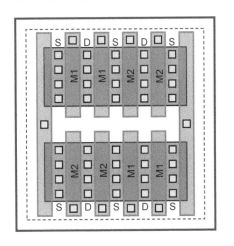

This is a typical common centroid NPN array for a bandgap, typically an eight to one array. So what I did is take two pieces of the 8x transistor. One on the left, one on the right, each has four emitters, put the other transistor in the middle and align the centroids. Each having their common centroid.

7. LATCHUP

What we're talking about is a whole range of phenomena, but the one that most people think about is CMOS latch which is really parasitic bipolar transistors. Bipolar engineers didn't like those parasitic MOSFETs as you don't like the parasitic bipolars, but you have to live with them.

153　**CONTENTS**

- **Latchup: The Problem**
  - Cross section of an N-well epitaxial CMOS process
  - How transients can initiate latchup
  - Review of BJT action
  - The latchup mechanism
- **Latchup: Solutions**
  - Finding CMOS latchup
  - Four ways to stop latchup
  - Guard rings

I have a MOS transistor embedded inside. It is a hiding a bipolar and those bipolars, if you treat them poorly, they will latch up on you. So I want to talk briefly about the nature of the problem. Review a little about BJT action in CMOS circuits and then talk about the ways to stop it.

154　**N-WELL CMOS PROCESS CROSS SECTION**

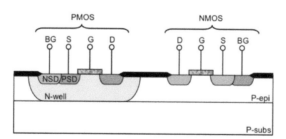

- **An NWell CMOS process fabricates NMOS and PMOS transistors on the same substrate.**
  - The NMOS transistors sit in the P-epi or in a P-well (not shown).
  - The PMOS transistors sit in an N-type well (N-well).

Here's a CMOS die you can see a PMOS on the left and an NMOS on the right, Here we're assuming a p-substrate. The NMOS transistor sits in the P-sub or perhaps in a P-well, the PMOS transistor sits in an N-well.

155 BACKGATE CONTACTS

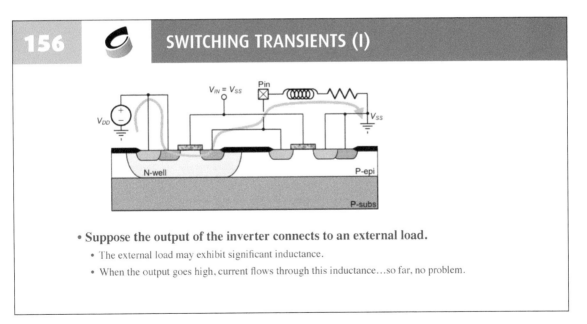

- **Suppose the NMOS and PMOS are connected to form an inverter.**
  - The PMOS backgate is connected to the positive supply, *VDD*.
  - The NMOS backgate is connected to the ground return, *VSS*

All right, imagine that we have this hooked up as an inverter. So the PMOS needs an insight back gate contact, hook it to the power supply that also feeds the source of the PMOS. We also need a back gate contact for the NMOS, hook it to the ground which also feeds the source of the NMOS.

156 SWITCHING TRANSIENTS (I)

- **Suppose the output of the inverter connects to an external load.**
  - The external load may exhibit significant inductance.
  - When the output goes high, current flows through this inductance…so far, no problem.

We have some sort of loadone which happens to include some inductance. There's the inductance and so I'm going to turn on the PMOS and current is flowing through the load to ground and of course the little inductor is now conducting current and everybody's happy.

157   SWITCHING TRANSIENTS (II)

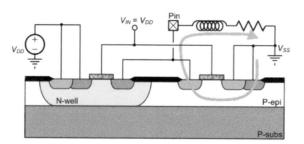

- **Now switch the inverter so that its output goes low.**
  - Current flow should stop, but the inductance momentarily continues to draw current.
  - This current pulls the NMOS drain below *VSS*, *forward-biasing the drain-backgate junction.*

Then I change the state of a circuit. The current flowing through the inductor will not instantly change, so that implies the current continues to circulate, obviously it flows through the NMOS but the NMOS is off, so it drag the Nwell below ground to this Junction forward bias, here is the alarm. There's a standard test for ICs, where you have to go around to every pin and deliberately pull it below ground and above supply and inject usually 100 mA of current and you can't latch up.

158   DIGRESSION: REVERSE-BIASED PN JUNCTIONS

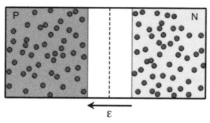

- **In a reverse-biased junction, the electric field across the depletion region holds the majority carriers on their respective sides of the junction.**
  - Holes stay on the P-type side of the junction (the *anode*).
  - Electrons stay on the N-type side of the junction (the *cathode*).
  - Little or no current flows across the junction.

Thank for reverse bias junctions and reverse bias. Junction the electric field across the junction holds the holes on the anode side and the electrons on the cathode side and current doesn't flow.

159   DIGRESSION: BIPOLAR JUNCTION TRANSISTORS

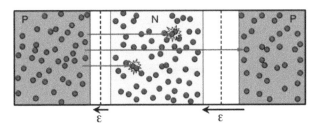

- **Minority carriers will typically recombine within a few microseconds.**
  - If in this time they can diffuse to a reverse-biased junctions, then they can diffuse across it and can again become majority carriers.
  - This is the essence of bipolar transistor action.

B ut when I forward bias the junction, the carriers flow across the Junction. Holes are injected into an end region, get through look like a bipolar. So when you forward bias one Junction on the region that has other reverse bias junctions of budding it. You've just formed a bipolar transistor.

160   MINORITY CARRIER INJECTION

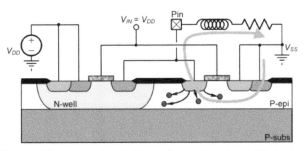

- **Pulling an N-type region below substrate potential injects electrons into the P-epi.**
  - These electrons become minority carriers in the P-epi.
  - They diffuse in all directions, although few enter the heavily doped substrate.

B ecause that's an N-type region, pulling it below substrate is a forward bias, and it spraying electron out into the P-epi. They're going to be captured by other junctions. That maybe not so harmless after all.

161   THE PARASITIC NPN

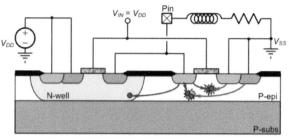

- **Most electrons recombine in the P-epi.**
  - This recombination consumes holes.
  - More holes flow in from substrate contacts.
  - A few electrons diffuse across to the N-well…

Let's imagine that some of the carriers are reached the N-well and enter into the N-well, where they become majority carriers again, they don't normally do much except a sort of tiny Junction leakages. But when I fire up a latch up test and suddenly shoot 100mA into the part. Suddenly the back gate begin to provide holes to support the recombination. Then other junctions will forward bias and begin to inject more minority carriers, like a nightmare.

162   SUBSTRATE DEBIASING

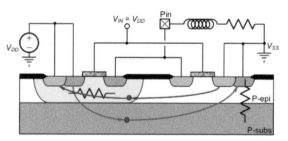

- **The hole current flowing to the substrate contact must cross the P-epi.**
  - A voltage gradient appears across the P-epi.
  - This voltage gradient forward-biases the NMOS source-backgate junction.

The P-epi parasitic resistance create a debiasing loop, holes flow through the resistance up to the contact, the current through the inductor stop flowing, but the latch up remains. The current flow just keeps on flowing, It's latched up.

163 DIGRESSION: LATCHUP TESTING

- **To test for latchup,**
  - Connect all power supplies to the *device under test* (DUT),
  - Record each supply current,
  - Inject a test current into/out of the pin under test,
  - Remove the test current,
  - If any supply current has significantly shifted (say by more than 10%), and the operation of the circuit cannot explain this shift, then latchup has occurred.

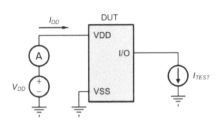

This is a standard test for latch up. I hook up a power supply to its power pin, put a current meter in the line, measure the current flow, then take some pin. When I pull it 100 mA, if this chip is behaving oddly, that's not failure. When I stop pulling the current, I expect the chip to resume normal operation, if it does so it passed the latch up test, if not, it fails the latch up test.

164 DIGRESSION: EMISSION MICROSCOPY

- **Minority carrier recombination generates low levels of photoemission, allowing an IR microscope to image latchup.**

The usual way you check is by monitoring the current to see if it shifts by 10%. This is a subtle and difficult test, as you may wake up some sleeping circuit like flip-flop, So, the 10% current is just an indicator suggesting The gold standard is you try to make the part work and something doesn't work now.

165 CONDITIONS FOR LATCHUP

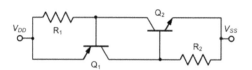

- **Parasitic PNP Q1, parasitic NPN Q2, well resistance *R1*, and substrate resistance *R2* form a positive-feedback device called a *silicon-controlled rectifier* (SCR).**

- **In order for latchup to occur,**
  - $V(R1) > VBE1$
  - $V(R2) > VBE2$
  - $\beta 1 \beta 2 > 1$

L et's talk about conditions for latch up. We basically have an NPN and a PNP pair of sitting that are cross coupled. There are also turn off resistors R1 and R2. So for latch up to occur, Q1 must be biased into the forward active region, which means the voltage across R1 must be large enough to give a VBE across Q1. Q2 must be biased into the forward active region which means the voltage across R2 must be big enough to forward bias.

166 FOUR WAYS TO STOP LATCHUP (I)

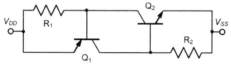

- **Reduce well resistance *R1*.**
  - Add more well contacts.
  - Use a retrograde well.
  - Add an N-type buried layer (NBL) to the N-well.

- **Reduce substrate resistance *R2*.**
  - Add more substrate contacts.
  - Increase substrate doping.

To reduce R1, the well resistance, that would prevent transistor Q1 from biasing forward. Add more well contacts, put them between the part of the transistor. Reduce the lateral resistance in the bottom of the well by making it more heavily doped, another possibility is NBL. To reduce R2, add more substrate contacts, the process people can help you increasing the substrate doping.

167 FOUR WAYS TO STOP LATCHUP (II)

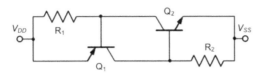

- **Reduce PNP Q1's beta.**
  - Use a retrograde well or add NBL.
  - Increase overlap of well over devices.
  - Add a hole-blocking guard ring (HBGR).
  - Add a hole-collecting guard ring (HCGR).

We could reduce the PNP beta by process variations in retrograde Wells. We could increase the overlap of the well over the device which requires lots of space, or add a hole collecting guard ring, these are the approach.

168 FOUR WAYS TO STOP LATCHUP (III)

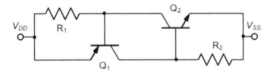

- **Reduce NPN Q2's beta.**
  - Increase spacing between NMOS and N-well.
  - Employ an electron-collecting guard ring (ECGR).

We could also reduce the NPN beta, increased spacings sometimes, but sometimes using an electron collecting guard ring. If you're in a CMOS process. It's much harder and probably your only hope is to add more back gate contacts.

169   THE HOLE-BLOCKING GUARD RING

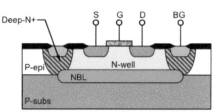

- **A hole-blocking guard ring (HBGR) blocks hole injection into the P-epi and P-substrate.**
  - Holes cannot surmount the electric field caused by the built-in potential of an N+/N− interface.
  - The HBGR uses a deep-N+ sinker and NBL to construct the N+/N− interface.

The hole-blocking guard ring looks like this. It must have a complete ring of deep end around the offending device and have a complete floor of NBL beneath it. Now the offending device spews out holes that want to seek out the substrate, but they can't get through the NBL because the N+N-interface repels the. The rule is 100:1, you can get 100 times more concentration at the deep end than in the adjacent end well, and 100 times more in the NBL than in the end well.

170   THE HOLE-COLLECTING GUARD RING

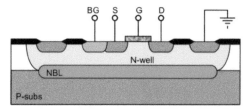

- **The hole-collecting guard ring (HCGR) collects some fraction of the holes before they can reach the substrate.**
  - NBL or a strong retrograde profile is necessary to block vertical hole injection.
  - The limited depth of PSD implants make them poor HCGR's.
  - P-well or P-epi can be used as an HCGR, but its low doping renders it vulnerable to debiasing.

A hole-collecting guard ring collects some fraction of the holes before they can reach the substrate, but it won't work well because the limited depth of PSD and the low doping of P-epi.

171 THE ELECTRON-COLLECTING GUARD RING

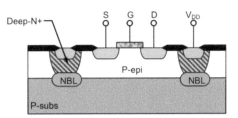

- **The electron-collecting guard ring (ECGR) collects a fraction of the electrons before they can reach an N-well or NSD region.**
  - A P+ substrate is required to constrain the electrons within the P-epi.
  - The best ECGR structures use deep heavily doped diffusions, but N-well alone will often suffice if it is connected to a supply.

I could create electron collecting rings by taking an n-type ring, but it is not very deep, so I might make it wide. The Nwell, it's plenty deep, but the problem is it's very resistive and since biasing is controlled, so I might hook it to a power supply, which gives it quite a lot of voltage before the biases to the point where it begins to re-inject.

 172 SUMMARY

- **All junction-isolated integrated circuits may experience latchup.**
  - All diffusions that hook either directly to pins, or to pins through less than about 50 kilohms of resistance, can potentially trigger latchup.
  - Capacitors switching within a circuit can sometimes cause minority carrier injection that triggers latchup.

- **Proper layout practices can minimize the risk of latchup.**
  - Adequate well and substrate contacts.
  - Maintain recommended spacings.
  - Add guard rings where feasible.

All junction isolated integrated circuits may experience latch up, and you can use below method to minimize the risk of latch up.

About
the Editors

Milin Zhang

Milin Zhang received the B.S. and M.S. degrees in electronic engineering from Tsinghua University, Beijing, China, in 2004 and 2006, respectively, and the Ph.D. degree in the Electronic and Computer Engineering Department, Hong Kong University of Science and Technology (HKUST), Hong Kong. After finishing her doctoral studies, she worked as a postdoctoral researcher at the University of Pennsylvania (UPenn). She joined Tsinghua University as an Assistant Professor in the Department of Electronic Engineering in 2016. Her research interests include designing of traditional and various non-traditional imaging sensors, such as polarization imaging sensors and focal-plane compressive acquisition image sensors. She is also interested in brain-machine-interface (BMI) and relative biomedical sensing applications and new sensor designs.

She has received the Best Paper Award of the BioCAS Track of the 2014 International Symposium on Circuits and Systems (ISCAS), and the Best Paper Award (1st place) of the 2015 Biomedical Circuits and Systems Conference (BioCAS). She received the Thousand Youth Talents Award in 2016.

Zhihua Wang

Zhihua Wang received the B.S., M.S., and Ph.D. degrees in Electronic Engineering in 1983, 1985 and 1990, respectively, from Tsinghua University, Beijing, China, where he has served as full professor and Deputy Director of the Institute of Microelectronics since 1997 and 2000. He was a visiting scholar at CMU (1992-1993) and KU Leuven (1993-1994), and was a visiting professor at HKUST (2014.9-2015.3). His current research mainly focuses on CMOS RFIC and biomedical applications, involving RFID, PLL, low-power wireless transceivers, and smart clinic equipment combined with leading edge RFIC and digital image processing techniques. He has co-authored 11 books/chapters, over 160 (439) papers in international journals (conferences), and holds 121 Chinese and 7 US patents.

Prof. Wang has served as the chairman of IEEE SSCS Beijing Chapter (1999-2009), an AdCom Member of the IEEE SSCS (2016-2019), a technology program committee member of the IEEE ISSCC (2005-2011), a steering committee member of the IEEE A-SSCC (2005-), the technical program chair for A-SSCC 2013, a Guest Editor for IEEE JSSC Special Issues (2006.12, 2009.12 and 2014.11), an associate editor of IEEE Trans on CAS-I, II and IEEE Trans on BioCAS, and other administrative/expert committee positions in China's national science and technology projects.

Jan Van der Spiegel

Jan Van der Spiegel is a Professor of the Electrical and Systems Engineering Department, and Associate Dean for Professional Programs of the School of Engineering and Applied Science at the University of Pennsylvania. He is the former chair of the Electrical Engineering and interim chair of the Electrical and Systems Engineering Departments. Dr. Van der Spiegel received his Masters degree in Electro-Mechanical Engineering and his Ph.D. degree in Electrical Engineering from the University of Leuven, Belgium. His primary research interests are in mixed-mode VLSI design, CMOS vision sensors for polarization imaging, biologically based image sensors and sensory information processing systems, low-power brain-machine-interfaces, and micro-sensor technology. He is the author of over 180 journal and conference papers and holds 4 patents.

He is a fellow of the IEEE and recipient of the 2007 EAB Major Educational Innovation Award. He received the IEEE Third Millennium Medal, the UPS Foundation Distinguished Education Chair and the Bicentennial Class of 1940 Term Chair. He was awarded the Christian and Mary Lindback Foundation, and the S. Reid Warren Award for Distinguished Teaching, and the Presidential Young Investigator Award.

He has served on several IEEE program committees (IEDM, ICCD, ISCAS and ISSCC) and was the technical program chair of the 2007 International Solid-State Circuit Conference (ISSCC 2007). He is an associate editor of the IEEE Transactions on Biomedical Circuits and Systems (TBCAS), and is on the Editorial Board of the Proceedings of the IEEE and Sensors and Actuatorss A. He is a member of the IEEE Solid-State Circuits Society where he served as the chapters Chairs coordinator. He is the 2016-2017 President of the IEEE Solid-State Circuits Society. He is also a member of Phi Beta Delta and Tau Beta Pi.

Franco Maloberti

Franco Maloberti received the Laurea Degree in Physics (Summa cum Laude) from the University of Parma, Parma Italy, in 1968 and the Dr. Honoris Causa degree in Electronics from the Instituto Nacional de Astrofisica, Optica y Electronica (Inaoe), Puebla, Mexico in 1996. He was a Visiting Professor at ETH-PEL, Zurich in 1993 and at EPFL-LEG, Lausanne in 2004. He was Professor of Microelectronics and Head of the Micro Integrated Systems Group University of Pavia, Pavia, Italy and the TI/J.Kilby Analog Engineering Chair Professor at the Texas A&M University. He was also the Distinguished Microelectronic Chair Professor at the University of Texas at Dallas. Currently, he is Professor at the University of Pavia, Italy and Honorary Professor, University of Macau, China SAR. His professional expertise is in the design, analysis, and characterization of integrated circuits and analog digital applications, mainly in the areas of switched capacitor circuits, data converters, interfaces for telecommunication and sensor systems, and CAD for analog and mixed A-D design. He has written more than 550 published papers, five books and holds 34 patents.

He was in 1992 recipient of the XII Pedriali Prize for his technical and scientific contributions to national industrial production. He was co-recipient of the 1996 Institute of Electrical Engineers (U.K.) Fleming Premium for the paper "CMOS Triode Transistor Transconductor for high-frequency continuous-time filters." The research activity of Prof. Franco Maloberti is a proper balance of scientific innovation and technical advance for the industry. He received the 1999 IEEE CAS Society Meritorious Service Award, the 2000 CAS Society Golden Jubilee Medal, and the IEEE Millenium Medal. He received the ESSCIRC 2007 Best Paper Award and the IEEJ Workshop 2007 and 2010 Best Paper Award. He received the IEEE CAS Society 2013 Mac Van Valkenburg Award. He is a Fellow of IEEE.

About

the Authors

Minoru Fujishima

Minoru Fujishima (S'89–M'93) received the B.E., M.E., and Ph.D. degrees in electronics engineering from the University of Tokyo, Japan in 1988, 1990, and 1993, respectively. He joined the Faculty of the University of Tokyo as a Research Associate in 1988, and has been an Associate Professor with the School of Frontier Sciences, University of Tokyo, since 1999. He was a Visiting Professor with the ESAT-MICAS Laboratory, Katholieke Universiteit Leuven, Belgium, from 1998 to 2000. Since 2009, he has been a Professor with the Graduate School of Advanced Sciences of Matter, Hiroshima University. He studied design and modeling of CMOS and BiCMOS circuits, nonlinear circuits, single-electron circuits, and quantumcomputing circuits. He has co-authored over 40 journal papers and 100 conference papers, and a book entitled Design and Modeling of Millimeter-Wave CMOS Circuits for Wireless Transceivers: Era of Sub-100nm Technology (Springer, 2008). His current research interests include the designs of low-power millimeterand short-millimeter-wave wireless CMOS circuits. Dr. Fujishima is a member of IEICE and JSAP. He is currently serving as a technical committee member of several international conferences. He was a Distinguished Lecturer of SSCS from 2011 to 2012. He is the Chair of the IEEE SSCS Kansai Chapter from 2013 to 2014.

[Chapter 5]

Alan Hastings

Alan Hastings has been a power integrated circuit designer for thirty years. During this time, he has helped design analog and power integrated circuits for custom automotive applications, cellphones, Power-over-Ethernet power device controllers, and USB type-C power delivery controllers. He has a particular interest in the effects of layout upon circuit performance. His textbook entitled The Art of Analog Layout has sold over 15,000 copies since 2000. A second edition was published in 2009, and a third edition is in preparation. Alan holds 32 patents in the areas of analog and power integrated circuit design. He is currently employed by Texas Instruments, where he holds the title of TI Fellow. He received his BSEE from the University of Florida in 1985. He was inducted into the University of Florida Electrical and Computer Engineering Academy in 2009 as one of the first 25 inductees honored for "having made a significant mark in engineering." He is also a senior member of the IEEE.

[Chapter 6]

Rakesh Kumar

Dr. Rakesh Kumar is a 43 year semiconductor industry veteran, an entrepreneur, and is now an educator. As President and CEO of TCX Technology Connexions he provides management, business and technical 'bridging the gaps' consulting services in advanced semiconductor technology and virtual operations areas. Clients have included over 20 emerging fabless IC companies, mid-size, a few Fortune 500 IC companies, and two leading research organizations. Currently he provides successful startup education to engineering students, and serves as a Business and Technology Advisor at the University of California San Diego's Entrepreneurism Center. Dr. Kumar authored the book "Fabless Semiconductor Implementation", published by McGraw Hill in addition to numerous publications. He is an IEEE Life Fellow and an active volunteer. He is a Past President (2012-13) of the IEEE Solid-State Circuits Society. During his semiconductor industry career Dr. Kumar has also been the VP&GM of Cadence Design's worldwide Silicon Technology Services business unit, and has held various technical and management positions at Unisys and Motorola. He received the M.S. and Ph.D. degrees from the University of Rochester in 1971 and 1974, the B.Tech from IIT Delhi in 1969, and an executive MBA from UCSD in 1989.

[Chapter 1]

Seng Pan U, Ben

Seng-Pan U (Ben) (S'94 - M'00 - SM'05 - F'16) received the B.Sc. and M.Sc degree in 1991 and 1997, respectively, and the dual Ph.D. (Hons.) degrees from the University of Macau (UM) and the Instituto Superior Técnico (IST), Portugal in 2002 and 2004 respectively.

Prof. U has been with Faculty of Science & Technology, UM since 1994, where he is currently Professor and Deputy Director of State-Key Laboratory of Analog & Mixed-Signal (AMS) VLSI. From 1999 to 2001, he was on leave to the Center of Microsystems in IST, as a Visiting Research Fellow.

In 2001, Prof. U co-founded the Chipidea Microelectronics (Macau), Ltd. as the Engineering Director, and later the corporate VP-IP Operations Asia Pacific devoting to advanced AMS Semiconductor IP product development. The company was acquired in 2009 by the world leading EDA & IP provider Synopsys Inc. (NASDAQ: SNPS), currently as Synopsys Macau Ltd. He is also the corporate R&D director and site general manager.

Prof. U has co-authored about 200 publications, four books (Springer and China Science Press) in the area of VHF SC filters, analog baseband for Multi-standard wireless transceivers and very high-speed TI ADCs. He also co-holds 14 US patents.

(cont.)

(cont.)

Prof. U received 30+ research & academic/teaching awards and is co-recipient of the 2014 ESSCIRC Best Paper Award. He is also the Advisor for 30+ various student research award recipients, e.g. SSCS Pre-Doctoral Achievement Award, ISSCC Silk-Road Award, A-SSCC Student Design Contest, IEEE DAC/ISSCC Student Design Contest, ISCAS, MWSCAS, PRIME and etc. As the Macau founding Chairman, he received the 2012 IEEE SSCS Outstanding Chapter Award. Both at the 1st time from Macau, he received the Science & Technology (S&T) Innovation Award of Ho Leung Ho Lee Foundation in 2010, and also The National State S&T Progress Award in 2011. He also received both the 2012, 2014 and 2016 Macau S&T Invention Award and Progress Award. In recognition of his contribution in academic research & industrial development, he was awarded by Macau SAR government the Honorary Title of Value in 2010. He was also elected as the "Scientific Chinese of the Year 2012".

Prof. U is currently IEEE Fellow, Honorary Chairman of IEEE SSCS (as founder) and CAS/COMM Macau chapter. He is appointed a member of the S&T Commission of China Ministry of Education and also S&T Committee of Macau SAR. He was IEEE SSCS Distinguished Lecturer (2014-2015), A-SSCC 2013 and ISSCC 2018 Tutorial Speaker. He has also been on the technical review committee of various IEEE journals and the program committee/chair of IEEJ AVLSIWS, IEEE APCCAS, ICICS, PRIMEAsia and IEEE ASP-DAC'16. He is currently TPC of ISSCC, Sub-Committee Chair (Data Converters) of A-SSCC, the Analog Sub-Committee Chair of VLSI-DAT and Editorial Board member of the Journal AICSP.

[Chapter 2]

Alice Wang

Alice Wang received her Bachelors, Masters, and Ph.D. degrees in electrical engineering and computer science from the Massachusetts Institute of Technology, in 1997, 1998, and 2004, respectively. She wrote the paper "A 180-mV Subthreshold FFT Processor Using a Minimum Energy Design Methodology" with Professor Anantha Chandrakasan which inspired a new research field in ultra-low power technology. After her PhD, she spent 8 years at Texas Instruments developing low-power circuit and system technology for mobile, application processors and radios. Her work on low-power technology has been showcased in 30+ IEEE publications and she has co-authored two books. She is a Senior Member of IEEE.

Currently, Alice is a Assistant General Manager in High-Performance Processor Technology at MediaTek working on advanced processors for consumer electronics including Smartphones, Tablets and Smart TV's and managing the Foundation IP teams (Standard cell library, Memory and I/O). She was elected to the Advisory Committee for the Solid State Circuits Committee (2017-2019) and heads up the Women in Circuits committee.

[Chapter 3]

Hoi-Jun Yoo

Prof. Hoi-Jun Yoo is the full professor of Department of Electrical Engineering at KAIST and the director of SDIA(System Design Innovation and Application Research Center). From 2003 to 2005, he was the full time Advisor to the Minister of Korean Ministry of Information and Communication for SoC and Next Generation Computing. His current research interests are Bio Inspired Intelligent SoC Design, Wearable Computing and Wearable Healthcare. He published more than 200 papers, and wrote 5 books including "Mobile 3D Graphics SoC"((2010, Wiley) and "Biomedical CMOS ICs"(2011, Springer).

Dr. Yoo received the National Medal for his contribution to Korean Memory Industry in December of 2011, the Korean Scientist of the Month award in Dec. 2010, Best Research of KAIST Award in 2007, Design Award of 2001 ASP-DAC, and Outstanding Design Awards 2005, 2006, 2007, 2010, 2011, 2014 A-SSCCs, and Best Demo Award of ISSCC 2016. He is an IEEE Fellow, a member of the executive committee of Symposium on VLSI, and A-SSCC. He was the TPC Co-Chair of ISWC 2010, IEEE Distinguished Lecturer('10-'11), and Asia Chair of ISSCC('10-'11). He was TPC Chair of ISSCC 2015, Vice Chair of ISSCC 2014, Technology Direction Sub-Committee Chair of ISSCC 2013, a member of Executive Committee of ISSCC 2008-2015 and recognized as the top 4 paper-contributor for 2004-2013 ISSCCs and top 10 paper contributor for 1954-2013 ISSCCs.

[Chapter 4]